# Concise Textbook of
# BIOCHEMISTRY
## for Paramedical Students

# Concise Textbook of
# BIOCHEMISTRY
## for Paramedical Students
*(for BSc Nursing, BPT and Other Paramedical Courses)*

**Second Edition**

**DM Vasudevan**
MBBS MD FRCPath (London)
Distinguished Professor
Department of Biochemistry
Chairman, PG Programs and Research
College of Medicine
Amrita Institute of Medical Sciences
Kochi, Kerala, India

*Formerly*

Principal
Amrita College of Medicine
Kochi, Kerala, India

Dean
Sikkim Manipal Institute of Medical Sciences
Gangtok, Sikkim, India

**Sukhes Mukherjee** MSc PhD
Additional Professor
Department of Biochemistry
All India Institute of Medical Sciences
Bhopal, Madhya Pradesh, India

**JAYPEE BROTHERS MEDICAL PUBLISHERS**
*The Health Sciences Publisher*
New Delhi | London

 **Jaypee Brothers Medical Publishers (P) Ltd.**

#### Headquarters
Jaypee Brothers Medical Publishers (P) Ltd
EMCA House
23/23-B, Ansari Road, Daryaganj
New Delhi - 110 002, India
Landline: +91-11-23272143, +91-11-23272703
+91-11-23282021, +91-11-23245672
Email: jaypee@jaypeebrothers.com

#### Corporate Office
Jaypee Brothers Medical Publishers (P) Ltd
4838/24, Ansari Road, Daryaganj
New Delhi 110 002, India
Phone: +91-11-43574357
Fax: +91-11-43574314
Email: jaypee@jaypeebrothers.com

#### Overseas Office
J.P. Medical Ltd
83 Victoria Street, London
SW1H 0HW (UK)
Phone: +44 20 3170 8910
Fax: +44 (0)20 3008 6180
Email: info@jpmedpub.com

Website: www.jaypeebrothers.com

Website: www.jaypeedigital.com

© 2021, Jaypee Brothers Medical Publishers

The views and opinions expressed in this book are solely those of the original contributor(s)/author(s) and do not necessarily represent those of editor(s) of the book.

All rights reserved. No part of this publication may be reproduced, stored or transmitted in any form or by any means, electronic, mechanical, photocopying, recording or otherwise, without the prior permission in writing of the publishers.

All brand names and product names used in this book are trade names, service marks, trademarks or registered trademarks of their respective owners. The publisher is not associated with any product or vendor mentioned in this book.

Medical knowledge and practice change constantly. This book is designed to provide accurate, authoritative information about the subject matter in question. However, readers are advised to check the most current information available on procedures included and check information from the manufacturer of each product to be administered, to verify the reco mmended dose, formula, method and duration of administration, adverse effects and contraindications. It is the responsibility of the practitioner to take all appropriate safety precautions. Neither the publisher nor the author(s)/editor(s) assume any liability for any injury and/or damage to persons or property arising from or related to use of material in this book.

This book is sold on the understanding that the publisher is not engaged in providing professional medical services. If such advice or services are required, the services of a competent medical professional should be sought.

Every effort has been made where necessary to contact holders of copyright to obtain permission to reproduce copyright material. If any have been inadvertently overlooked, the publisher will be pleased to make the necessary arrangements at the first opportunity. The **CD/DVD-ROM** (if any) provided in the sealed envelope with this book is complimentary and free of cost. **Not meant for sale.**

**Inquiries for bulk sales may be solicited at:** jaypee@jaypeebrothers.com

*Concise Textbook of Biochemistry for Paramedical Students*

*First Edition: 2016*

*Second Edition:* **2021**

ISBN: 978-93-90281-34-3

# Preface to the Second Edition

I am glad to see that this *Concise Textbook of Biochemistry for Paramedical Students* is now in the second edition. The students of BSc Nursing and other paramedical students find biochemistry as the most difficult subject. It is mainly due to the fact that they are following textbooks written for medical or dental courses. In this textbook, I have tried to keep a balance between the basic knowledge and the academic requirements. I also emphasized the clinical applications of biochemical knowledge. In this second edition, a number of new figures and tables were added, which will aid the students to understand the facts easily. Few question papers of BSc Nursing of various universities are included for giving a general idea to the students. This may enable them to prepare for the university examinations.

This is to report the sad and untimely demise of Dr T Vijayakumar, the second author of the first edition. He was my colleague for the past 40 years. He is fondly remembered by all his students. May his soul rest in peace.

Dr Sukhes Mukherjee has now taken up the responsibility of the second author. I am sure that this short textbook will be of immense help to the nursing students and teachers alike. Any suggestions to improve this book are appreciated.

**DM Vasudevan**

# Preface to the First Edition

The students of paramedical students [BSc (Nursing), BPharm, BPT, MLT] find biochemistry as the most difficult subject. It is mainly due to the fact that they are following textbooks written for medical or dental courses. In this book, we have tried to keep a balance between the basic knowledge and the academic requirements. We have also given importance to the clinical applications of biochemical knowledge.

A few questions are given in the Appendix 4. This will certainly enable the students to prepare for the university examinations.

We expect that the short textbook will be of immense help to the paramedical students and teachers alike. Suggestion to improve the book is appreciated. All are welcome to communicate with the first author in his email address <dmvasudevan@yahoo.co.in>

**DM Vasudevan**
**T Vijayakumar**

# Acknowledgments

I am indebted to my postgraduate students in compiling this textbook. A few tables, figures and sentences were taken from the *Textbook of Biochemistry for Medical Students*, authored by Dr DM Vasudevan, Sreekumari S and Kannan Vaidyananthan published by M/s Jaypee Brothers Medical Publishers (P) Ltd, New Delhi, India.

Last but not the least, we would like to thank Shri Jitendar P Vij (Group Chairman), Mr Ankit Vij (Managing Director), Mr MS Mani (Group President), Dr Madhu Choudhary (Publishing Head–Education), Ms Pooja Bhandari (Production Head), Ms Sunita Katla (Executive Assistant to Group Chairman and Publishing Manager), Ms Samina Khan (Executive Assistant to Director–Content Strategy), Mr Rajesh Sharma (Production Coordinator), Ms Seema Dogra (Cover Visualizer), Mr Laxmidhar Padhiary (Proofreader), Mr Kapil Dev Sharma (Typesetter), Mr Gopal Singh Kirola (Graphic Designer), and their team of M/s Jaypee Brothers Medical Publishers (P) Ltd, New Delhi, India, for making my dream come true!

# Contents

1. **Introduction to Biochemistry** 1
   - Study of Metabolic Processes 2
2. **The Cell** 4
   - Cell Theory 4
   - Structure of a Typical Cell 5
   - Subcellular Organelles 5
   - Transport Mechanisms 9
3. **Enzymes** 14
   - Coenzymes 14
   - Classification of Enzymes 15
   - Factors Influencing Enzyme Activity 16
   - Mechanisms of Enzyme Reaction 19
   - Active Site or Active Center of Enzyme 20
   - Enzyme Inhibition 21
   - Zymogens 23
   - Isoenzymes 25
4. **Carbohydrate Chemistry** 29
   - Functions of Carbohydrates 29
   - Classification of Carbohydrates 29
   - Properties of Monosaccharides 31
   - Reactions of Carbohydrates 33
   - Modified Monosaccharides 34
   - Disaccharides 34
   - Polysaccharides 37
5. **Carbohydrate Metabolism** 41
   - Digestion 41
   - Absorption 41
   - Glucose Metabolism 42
   - Glycolysis (Embden–Meyerhof Pathway) 42
   - Gluconeogenesis 46
   - Cori Cycle 48
   - Glycogenolysis 48
   - Glycogen Synthesis (Glycogenesis) 49
   - Glycogen Storage Diseases 50
   - Hexose Monophosphate Shunt Pathway/Pentose Phosphate Pathway/Phosphogluconate Oxidative Pathway/Dickens–Horecker Pathway 50
   - Glucuronic Acid Pathway 51
   - Fructose Metabolism 52
   - Galactose Metabolism 53
6. **Regulation of Blood Glucose and Diabetes Mellitus** 56
   - Blood Glucose Level 56
   - Insulin 57
   - Diabetes Mellitus 58
   - Reducing Substances in Urine 63
7. **Lipid Chemistry** 67
   - Functions of Lipids 67
   - Clinical Applications of Lipids 67
   - Classification of Lipids 68
   - Fatty Acids 69
   - Essential Fatty Acids 71
   - Simple Lipids 71
   - Compound Lipids 72
8. **Lipid Metabolism** 80
   - Digestion of Lipids 80
   - Absorption of Long-Chain Fatty Acids 80
   - Beta-Oxidation of Fatty Acids 82
   - Synthesis of Fatty Acids 85
   - Ketogenesis (Formation of Ketone Bodies) 88
   - Ketolysis 88
   - Metabolism of Triglycerides 89
   - Fatty Liver 90
   - Essential Fatty Acids 91
9. **Cholesterol and Lipoproteins** 96
   - Structure of Cholesterol 96
   - Biosynthesis of Cholesterol 96
   - Importance of Cholesterol 97
   - Plasma Lipids 99
   - Cardiac Biomarkers 104
10. **Amino Acids and Proteins: Chemistry** 108
    - General Functions of Proteins 108
    - Amino Acids 108
    - Classification of Proteins 115
    - Levels of Organization of Proteins 117
    - Denaturation of Proteins 117

## 11. Amino Acids and Proteins: Metabolism — 121

- Digestion of Proteins 121
- Absorption of Amino Acids 122
- Amino Acid Pool 122
- Urea Cycle 123
- General Reactions of Amino Acid Metabolism 124
- Metabolism of Individual Amino Acids 126
- Nitric Oxide 129
- Branched-Chain Amino Acids 130

## 12. Inborn Errors of Metabolism — 136

- Salient Features of Inborn Errors of Metabolism 136
- Inborn Errors Associated with Amino Acid Metabolism 136
- Inherited Disorders of Lipid Metabolism 139
- Inborn Errors of Carbohydrate Metabolism 140
- Inborn Errors of Nucleic Acid Metabolism 141
- Important Groups of Inborn Errors of Metabolism 141

## 13. Citric Acid Cycle, Biological Oxidation, and Free Radicals — 143

- Citric Acid Cycle or Tricarboxylic Acid Cycle or Krebs Cycle 143
- Biological Oxidation 148
- Free Radicals or Reactive Oxygen Species 150

## 14. Plasma Proteins, Immunoglobulins, and Tissue Proteins — 156

- Normal Values of Proteins in Plasma 156
- Separation of Plasma Proteins 156
- Normal Values and Interpretations 157
- Abnormal Patterns in Clinical Diseases 157
- Albumin 157
- Globulins 158
- Transport Proteins 159
- Functions of Plasma Proteins 159
- Acute Phase Proteins 159
- Gamma-Globulins (Immunoglobulins) 159
- General Immunity 162
- Enzyme-Linked Immunosorbent Assay Test 165
- Special Proteins 166
- Muscle Proteins 168
- Lens Proteins 168

## 15. Hemoglobin — 171

- Heme Metabolism 171
- Hemoglobin 176
- Abnormal Hemoglobins (Hemoglobin Variants or Hemoglobinopathies) 177
- Transport of Oxygen by Hemoglobin 178
- Transport of Carbon Dioxide 179

## 16. Vitamins — 182

- Vitamins in General 182
- Vitamin A 183
- Vitamin D (Cholecalciferol) 185
- Vitamin E 188
- Vitamin K 188
- Water-Soluble Vitamins 188
- Pantothenic Acid 193
- Cobalamin 195

## 17. Minerals — 200

- Calcium 200
- Phosphorus 203
- Magnesium 204
- Sodium 204
- Potassium 205
- Chloride 205
- Sulfur 206
- Trace Elements 206
- Iron 206
- Copper 209
- Iodine 210
- Zinc 210
- Fluoride 211
- Selenium 211
- Heavy Metal Poisoning 211

## 18. Nutrition — 215

- Calorific Value 215
- Respiratory Quotient 215
- Energy Requirements of a Normal Person 216
- Proximate Principles 217
- Obesity 219
- Balanced Diet 219

## 19. Acid-Base and Electrolyte Balance — 223

- Acids and Bases 223
- Mechanism of Regulation of pH in the Body 225

- Acid–Base Imbalance  227
- Electrolyte Balance  230
- Sodium  231
- Potassium  232
- Chloride  233
- Magnesium  233

## 20. Liver and Renal Function Tests  237
- Liver Function Tests  237
- Clinical Manifestations of Liver Dysfunction  237
- Tests Based on Bilirubin Metabolism  239
- Tests Based on Plasma Protein Estimation  240
- Tests Based on Serum Enzyme Activity  241
- Markers of Obstructive Liver Disease  241
- Renal Function Tests  241
- Normal Constituents of Urine  242
- Abnormal Constituents of Urine  242
- Clearance Tests (Markers for Glomerular Filtration Rate)  243
- Gastric Function Tests  246
- Fractional Test Meal  247
- Cerebrospinal Fluid  248

## 21. Molecular Biology  251
- Composition of Nucleotides  251
- Biosynthesis of Purine Nucleotides  253
- Uric Acid  255
- Gout  255
- De Novo Synthesis of Pyrimidine  256
- Structure of DNA  256
- Replication of DNA  259
- RNA  259
- Transcription Process  260
- Protein Biosynthesis  262
- Translation Process  263

## 22. Biophysical and Biomedical Techniques  268
- Osmosis  268
- Dialysis  270
- Filtration  270
- Microscopes  271
- Electrophoresis  272
- Chromatography  276
- Recombinant DNA Technology  277
- Gene Therapy  278
- Polymerase Chain Reaction  280

## Appendices  285
- Appendix A: Normal Values (Reference Values)  285
- Appendix B: Recommended Daily Allowance of Essential Nutrients  287
- Appendix C: Nutritional Value of Food Items  289

*Index*  291

# CHAPTER 1

# Introduction to Biochemistry

Biochemistry is the language of biology. The tools for research in all the branches of medical science are mainly biochemical in nature. The study of biochemistry is essential to understand the basic functions of the body. This study will give information regarding the functioning of cells at the molecular level. How is the food that we eat digested, absorbed, and used to make the ingredients of the body? How does the body derive energy for the normal day-to-day work? How are the various metabolic processes interrelated? What is the function of genes? What is the molecular basis for immunological resistance against invading organisms? Answers to such basic questions can only be derived by a systematic study of biochemistry.

Modern-day medical practice is highly dependent on the laboratory analysis of body fluids, especially the blood. The disease manifestations are reflected in the composition of blood and other tissues. Hence the demarcation of abnormal from normal constituents of the body is another aim of the study of biochemistry.

The word *chemistry* is derived from the Greek word *chemi* (the black land), the ancient name of Egypt. Indian medical science, even from ancient times, had identified the metabolic and genetic basis of diseases. Charaka, the great master of "Ayurveda"—the ancient Indian medicine, in his treatise (circa 400 BC) observed that *madhumeha* (diabetes mellitus) is produced by the alterations in the metabolism of carbohydrates and fats; the statement still holds good. Some of the milestones in the development of the science of biochemistry are given in **Table 1.1**.

**Table 1.1:** Milestones in the history of biochemistry.

| Scientists | Year | Landmark discoveries |
|---|---|---|
| Wohler | 1828 | Synthesis of urea |
| Berzelius | 1835 | Enzyme catalysis theory |
| Louis Pasteur | 1860 | Fermentation process |
| Fiske & Subbarow | 1926 | Isolated adenosine triphospchate (ATP) from muscle |
| Lohmann | 1932 | Creatine phosphate |
| Hans Krebs | 1937 | Citric acid cycle |
| Avery & Macleod | 1944 | Deoxyribonucleic acid (DNA) is genetic material |
| Watson & Crick | 1953 | Structure of DNA |
| Khorana | 1965 | Synthesized the genes |
| James Watson | 1990 | Human genome project started |
| Aristides Patrinos | 2003 | Human genome project completed |

The practice of medicine is both an art and a science. The word *doctor* is derived from the Latin root *docere*, which means "to teach." Knowledge devoid of ethical background may sometimes be disastrous! Hippocrates (460 BC–377 BC), the father of modern medicine, articulated "the Oath." About one century earlier, Sushrutha (approx. 500 BC), the great Indian surgeon, enunciated a code of conduct for the medical practitioners, which is still valid. He proclaims, "You must speak only truth; care for the good of all living beings; devote yourself to the healing of the sick even if your life be lost by your work; be simply clothed and drink no intoxicant; always seek to grow in knowledge; in face of God, you can take upon yourself these vows."

Biochemistry is perhaps the most rapidly developing discipline in medicine. No wonder, the major share of Nobel prizes in medicine has gone to research workers engaged in biochemistry. Thanks to the advent of DNA recombination technology, genes can now be transferred from one person to another, so that many of the genetically determined diseases are now amenable to gene therapy. Many genes (e.g., human insulin gene) have already been transferred to microorganisms for large-scale production of human insulin. Studies on oncogenes have identified molecular mechanisms of control of normal and abnormal cells. With the help of the human genome project (HGP) the sequences of whole human genes are now available; it has already made a great impact on medicine and related health sciences.

## STUDY OF METABOLIC PROCESSES

Our food contains carbohydrates, fats, and proteins as principal ingredients. These macromolecules are to be first broken down to smaller units; carbohydrates to monosaccharides and proteins to amino acids. This process takes place in the gastrointestinal tract and is called digestion or **primary metabolism**. After absorption, these smaller molecules are further broken down and oxidized to carbon dioxide. In this process, nicotinamide adenine dinucleotide (NADH) or flavin adenine dinucleotide ($FADH_2$) are generated. This process is named as **secondary or intermediary metabolism**. Finally, these reducing equivalents enter the electron transport chain in the mitochondria, where they are oxidized to water; in this process, energy is trapped as adenosine triphosphate (ATP). This process is termed as **tertiary metabolism**. Metabolism is the sum of all chemical changes of a compound inside the body, which includes synthesis (anabolism) and breakdown (catabolism) (in Greek, kata = down; ballein = change). We shall discuss the details of these metabolic processes in the next chapters.

## Importance of Biochemistry for Nursing and Allied Health Professionals

Biochemistry involves the chemical processes that occur in all living cells and organisms and is crucial to nurses and allied health professionals in understanding how the human body functions in both normal state and various disease states in bedside clinics.

Nursing and allied health professional therapeutic interventions are based on this understanding of various biochemical pathways and cellular environment. In critical care, for instance, nurses learn how to preserve patients' energy. Blood estimations are performed to ensure that patients' acid–base balance and oxygenation levels are maintained to promote aerobic metabolism.

How medications work is directly related to biochemistry? Antibiotics, for example, work on the various microorganisms differently. Some antibiotics kill bacteria outright, by making holes in their cell walls. Others keep bacteria from replicating by disrupting intracellular processes. The aim of this book is to make nursing and allied health professionals understand the basic knowledge of biochemical pathways and their applications to the bedside medicine.

Nurses work as part of patient care teams and interact closely with patients' families. Nurses would be able to help patients' family members understand what is going on with their patients, only if they have a thorough understanding of biochemical processes and principles. Thus, it is essential to learn biochemistry and its applications in medicine.

# CHAPTER 2

# The Cell

At the completion of this chapter, the reader will be able to answer questions on the following topics:

- Cell theory
- Structure of cell
- Subcellular organelles
- Nucleus
- Endoplasmic reticulum
- Ribosomes
- Mitochondria
- Golgi apparatus
- Lysosomes
- Peroxisomes
- Plasma membrane
- Fluid mosaic model
- Diffusion
- Facilitated transport
- Active transport

The cell is the basic unit of life. It is the smallest part of an organism that displays the characteristics of life. Hence it is the unit of structure and function. The term *cell* was coined by Robert Hooke.

## CELL THEORY

- All animals and plants are made up of cells and cell products.
- The cell is the unit of structure and function of all living things.
- The cell can arise only from preexisting cells.

The cells differ much in their shape, size, and structure. A major difference between a plant cell and an animal cell is that all plant cells have cell walls formed of cellulose, which is absent in animals. Structurally, the cells fall under two broad categories, the **prokaryotic cells** and the **eukaryotic cells.** The important differences between the prokaryotic and eukaryotic cells are given in **Table 2.1**.

**Table 2.1:** Differences between prokaryotic and eukaryotic cells.

| | Prokaryotic cell | Eukaryotic cell |
|---|---|---|
| 1. | Primitive in structure and organization | More evolved cells |
| 2. | Nuclear materials without nuclear envelope and hence no separation of protoplasm into cytoplasm and nucleoplasm | A proper nucleus present with nuclear membrane separating chromosome and cytoplasm |
| 3. | A single chromosome composed of circular deoxyribonucleic acid (DNA) | Chromosome composed of mostly linear DNA and proteins (histones) |
| 4. | Lacks many cellular organelles such as Golgi complex, mitochondria, etc. | Mitochondria, endoplasmic reticulum, Golgi complex, etc., are present |
| 5. | Plasma membrane carries out cellular respiration | Cellular respiration by mitochondria |
| 6. | Found in bacteria, blue–green algae, etc. | Found in all plants and animals |

## STRUCTURE OF A TYPICAL CELL

All eukaryotic cells have a basic uniform structure **(Fig. 2.1)**. They have a definite nucleus bounded by the nuclear membrane. Outside the nucleus is the cytoplasm, which is bounded by the plasma membrane. For all animal cells the plasma membrane is the limiting membrane of the cell.

## SUBCELLULAR ORGANELLES

Cells contain various organized structures collectively called as cell organelles.

### Nucleus

It is the most prominent organelle of the cell **(Fig. 2.1)**. All cells in the body contain nucleus, except mature red blood cells (RBCs) in circulation. The uppermost layer of the skin also may not possess a readily identifiable nucleus. In some cells, nucleus occupies most of the available space, e.g., small lymphocytes.

Nucleus is surrounded by a membrane called nuclear membrane. It is made up of a double layer of membrane; the inner one is called pronuclear membrane with numerous pores, while the outer membrane is continuous with the membrane of endoplasmic reticulum (ER). Nucleus contains the deoxyribonucleic acid (DNA); the very long DNA molecules are organized into chromosomes. There is a spherical body lying within the nuclear sap called nucleolus. This is the area for RNA processing and ribosome synthesis.

### Endoplasmic Reticulum

Endoplasmic reticulum is an elaborate network of membrane-bound vesicles, tubules, and flattened sacs spread throughout cytoplasm. The flattened sacs are arranged in a parallel manner; others are irregularly distributed in cytoplasm **(Fig. 2.1)**. In many places of the cell this system is continuous with the plasma membrane and the nuclear membrane. The outer surface of the ER may be studded with ribosome; these regions are called rough ER. This will be very prominent in cells actively synthesizing proteins, e.g., immunoglobulin-secreting plasma cells. The ribosome-free regions of ER are called smooth ER. George Palade was awarded Nobel Prize in 1974 for identifying the ER.

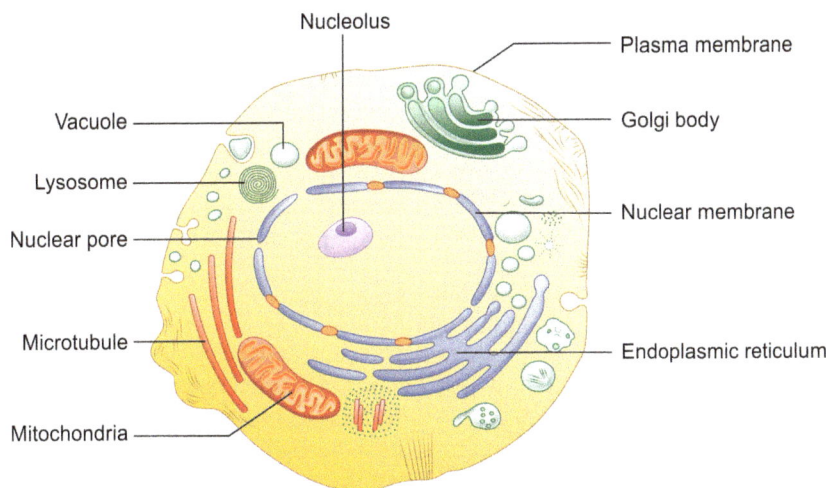

**Fig. 2.1:** Typical cell.

*Functions of Endoplasmic Reticulum*

- Synthesizing proteins and glycoproteins by rough ER.
- Synthesizing lipids, glycogen, and hormones by smooth ER.
- Transporting substances from one part of the cell to another.
- Giving mechanical support to the cell.
- Being involved in the detoxification process.

## Ribosome

Ribosomes are spherical bodies composed of RNA and proteins. They are formed of two subunits, one slightly smaller than the other. Ribosomes are sites of protein synthesis. During protein synthesis, ribosomes get attached to a specific mRNA molecule. Sometimes many ribosomes involved in the protein synthesis are clustered together, get attached to the mRNA molecule, and are called **polyribosomes** or polysomic.

## Mitochondria

Mitochondria are minute rod-like or spherical bodies **(Fig. 2.1)**. Each mitochondrion is bounded by two membranes. The inner membrane bears several folds projecting into the interior. These folds are called **cristae**. They increase the internal surface area of the mitochondria. The interior of a mitochondrion is filled with a dense fluid called **matrix**. The mitochondrial membrane and matrix contain respiratory enzymes. The inner surface of the inner membrane and cristae carry small particles called $F_1$ particles or elementary particles. They contain enzymes that catalyze the synthesis of adenosine triphosphate (ATP). Mitochondria are the **powerhouse of the cell**, where energy released from the oxidation of food stuffs is trapped as chemical energy in the form of ATP. Beta oxidation of fatty acid also takes place inside the mitochondria. The metabolic functions of subcellular organelles are shown in **Table 2.2**.

**Table 2.2:** Metabolic functions of subcellular organelles.

| Organelle | Functions |
|---|---|
| Nucleus | DNA replication, transcription |
| Endoplasmic reticulum | Biosynthesis of proteins, drug metabolism, and synthesis of cholesterol (partial) |
| Lysosome | Degradation of proteins, carbohydrates, lipids, and nucleotides |
| Mitochondria | Electron transport chain, ATP generation, TCA cycle, beta oxidation of fatty acids, and ketone body production |
| Cytosol | Protein synthesis, glycolysis, glycogen metabolism, HMP shunt pathway, transamination, fatty acid synthesis, and cholesterol synthesis |

(DNA: deoxyribonucleic acid; ATP: adenosine triphosphate; HMP: hexose monophosphate; TCA: tricarboxylic acid)

## Golgi Apparatus

Camillo Golgi described the structure in 1898 **(Fig. 2.1)**. The Golgi organelle is a network of flattened smooth membranes and vesicles. It may be considered as the converging area of endoplasmic reticulum. Associated with the cisternae are vesicles and tubules of different sizes. During protein biosynthesis, while moving through ER, carbohydrate groups are successively added to the nascent proteins. These glycoproteins reach the Golgi area. The main function of the Golgi apparatus is protein sorting, packaging, and secretion. Primary lysosomes are formed from the Golgi apparatus.

## Lysosomes

Lysosomes, discovered in 1950 by René de Duve, are tiny organelles **(Fig. 2.1)**. As solid wastes of a township are usually decomposed in incinerators, inside a cell, such a process takes place within the lysosomes. They are bags of hydrolytic enzymes capable of breaking down complex carbohydrates, proteins, lipids, and nucleic acids. Primary lysosome vesicles contain hydrolytic enzymes only. They are formed from the Golgi apparatus. The primary lysosomes fuse with endocytic vesicles containing food particles or other materials to form phagosome. The endocytic vesicles and phagosome with primary lysosome form the secondary lysosome.

*Functions of Lysosome*

- Intracellular digestion of food particles
- Destruction of harmful foreign particles, bacteria, viruses, and so on
- Disposition of worn out cellular organelles through autophagy.

## Peroxisomes

It is the small membranous bag resembling lysosome. They are prominent in leukocytes and platelets. They contain enzymes like peroxidase and catalase. These enzymes will destroy the unwanted peroxides and other free radicals. Comparison of the activities of cell organelles is shown in **Table 2.3**.

## Plasma Membrane

The plasma membrane or the cell membrane is the limiting membrane or the outer boundary of the cell. Chemically, it is made up of proteins, phospholipids, and small amounts of carbohydrate. The content of these compounds vary according to the nature of the membrane.

*Fluid Mosaic Model of Plasma Membrane*

This model was proposed by Singer and Nicolson in 1972. According to this model, the phospholipids are arranged in bilayers with the polar head groups oriented toward the extracellular side and the cytoplasmic side with a hydrophobic core (**Fig. 2.2**). The polar head or the hydrophilic core is made of phosphate and glycerol of phospholipids. The hydrophobic core or nonpolar tail is made of fatty acids.

The lipid bilayer shows free lateral movements of its component; hence the membrane is said to be fluid in nature. Fluidity enables the membrane to perform endocytosis and exocytosis. The cholesterol content and the nature of fatty acid of the membrane alters the fluidity of the membrane. The proteins are seen embedded in the lipid layer. There are two types of proteins, namely **integral** (intrinsic) and **peripheral** (extrinsic) proteins. The integral proteins are embedded at varying degrees in the lipid bilayer with free ends peeping out at the surfaces (**Fig. 2.2**). The integral proteins are tightly bound to the membrane and are not easily removed. The peripheral proteins are distributed at the inner surface of the plasma membrane, which faces the cytoplasm. They can be removed easily. The carbohydrates are seen as short-branched chains attached to the proteins or lipids at the outer surface.

The cell membrane is dynamic in nature. It has great selective properties to regulate the flow of materials into and out of the cell. The cell membrane allows the passage of only some molecules, and hence it is called **selectively permeable** or differentially permeable.

## Tight Junction

When two cells are in close approximation, in certain areas, instead of four layers, only three layers of plasma membranes are seen. This tight junction permits calcium and other

**Table 2.3:** Comparison of activities of cellular components.

| | |
|---|---|
| Plasma membrane | Fence with gates; gates open when message is received |
| Nucleus | Manager's office |
| Endoplasmic reticulum | Conveyer belt of production units |
| Golgi apparatus | Packing units |
| Lysosomes | Incinerators |
| Vacuoles | Lorries carrying finished products |
| Mitochondria | Power-generating units |

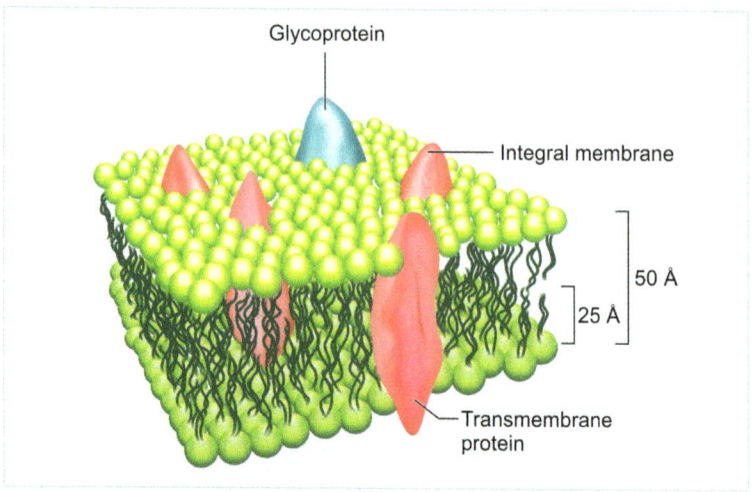

**Fig. 2.2:** Fluid mosaic model of plasma membrane.

small molecules to pass through from one cell to another through narrow pores. Some sort of communication between cells thus results (**Fig. 2.3**). Absence of tight junction is implicated in the loss of contact inhibition in cancer cells. Tight junctions also seal off subepithelial spaces of organs from the lumen.

## Gap Junction

Most eukaryotic cells are in contact with their neighboring cells. Neighboring cells are in metabolic contact through **gap junctions.** Gap junctions are intercellular channels to ensure a supply of nutrients to the cells of an organ that are not in direct contact with the blood supply.

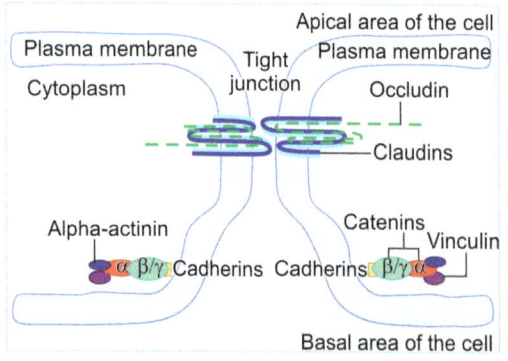

**Fig. 2.3:** Tight junction.

## Cytoskeleton

Human body is supported by the skeletal system; similarly the structure of a cell is maintained by the cytoskeleton present underneath the plasma membrane. The cytoskeleton is responsible for the shape of the cell, its motility, and chromosomal movements during cell division. The cytoskeleton is composed of microfilaments, intermediate filaments, and microtubules, forming a network within the cell. These filaments form rod-like elongated structures, which are stable components of the cytoskeleton.

### Functions of a Cell Membrane

- Plasma membrane acts as a limiting membrane for all the cells. It keeps the integrity of the cell by separating the intracellular medium from the extracellular medium.
- It is a protective envelope for the cell.
- Various enzymes, antigens, and so forth are present in the plasma membrane.
- In muscle cells, the plasma membrane (sarcolemma) is excitable and conducts impulse for muscle contraction.
- The most important function of the plasma membrane is transporting materials across

it. The important transport mechanisms are described in the next section.

## TRANSPORT MECHANISMS

The permeability of substances across the cell membrane is dependent on their solubility in lipids and not on their molecular size. Water-soluble compounds are generally impermeable and require carrier-mediated transport. An important function of the membrane is to withhold unwanted molecules, while permitting entry of molecules necessary for cellular metabolism. Transport mechanisms are classified into:
1. Passive transport
    - Simple diffusion
    - Facilitated diffusion
2. Active transport

### Passive Transport

*Simple Diffusion*

Solutes and gases enter into the cells passively. They are driven by the concentration gradient. The rate of entry is proportional to the solubility of that material in lipids. In this type of solute movement there is no active participation of any membrane proteins.

**Diffusion of electrolytes**

The ions are influenced by the electric field, and the net movement is according to the sign of their charge. One ion attracts other ions of opposite sign around itself and tends to drag them along with itself. So, ions of opposite charge do not move singly but rather in pairs. A faster ion is thus slowed down by its sluggish partner, while its association with the former speeds up the latter.

The anions and cations of electrolyte solution move at different rates. As a result, a potential is set up across the membrane through which these ions diffuse. It can happen when a membrane separates two solutions of the same electrolyte at different concentrations, and also two different electrolytes of the same concentration are kept separate by a membrane. These potentials are called **diffusion potentials**.

**Biochemical importance of diffusion**

- Mechanism of respiratory gas exchange: Oxygenation of venous blood, removal of excess $CO_2$ from blood, and distribution of oxygenated blood to the tissues are governed by simple diffusion.
- Intestinal absorption of pentoses, minerals, and water-soluble vitamins.
- Ions and small molecules pass largely by diffusion through plasma membrane.
- Diffusion of ions across cell membrane influences polarization of membrane and produces membrane potential.

**Factors affecting diffusion**

- Pressure gradient: A gas diffuses down its own pressure gradient.
- Concentration gradient: Diffusion of solute is directly proportional to its concentration gradient.
- Electrical gradient: When a potential difference exists between two sides of the membrane, anions and cations diffuse through electropositive and electronegative sides.
- Temperature: Diffusion of gas or solute is directly proportional to temperature.
- Pore size of the membrane.
- Surface area of the membrane.
- Thickness of the membrane.
- Size and shape of the solute particle.
- Viscosity of the solvent; more viscous solvent resists diffusion.

*Facilitated Diffusion*

Facilitated diffusion occurs with the help of a carrier molecule, usually a membrane protein **(Fig. 2.4)**. Facilitated diffusion is performed through any one of the following ways:
- Carrier molecule may move across the membrane with the substance to release it on the other side and then return to its original position to continue the cycle.

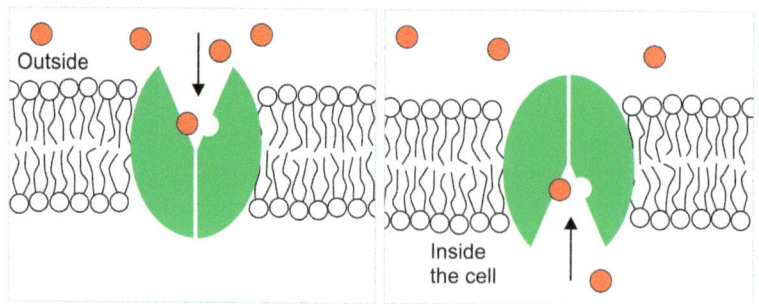

**Fig. 2.4:** Facilitated diffusion. The carrier molecules exist in two conformations.

- A conformational change may take place in the transport protein on binding the solute, which may help in the translocation of the latter to the opposite side.
- The solute may be shuttled from one membrane protein to another till it reaches its destination.

## Active Transport

The salient features of active transport are:
- This form of transport **requires energy** (**Fig. 2.5**). About 40% of the total energy expenditure in a cell is used for the active transport system.
- The active transport is unidirectional.
- It requires specialized integral proteins called **transporters**.
- The transport system is saturated at higher concentrations of solutes.
- The transporters are susceptible to inhibition by specific organic or inorganic compounds.

*Sodium Pump*

It is the best example for active transport. **Cell has low intracellular sodium**; but concentration of potassium inside the cell is very high. This is maintained by the **sodium–potassium activated ATPase**, generally called as sodium pump. The ATPase is an integral protein of the membrane. It has binding sites for ATP and sodium on the inner side, and the potassium-binding site is located outside the membrane (**Fig. 2.6**).

The hydrolysis of one molecule of ATP can result in the expulsion of three Na⁺ ions and influx of two K⁺ ions. The ion transport and ATP hydrolysis are tightly coupled.

**Clinical applications**

Cardiotonic drug digoxin inhibits the sodium–potassium pump. This leads to an increase in Na⁺ level inside the cell and the extrusion of Ca⁺ from the myocardial cell. This enhances the contractility of the cardiac muscle and so improves the function of the heart.

## Uniport, Symport, and Antiport

In general, transport systems are classified into uniport, symport, and antiport systems (**Figs. 2.7A to C**).

**Uniport system (Fig. 2.7A)** carries single solute across the membrane, e.g., **glucose** transporter in most of the cells. Calcium

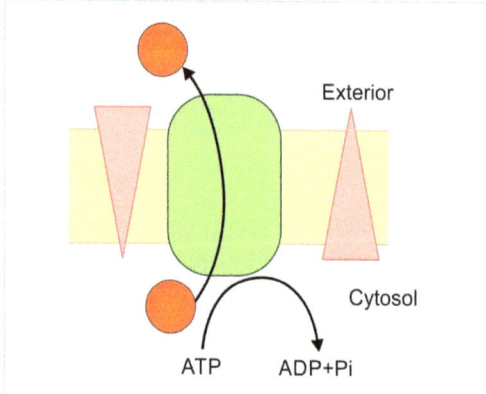

**Fig. 2.5:** Active transport.

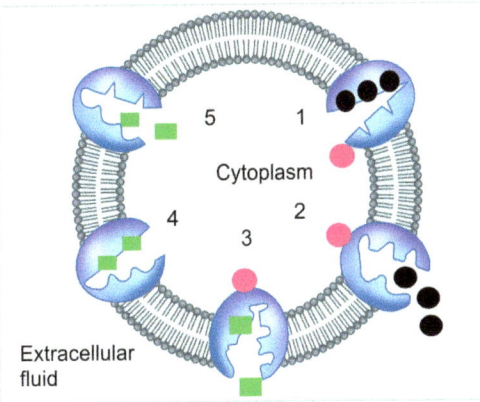

**Fig. 2.6:** The sodium–potassium pump. It brings sodium ions out of the cells and potassium ions into the cells. Black circle = sodium ion; green square = potassium ion; pink circle = phosphate. (1) Cytoplasmic sodium ions (three numbers) bind to the channel protein. This favors phosphorylation of the protein along with hydrolysis of ATP. (2) Phosphorylation causes the protein to change conformation, expelling the sodium ions across the membrane. (3) Simultaneously, extracellular potassium ions (two numbers) move to the carrier protein. Potassium binding leads to the release of phosphate group. (4) So, original conformation is restored. (5) Potassium ions are released into the cytoplasm. The cycle repeats state; the active sites are exposed to the exterior, when the solutes bind to the specific sites.

pump is another example. If the transfer of one molecule depends on simultaneous or sequential transfer of another molecule, it is called the **cotransport** system.

The co-transport system may either be a symport or an antiport. In **symport (Fig. 2.7B)**, the transporter carries two solutes in the same direction across the membrane. Amino acid transport is an example for symport. Sometimes the transport may be coupled with energy indirectly. Here, movement of the substance against a concentration gradient is coupled with movement of a second substance down the concentration gradient, the second molecule being already concentrated within the cell by an energy requiring process. The co-transport of glucose in the intestinal cell is

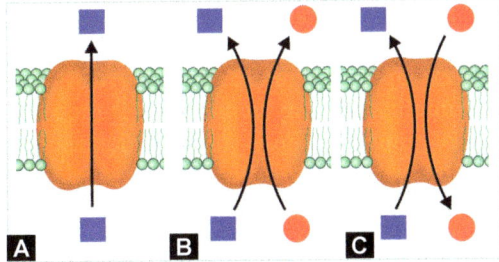

**Figs. 2.7A to C:** (A) Uniport, (B) symport, and (C) antiport transport systems.

coupled with the action of sodium–potassium ATPase to extrude sodium from the cell and take in potassium; this is secondary mediated transport.

The **antiport** system **(Fig. 2.7C)** carries two solutes or ions in the opposite direction, e.g., **sodium pump (Fig. 2.6)**. Transport mechanisms are summarized in **Table 2.4**.

Osmosis, filtration, and microscopy are described in **Chapter 22**.

**Table 2.4:** Summary of transport mechanisms.

| Type | Carrier | Energy required | Examples |
| --- | --- | --- | --- |
| Simple diffusion | Nil | Nil | Water |
| Facilitated diffusion | Yes | Nil | Glucose to RBCs |
| Primary active | Yes | Direct | Sodium pump |
| Secondary active | Yes | Indirect | Glucose to intestine |
| Ion channels | Yes | No | Sodium channel |

(RBCs: red blood cells)

## SUMMARY OF THE CHAPTER

- Marker enzymes are present only in particular organelles and are used to identify these organelles during cell fractionation.
- Nucleus is the storehouse of genetic information containing DNA organized into 23 pairs of chromosomes. All cells in

the body contain nucleus except mature erythrocytes.
- Endoplasmic reticulum, a network of interconnecting membranes, is the site of protein synthesis (rough endoplasmic reticulum).
- Lysosomes are bags of hydrolytic enzymes responsible for autophagy, postmortem autolysis, and phagocytosis.
- Mitochondria, the powerhouse of the cell, has its own DNA-encoded mitochondrial proteins and has a role in triggering apoptosis.
- Plasma membrane is mainly made up of phospholipid bilayers, interspersed with proteins and carbohydrate residues attached to proteins and lipids.
- Cholesterol is also a component of animal cell membranes.
- Phospholipid bilayers are oriented in such a way that there is a hydrophobic core with hydrophilic interior (cytoplasmic side) and outer layer (extracellular side).
- Transmembrane proteins serve as receptors, tissue specific antigens, ion channels, and so forth.
- Membrane structure is described as the fluid mosaic model.
- Transport of molecules across the plasma membrane may be energy dependent (active) or energy independent (passive).
- Active transport involves expenditure of energy and occurs against a concentration gradient.
- $Na^+/K^+$ ATPase (sodium pump) is an example of active transport. Cardiotonic drugs like digoxin and ouabain competitively inhibit $K^+$ ion binding.
- Glucose transporters are examples of uniport transport by facilitated diffusion.

## QUESTIONS FOR SELF-ASSESSMENT

- Briefly describe the structure of a cell with all its components.
- Add a note on endoplasmic reticulum and mention its functions.
- Describe the structural aspects of mitochondria and mention its major functions.
- Briefly describe the different types of transport occurring inside the cell.
- Schematically describe the functions of different cell organelles.
- Highlight the importance of endocytosis and exocytosis.
- Describe what are the different transport mechanisms occurring inside the cell.
- Add a note on fluid mosaic model of cell membrane with a neat diagram.
- Mention the clinical importance of $Na^+/K^+$-ATPase (sodium pump).
- Explain with a diagram uniport, antiport, and symport transport mechanisms with examples.

## MULTIPLE CHOICE QUESTIONS (MCQs)

**2-1.** Mention the incorrect statement regarding mitochondria:
   a. Glycolysis occurs here
   b. ETC takes place in mitochondria
   c. It is considered as the power house of the cell
   d. Oxidation of pyruvate takes place here

**2-2.** The main function of the Golgi apparatus is:
   a. RNA synthesis
   b. DNA synthesis
   c. Protein synthesis
   d. Production of ATP

**2-3.** Lysosomes are known as "suicidal bags" because of the:
   a. Parasitic activity
   b. Presence of food vacuole
   c. Hydrolytic activity
   d. Catalytic activity

**2-4.** Which cell organelle is involved in apoptosis?
   a. Lysosome
   b. Endoplasmic reticulum
   c. Golgi body
   d. Mitochondria

2-5. **The resting potential membrane is determined by:**
 a. Potassium-ion gradient
 b. Sodium-ion gradient
 c. Bicarbonate-ion gradient
 d. None

2-6. **The oxygen and carbon dioxide cross the plasma membrane by the process of:**
 a. Active diffusion
 b. Facilitated diffusion
 c. Passive diffusion
 d. Random diffusion

2-7. **The fluidity of the plasma membrane increases with:**
 a. Increase in unsaturated fatty acids in the membrane
 b. Increase in saturated fatty acids in the membrane
 c. Increase in glycolipid content in the membrane
 d. Increase in phospholipid content in the membrane

2-8. **The term cell is not applied to:**
 a. Algae
 b. Bacteria
 c. Virus
 d. Fungi

2-9. **Distribution of intrinsic proteins in the plasma membrane is:**
 a. Random
 b. Symmetrical
 c. Asymmetrical
 d. Circular

## ANSWER KEYS TO MCQs

2.1 (a)  2.2 (c)  2.3 (c)  2.4 (d)  2.5 (a)  2.6 (c)  2.7 (a)  2.8 (c)  2.9 (c)

# CHAPTER 3

# Enzymes

At the completion of this chapter, the reader will be able to answer questions on the following topics:

- Apoenzyme, coenzyme, and holoenzyme
- Classification of enzymes
- Factors influencing enzyme activity
- Michaelis constant
- Effect of temperature and pH
- Michaelis–Menten theory
- Fischer's template theory
- Koshland's induced fit theory
- Competitive and noncompetitive inhibitions
- Allosteric inhibition
- Therapeutic and diagnostic enzymes
- Cardiac and liver enzymes
- Isoenzymes
- Zymogens

Enzymes may be defined as biocatalysts synthesized by living cells. They are proteins, with high molecular weight, colloidal, thermolabile in character and specific in their action. Several enzymes occur in the form of multienzyme complexes.

## COENZYMES

Most of the enzymes are simple proteins; while others require the presence of certain additional organic substances called coenzymes (**Table 3.1**). Protein part of an enzyme is called apoenzyme. When apoenzyme and coenzyme combine, it is called holoenzyme. The nonprotein part of the holoenzyme is known as the prosthetic group.

Apoenzyme (protein) + coenzyme (prosthetic group) = holoenzyme

If the prosthetic group is removed, the biological activity of the enzyme is lost. The prosthetic group may be an organic compound (coenzyme) or inorganic metal ions (cofactor). Some enzymes require the presence of certain metallic ions such as $Mg^{2+}$, $Mn^{2+}$, $Ca^{2+}$, etc. for their activity (**Table 3.2**). They are called cofactors.

**Table 3.1:** Examples of coenzymes.

| Coenzyme | Group transferred |
| --- | --- |
| Thiamine pyrophosphate (TPP) | Hydroxyethyl |
| Pyridoxal phosphate (PLP) | Amino group |
| Biotin | Carbon dioxide |
| Coenzyme-A (Co-A) | Acyl groups |
| Tetrahydrofolate (FH4) | One-carbon groups |
| Adenosine triphosphate (ATP) | Phosphate |

## Salient Features of Coenzymes

- The protein part of the enzyme gives the necessary three-dimensional infrastructure for chemical reaction, but the group is transferred from or accepted by the coenzyme.

**Table 3.2:** Enzymes containing metals.

| Metal | Enzyme-containing the metal |
|---|---|
| Zinc | Carbonic anhydrase, carboxypeptidase |
| Magnesium | Hexokinase, hosphofructokinase, enolase |
| Manganese | Phosphoglucomutase, hexokinase, enolase |
| Copper | Tyrosinase, cytochrome oxidase |
| Iron | Cytochrome oxidase, catalase, xanthine oxidase |
| Calcium | Lecithinase, lipase |
| Molybdenum | Xanthine oxidase |

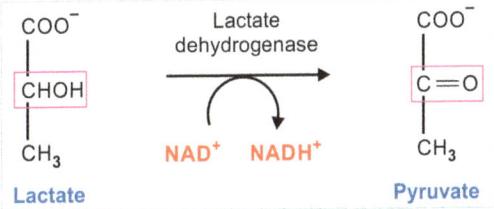

**Fig. 3.1:** Reaction of lactate dehydrogenase.

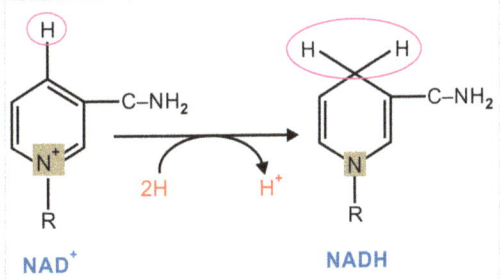

**Fig. 3.2:** Nicotinamide adenine dinucleotide (NAD$^+$) accepts hydrogen.

- Coenzyme is essential for the biological activity of the enzyme.
- Coenzyme is a low-molecular-weight organic substance and is heat stable.
- Generally, the coenzymes combine loosely with the enzyme molecules. The enzyme and coenzyme can be easily separated by dialysis.
- Inside the body, when the reaction is completed, the coenzyme is released from the apoenzyme, which can bind to another enzyme molecule.
- One molecule of a coenzyme is able to convert a large number of substrate molecules with the help of enzymes.
- Most of the coenzymes are derivatives of vitamin B complex group of substances.

## Nicotinamide Adenine Dinucleotide

This is a coenzyme synthesized from nicotinamide, a member of vitamin B complex. The structure of nicotinamide adenine dinucleotide (NAD$^+$) is nicotinamide–Ribose-P-P-Ribose-Adenine (Chapter 16). The reversible reaction of lactate to pyruvate is catalyzed by the enzyme lactate dehydrogenase (LDH), but the actual transfer of hydrogen is taking place on the coenzyme NAD$^+$ **(Fig. 3.1)**. In this case, two hydrogen atoms are removed from lactate, of which one hydrogen and two electrons are accepted by the NAD$^+$ to form NADH, and the remaining H$^+$ is released into the surrounding medium. The hydrogen is accepted by the **nicotinamide** group as shown in **Figure 3.2**.

## CLASSIFICATION OF ENZYMES

Earlier, enzymes were classified on the basis of the substrate on which they are acting. But there may be more than one enzyme acting on the same substrate. The International Union of Biochemistry and Molecular Biology (IUBMB) suggested a new system of enzyme classification. It is a complex and difficult method but very accurate. Per the system, the name starts with enzyme class (EC) followed by four digits. The first digit represents the class. The second digit stands for subclass. The third digit represents subgroup and the fourth indicates the number of that particular enzyme in the list of enzymes. Enzymes are classified into the following six major classes **(Table 3.3)**.

**Table 3.3:** Classification of enzymes.

| Class | Name | Function | Example |
|---|---|---|---|
| Class 1 | Oxidoreductases | Transfer of hydrogen | Lactate dehydrogenase |
| Class 2 | Transferases | Transfer of groups other than hydrogen | Aminotransferases (subclass: kinase, transfer of phosphoryl group from ATP; e.g., hexokinase) |
| Class 3 | Hydrolases | Hydrolysis of bond, cleave bond, and add water | Acetylcholinesterase |
| Class 4 | Lyases | Cleave without adding water | Aldolase (subclass: hydratase; add water to double bond) |
| Class 5 | Isomerases | Intramolecular transfers | This class includes racemases and epimerases, e.g., triose phosphate isomerase |
| Class 6 | Ligases | ATP-dependent condensation of two molecules | Acetyl CoA carboxylase |

## Class 1: Oxidoreductases

**This group of enzymes will catalyze the oxidation of one substrate with the simultaneous reduction of another substrate or coenzyme.**

For example:

Alcohol + NAD $\longrightarrow$ aldehyde + NADH + H$^+$

The enzyme is alcohol dehydrogenase.

Glucose-6-phosphate + NADP $\rightarrow$ 6-phosphogluconolactone + NADPH + H$^+$

Enzyme is glucose-6-phosphate dehydrogenase (GPD)

## Class 2: Transferases

This class of enzyme transfer any group, other than hydrogen from one substrate to another substrate.

Hexose + ATP $\longrightarrow$ Hexose-6-phosphate + ADP

Enzyme: Hexokinase.

## Class 3: Hydrolases

This class of enzymes can hydrolyze ester, ether, peptides, or glycoside bond by adding H$_2$O.

For example, Acetylcholine + H$_2$O $\longrightarrow$ Choline + Acetate

Enzyme is acetylcholinesterase

## Class 4: Lyases

These enzymes can remove group from substrates or break bonds by mechanism other than hydrolysis.

For example, Fructose-1,6-diphosphate $\longrightarrow$ Glyceraldehyde-3-phosphate + dihydroxyacetone phosphate

Enzyme: Aldolase.

## Class 5: Isomerases

These enzymes can produce optical, geometric, or positional isomerism of substrate. For example, racemases, epimerase, and cis-trans isomerase.

For example, Glucose-6-phosphate $\xrightarrow{\text{Phosphohexose isomerase}}$ Fructose-6-phosphate

## Class 6: Ligases

These enzymes link two substrates together usually with the simultaneous hydrolysis of ATP.

Acetyl Co-A + CO$_2$ + ATP $\longrightarrow$ Malonyl Co-A + ADP + Pi (inorganic phosphate)

Enzyme: Acetyl-CoA carboxylase.

## FACTORS INFLUENCING ENZYME ACTIVITY

Contact between enzyme and substrate is the most essential prerequisite for enzyme activity.

The most important factors that influence the enzyme activity are the following.

## Concentration of Enzymes

As the concentration of the enzyme increases, the velocity of the reaction also increases proportionately. The concentration of the serum enzyme measured through its activity is used for the diagnosis of diseases. If the enzyme activity is plotted against enzyme concentration, a graph as shown in **Figure 3.3** will be obtained.

## Substrate Concentration

If the velocity is plotted against the substrate concentration, a typical curve **(Fig. 3.4)** will be obtained. As substrate concentration is increased, the velocity is also correspondingly increased in the initial phases but the curve flattens afterward. Increase in the substrate

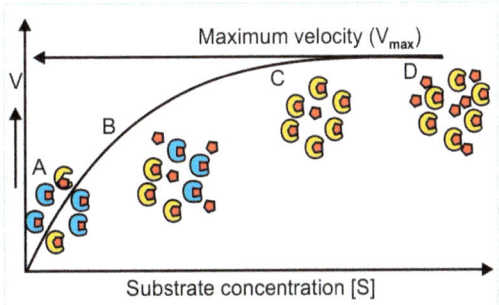

Fig. 3.5: Effect of substrate concentration on enzyme activity. Enzyme molecules are shown as half-circles. Substrate molecules are red dots. (A) Substrate molecules are low; so only a few enzyme molecules are working and velocity is less; (B) At half-maximal velocity ($K_m$), 50% enzyme molecules are bound with substrate; (C) As a lot of substrate molecules are available, all enzyme molecules are bound; (D) Further increase in the substrate will not increase the velocity further.

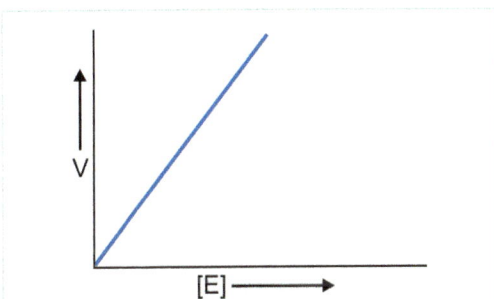

Fig. 3.3: As the concentration of enzyme increases, the enzyme activity also increases.

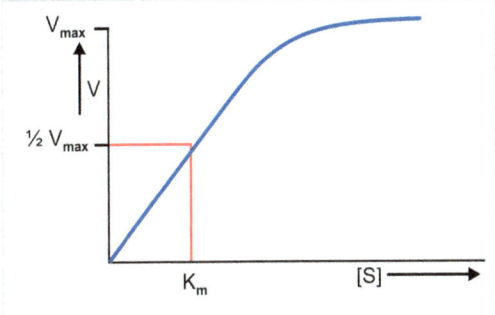

Fig. 3.4: Effect of substrate concentration on enzyme velocity (substrate saturation curve).

concentration gradually increases the velocity of the enzyme reaction within the limited range of substrate levels and then it will become second level (velocity will not change). This is explained in **Figure 3.5**. At lower concentrations of substrate (point A in the curve), some enzyme molecules remain idle. As the substrate increases, more and more enzyme molecules work. At half-maximal velocity, 50% enzymes attach with the substrate (point B in the curve). As more substrate is added, all enzyme molecules become saturated (point C). Further increase in substrate cannot cause any effect in the reaction velocity (point D). **The maximum velocity obtained is called $V_{max}$ (Fig. 3.5).** It represents the maximum reaction rate attainable in the presence of excess substrate (at **substrate saturation level**).

### Michaelis Constant ($K_m$ value)

The substrate concentration at half the maximum velocity ($\frac{1}{2}V_{max}$) is known as **$K_m$ value** or Michaelis constant **(Fig. 3.4)**. The maximum velocity obtained is called $V_{max}$ **(Fig. 3.4)**.

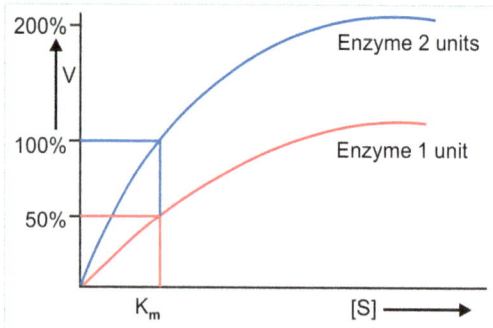

**Fig. 3.6:** Effect of enzyme concentration on $K_m$.

## Salient features of $K_m$

- $K_m$ value is the substrate concentration (expressed in moles/L) at half-maximal velocity.
- It denotes that 50% of enzyme molecules are bound with substrate molecules at that particular substrate concentration.
- $K_m$ is independent of enzyme concentration. If the enzyme concentration is doubled, the $V_{max}$ will be double. But $K_m$ will remain exactly the same **(Fig. 3.6)**. In other words, irrespective of enzyme concentration, 50% molecules are bound to substrate at that particular substrate concentration.
- $K_m$ is the signature of the enzyme. The $K_m$ value is thus a constant for an enzyme. It is the characteristic feature of a particular enzyme for a specific substrate.
- $K_m$ denotes the affinity of enzyme for substrate. The *lesser the numerical value of $K_m$, the affinity of the enzyme for the substrate is more.*

## Effect of Temperature

Velocity of an enzyme reaction increases with increase in temperature to a maximum and then declines. A bell-shaped curve is usually obtained **(Fig. 3.7)**. The optimum temperature for most of the enzymes is around 40°C. In general, when the enzymes are exposed to temperature of above 50°C, denaturation leading to derangement in the structure of proteins will take place. Most of the enzymes

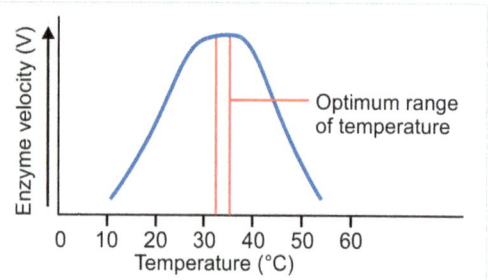

**Fig. 3.7:** Effect of temperature on velocity.

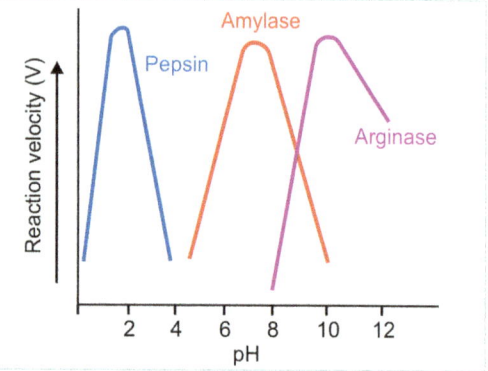

**Fig. 3.8:** Effect of pH on enzyme velocity.

become inactive above 70°C. But a few enzymes, such as muscle adenylcyclase and enzymes present in bacteria of thermal vents have an optimum temperature of about 100°C.

## Effect of pH

Increase in H$^+$ concentration considerably influences the enzyme activity and a bell-shaped curve is normally obtained **(Fig. 3.8)**. Each enzyme has got an optimum pH at which the velocity is maximum. Above or below this pH, the enzyme activity is much less; and at extreme pH, enzymes become totally inactive. Most of the enzymes of higher organisms show optimum activity around neutral pH. But there are many exceptions. Pepsin optimum pH is 1–2, acid phosphatase (ACP) pH 4–5, and alkaline phosphatase (ALP) pH 10–11 (liver and bone). Hydrogen ions influence the enzyme activity by altering the ion concentration on the amino acid.

### Effect of Product Concentration

The accumulation of products generally decreases the enzyme activity. For certain enzymes, the product combines with the active site of enzyme and forms a loose complex which inhibits the enzyme activity.

### Effect of Activators

Some enzymes require certain inorganic metallic cations, such as $Mg^{++}$, $Mn^{++}$, and $Zn^{++}$ are necessary for their optimum activity **(Table 3.2)**. Another type of activation is the conversion of an inactive **proenzyme** or **zymogen** to the active enzyme. By splitting a single peptide bond, and removal of a small polypeptide from **trypsinogen**, the active trypsin is formed. This results in unmasking of the active center (see zymogens described below).

**Coagulation factors** are seen in blood as zymogen forms. Their activities are needed only occasionally; but when needed, a large number of molecules are to be produced instantaneously. Hence, the cascade system of **chemical amplification** of such factors.

### Effect of Inhibitors

Please see the section of enzyme inhibition given below.

### Effect of Time

Under optimum conditions, the time needed for enzyme reaction is short. Variation in the time of reactions are generally related to change in pH and temperature.

## MECHANISMS OF ENZYME REACTION

Enzymes are biocatalysts. The catalysis taking place in biological system is similar to chemical reactions taking place in vitro (outside the biological system). The reactants have to be elevated to a particular activation stage or transition state. The following theories have been put forward to explain the mechanism of enzyme action.

### Lowering of Activation Energy

Energy required for the reactants to undergo the reaction is known as activation energy. When the reactants are heated, they attain the activation energy. The enzymes reduce the activation energy in the biological systems and this will result in the reaction at the lower energy. The enzymes cannot alter the equilibrium constant but only enhance the velocity of the reactions **(Fig. 3.9)**.

### Michaelis–Menten Theory

It is otherwise called the enzyme–substrate (ES) complex theory. Enzyme must combine with the substrate to form an ES complex, which results in the formation of products **(Fig. 3.10)**.

$$E + S \rightarrow [E\,S] \rightarrow E + P$$

This theory is more accepted than the other theories.

### Fischer's Template Theory (Lock and Key Model)

According to this theory, the conformation of the enzyme is rigid. The substrate fits into the active site, just as a key fits into a lock **(Fig. 3.11)**. The active site of the enzyme,

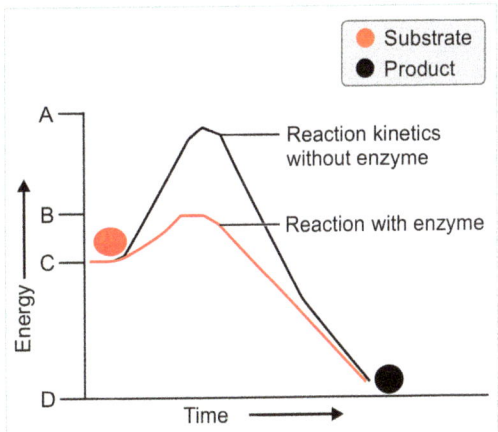

**Fig. 3.9:** Lowering of activation energy by enzymes. (C: Energy level of substrate; D: Energy level of product; C to A: Activation energy in the absence of enzyme; C to B: Activation energy in presence of enzyme; B to A: Lowering of activation energy by enzyme).

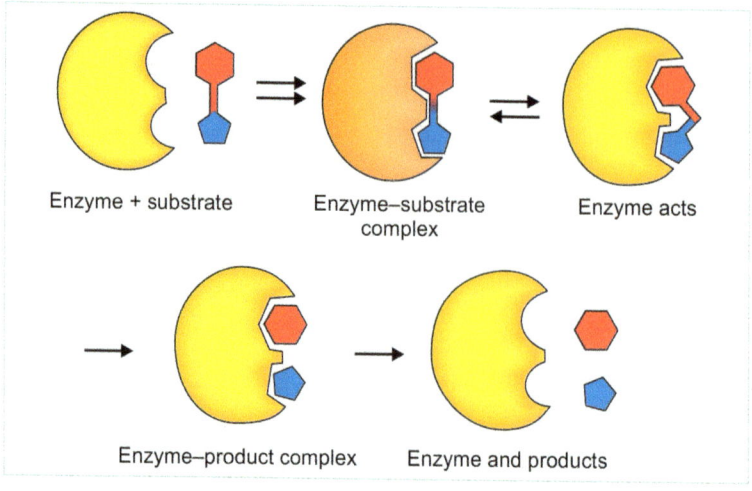

**Fig. 3.10:** Enzyme–substrate complex.

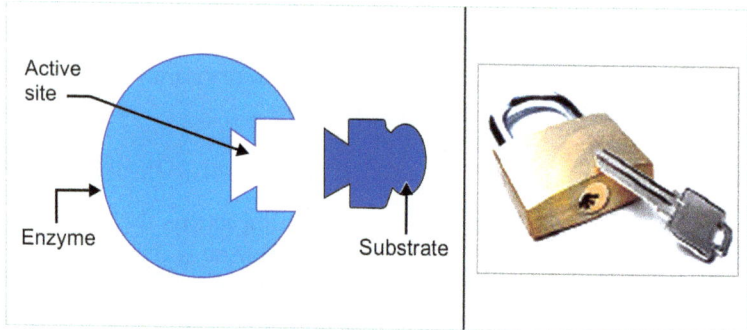

**Fig. 3.11:** Enzyme and substrate are specific to each other. This is similar to key and lock (Fischer's theory).

according this theory is a rigid and preshaped template, where only a specific substrate can bind. The defect of this theory is that it cannot give any explanation for the flexible model of the enzyme and cannot give explanation for the factors influencing enzyme action, especially for allosteric inhibition.

### Koshland's Induced Fit Model Theory

According to this theory, the active site is not rigid. The interaction of the substrate with the enzyme induces a conformational change in the enzyme, resulting in the formation of a strong substrate binding site. Due to the induced fit, the amino acids of the enzymes rearrange to form the active site and catalysis takes place **(Figs. 3.12A to D)**.

### ACTIVE SITE OR ACTIVE CENTER OF ENZYME

Catalysis occurs at the active center or active site. It is the area where substrate binding occurs. Salient features of the active centers are as follows:
- The region of the enzyme where substrate binding and catalysis occurs is referred to as active site or active center **(Figs. 3.13A and B)**.
- Although all parts are required for keeping the exact three-dimensional structure of the enzyme, the reaction is taking place at the active site. The active site occupies only a small portion of the whole enzyme.
- Generally active site is situated in a crevice or cleft of the enzyme molecule **(Figs. 3.13A and B)**. To the active site, the specific

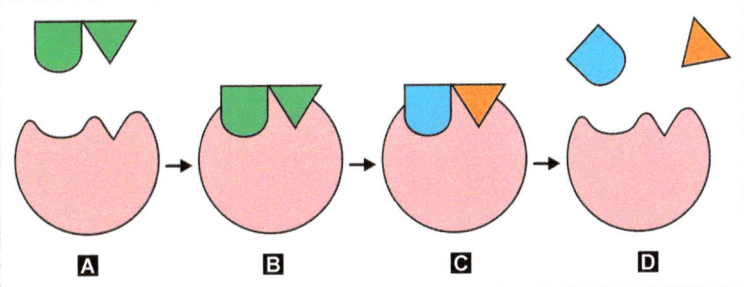

**Figs. 3.12A to D:** Koshland's induced fit theory. (A) Enzyme has shallow grooves; substrate alignment is not correct; (B) Fixing of substrate induces structural changes in enzyme; (C) Now substrate correctly fits into the active site of enzyme; (D) Substrate is cleaved into two products.

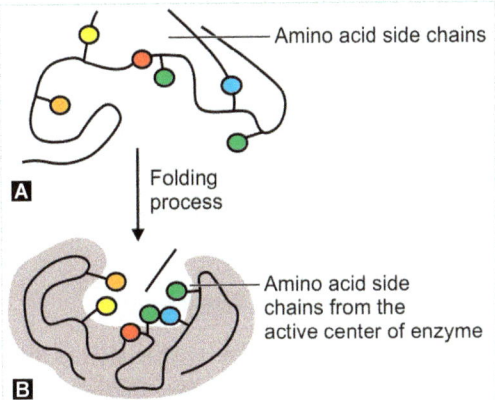

**Figs. 3.13A and B:** Correct alignment of amino acids in the active center of the enzyme. (A) Unfolded protein with no activity; (B) Folded protein with enzyme activity.

substrate is bound. The binding of the substrate to active site depends on the alignment of specific groups or atoms at the active site.
- During the binding, these groups may realign themselves to provide the unique conformational orientation, so as to promote exact fitting of the substrate to the active site **(Figs. 3.13A and B)**.
- The amino acids or groups that directly participate in making or breaking the bonds (present at the active site) are called catalytic residues or catalytic groups.

## ENZYME INHIBITION

All the reactions catalyzed by enzymes in the body are well controlled. Most of the enzymes in the metabolic pathways will be in abundance except one or two enzymes. Control of the whole pathway is achieved by inhibition of such **key enzymes** or **regulatory enzymes**. Different types of regulatory process are controlling these regulatory enzymes.

### Competitive Inhibition (Reversible Inhibition)

Here the inhibitor molecules are competing with the normal substrate molecules for attaching with the active site of the enzyme.

$$E + S \rightarrow [ES] \rightarrow E + P$$
$$E + I \rightarrow [EI] \rightarrow \text{No products}$$

Since the inhibitor can react with the enzyme to form enzyme-inhibitor complex, the number of enzymes available for (ES) complex formation is reduced. Since the effective concentration of the enzyme is reduced, the velocity of the reaction is less

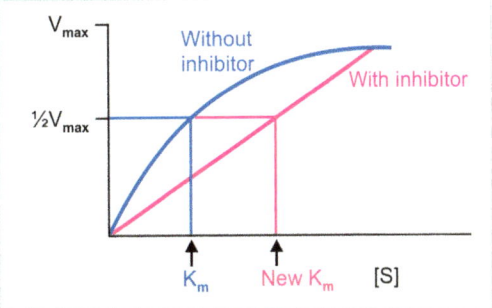

**Fig. 3.14:** Competitive Inhibition. Substrate saturation curve in the presence and absence of competitive inhibitor.

(**Fig. 3.14**). In this type of inhibition, the inhibitor will be a structural analog of the substrate. Competitive inhibition is usually reversible. Excess amount of substrate can abolish the inhibition.

Pharmacological actions of many drugs can be explained by competitive inhibition. Examples are: (a) sulfonamides employed as antibacterial agents function by competing with para-aminobenzoic acid (PABA), which is essential for the synthesis of folic acid; (b) methotrexate is a structural analog of folic acid and is used as an anticancer drug; (c) dicoumarol, a structural analog of vitamin K, is used as an anticoagulant; (d) isonicotinic acid hydrazide (INH), a structural analog to pyridoxine, functions as an anti-TB drug.

## Noncompetitive or Irreversible Inhibition

A variety of poisons, such as iodoacetate, heavy metallic ions (silver, mercury), and oxidizing agents acts as irreversible inhibitors. No competition is observed between the inhibitor and substrate. The inhibitor usually binds to a different site on the enzyme other than the binding site. The inhibitors will not have any structural resemblance to the substrate. A high concentration of the substrate normally cannot relieve the inhibition (**Fig. 3.15**).

Examples of noncompetitive inhibitions are: (a) cyanide inhibits the respiratory chain through cytochrome oxidase, (b) fluoride will remove $Mg^{2+}$ and $Mn^{2+}$ and inhibit the enzyme enolase in the glycolysis and hence used for the collection of blood samples for glucose estimation, (c) the British Anti-Lewisite (BAL) is an antidote for lead (Pb) poisoning. A comparison of competitive and noncompetitive inhibition is shown in **Table 3.4**.

**Table 3.4:** Comparison of the two types of inhibition.

|  | Competitive inhibition | Noncompetitive inhibition |
| --- | --- | --- |
| Structure of inhibitor | Substrate analogue | Unrelated molecule |
| Inhibition | Reversible | No effect |
| Excess substrate | Relieve inhibition | No effect |
| $K_m$ value | Increased | No change |
| $V_{max}$ | No change | Decreased |
| Significance | Drug action | Toxicity |

## Reversible Noncompetitive Inhibition

Some noncompetitive inhibitions are reversible. For example, trypsin inhibitors from soybean or ascaris will attach to trypsin and inactivate it. When the inhibitor is removed, trypsin regains its action.

## Uncompetitive Inhibition

The inhibitor has got no affinity for the free enzyme. But binds with the (ES) complex. For example, placental ALP is inhibited by phenylalanine.

## Suicidal Inhibition

Here the inhibitor, which is a structural analogue of the inhibitor, is converted to a more effective inhibitor with the help of the enzyme to be inhibited. This type of inhibition is also known as **mechanism-based inactivation**. Allopurinol, which is used as an inhibitor of uric acid synthesis, is oxidized by xanthine oxidase to alloxanthine which is a strong inhibitor of xanthine oxidase. Another

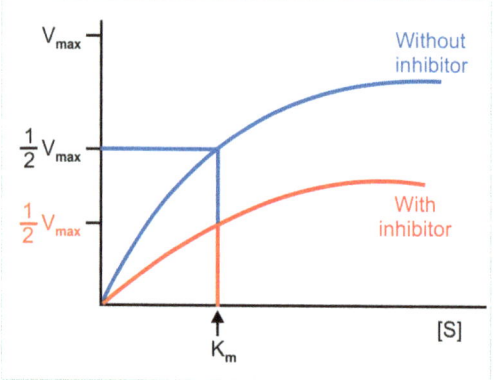

**Fig. 3.15:** Saturation curve of noncompetitive inhibition.

example is the anti-inflammatory action of **aspirin**. Arachidonic acid is converted to prostaglandin by the enzyme **cyclooxygenase**. Aspirin inhibits cyclooxygenase, thus prostaglandin synthesis is inhibited, thereby subsiding inflammation.

## Feedback Inhibition or End-product Inhibition

Here the enzyme activity is inhibited by the end-product. For example, in cholesterol biosynthesis the end product, i.e., cholesterol inhibits the key enzyme 3-hydroxy-3-methyl-glutaryl (HMG)-CoA reductase.

## Allosteric Inhibition

Allosteric enzymes have two binding sites: one for the substrate and the other (allosteric) for the modifier **(Figs. 3.16A and B)**. The binding of the modifier can activate the enzyme (allosteric activation) or inhibit the enzyme (allosteric inhibition). The binding of the modifier to the allosteric site modifies the binding sites to increase or decrease the activity of the enzyme as the case may be. Salient features of allosteric mechanism are as follows:
- The inhibitor is **not** a substrate analog.
- It is partially reversible, when excess substrate is added.

**Table 3.5:** Examples of allosteric enzymes.

| Enzyme | Allosteric inhibitor |
|---|---|
| 1. ALA synthase | Heme |
| 2. HMG-CoA reductase | Cholesterol |
| 3. Phosphofructokinase | ATP, citrate |
| 4. Acetyl-CoA carboxylase | Acyl-CoA |
| 5. Citrate synthase | ATP |

(ALA: aminolevulinic acid; HMG-CoA: β-hydroxy β-methylglutaryl-coenzyme A; ATP: adenosine triphosphate)

- When an inhibitor binds to the allosteric site, the configuration of catalytic site is modified, such that substrate cannot bind properly.

A few examples of allosteric inhibitors are shown in **Table 3.5**.

## ZYMOGENS

Zymogens are the inactive forms of enzymes. The active form is also known as zymase. Digestive enzymes are usually secreted in the inactive forms which are activated by H$^+$ ions or the active form of the enzyme. For example,

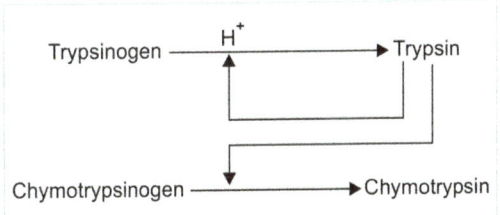

All the gastrointestinal enzymes are synthesized in the form of proenzymes, and only after secretion into the alimentary canal, they are activated. This prevents autolysis of the cellular structural proteins. Similarly coagulation factors are also present in the blood in the inactive form.

## Key Enzymes

Body uses allosteric enzymes for regulating the metabolic pathways. Such a **regulatory**

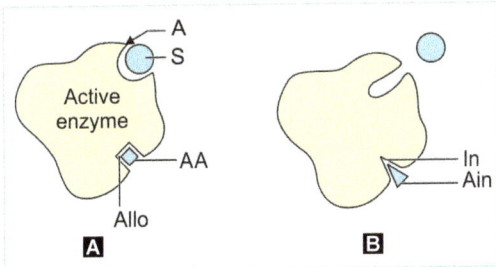

**Figs. 3.16A and B:** Action of allosteric enzymes. A = Active site; Allo = Allosteric site; S = Substrate; AA = Allosteric activator; In = Inhibitor site; Ain = Allosteric inhibitor. The enzyme has separate active site and allosteric site. (A) Depicts that the activator is fixed at the allosteric site, when active site has correct conformation, and the substrate is correctly fixed; (B) Shows that the inhibitor is fixed at the allosteric site when active site is deformed and the substrate could not fix.

**enzyme** in a particular pathway is called the **key enzyme** or **rate limiting** enzyme. The flow of the whole pathway is constrained as if there is a bottleneck at the level of the key enzyme. A typical example is the regulation of phosphofructokinase, the rate limiting step of glycolysis. Here the allosteric inhibitors are ATP and citrate. The whole glycolytic pathway is regulated by the phosphofructokinase reaction.

## Therapeutic Enzymes

Enzymes used as drugs are called therapeutic enzymes (**Table 3.6**). Streptokinase and urokinase are extremely useful as lifesaving drugs in acute myocardial infarction. They can lyse intravascular blood clots in the coronary arteries.

Similarly asparaginase is used as an anticancer drug to treat acute lymphoblastic leukemia. This enzyme hydrolyzes asparagine, a metabolite for leukemic cells, thereby inhibiting the proliferation of the cells.

Digestive enzymes, such as trypsin, chymotrypsin, etc., can be used for the treatment of indigestion and pancreatic insufficiency

## Diagnostic Enzymes

Functional enzymes are those enzymes which are secreted from tissue such as liver and kidney in significant amount into blood, where they catalyze reactions. Nonfunctional enzymes come into blood from different tissues in small amount due to normal cell death or damage. But the serum level of such nonfunctional enzyme in the blood will increase due to the secretion of the enzyme from damaged tissues in significantly higher amounts. Hence, the measurement of the level of such enzyme may be used to study the extend of tissue damage. These enzymes will function not only as diagnostic marker but also as prognostic markers.

## Biomarker

A biomarker is defined as one naturally occurring molecule by which a particular pathological process can be identified. In simple terms, a biomarker is a substance that can be used as an indicator of a particular disease state.

## Cardiac Biomarkers

Cardiac biomarkers are used to detect cardiac diseases. Commonly used biomarkers for early detection of acute myocardial infarction are: cardiac troponins (TnI and TnT) and creatine kinase isoenzyme [creatine kinase myocardial band (CK-MB)]. The brain natriuretic peptide (BNP) is a useful indicator of congestive cardiac failure.

## Cardiac Enzyme Markers

### Creatine Kinase

The level of this enzyme rises in the serum in myocardial infarction and also in muscular dystrophy. The normal value is 15–100 IU/L. In myocardial infarction, the value will rise within 3–6 hours. The isoenzymes of CPK, namely, CK-MB (heart) and CK-MM (muscle) will help in differential diagnosis.

### Lactate Dehydrogenase

It is an enzyme with four subunits and five isoenzymes. The normal value ranges from 100 to 200 IU/L. The value increases in many conditions including myocardial infarction, hepatocellular diseases, hemolytic anemia,

**Table 3.6:** Therapeutic use of enzymes.

| Enzyme | Therapeutic application |
|---|---|
| Asparaginase | Acute lymphoid leukemia |
| Streptokinase | To lyze intravascular clot |
| Urokinase | To lyze intravascular clot |
| Hyaluronidase | Enhances local anesthetics |
| Pancreatin | Pancreatic insufficiency |
| Papain | Anti-inflammatory |

muscular dystrophy, carcinomas, leukemia, and in any necrotic conditions. Nowadays the use of LDH as a cardiac marker is reduced because it is nonspecific and increases in various conditions. See also isoenzymes given below.

## Liver Enzymes

### Aspartate Aminotransferase

Aspartate aminotransferase (AST) or glutamate–oxaloacetate transaminase (GOT) increases in mycardian infarction. Normal value 10–20 U/L

### Alanine Aminotransferase

Alanine aminotransferase (ALT) or glutamate–pyruvate transaminase (GPT) levels in serum increases in viral and toxic hepatitis. The normal value ranges from 10 to 30 IU/L

### Alkaline Phosphatase

Level increases in all types of hepatitis, hepatocellular carcinoma, Hodgkin's lymphoma, and congestive cardiac failure. Very high levels are also encountered in bone diseases especially in carcinomas. The isoenzyme of ALP especially $\alpha$-2 heat-stable ALP known as Regan isoenzyme is of diagnostic and prognostic importance in the carcinoma of liver.

### Gamma Glutamyl Transferase

The normal serum levels vary from 10 to 30 IU/L. The serum levels of GGT increase in alcoholic cirrhosis, obstructive jaundice, and liver cancer. This enzyme is of help in detecting alcohol abuse.

## Other Diagnostic Enzymes

### Acid Phosphatase

Acid phosphatase is seen in prostate cells, red blood cells (RBCs), white blood cells (WBCs), and platelets. It is an important tumor marker as the serum level is found to be increased in prostate cancer especially in bone metastasis arising from prostatic cancer. The tartrate labile isoenzyme is of importance in the diagnosis and prognosis of prostate cancer.

### Glucose-6-Phosphate Dehydrogenase

Glucose-6-phosphate dehydrogenase value is reduced in drug-induced hemolytic anemia. The deficiency of this enzyme will result in decreased production of NADPH and hence RBC destruction.

### Amylase

Amylase is a pancreatic enzyme whose the level increases many fold in acute pancreatitis. The normal value is 50–120 IU/L. The urinary amylase value also increases in this condition from the normal level of 350–400 IU/L.

### Lipase

Lipase is another pancreatic enzyme, which is elevated in acute pancreatitis. The normal level of this enzyme in serum is 0.5–1.5 IU/L.

### Cholinesterase

Cholinesterase (ChE) is present in nerve endings and RBCs (normal levels 7–20 IU/L). The pseudo-ChE level in serum is reduced in viral hepatitis, cirrhosis, hepatocellular carcinoma, and metastatic cancer of liver and malnutrition. **Table 3.7** shows a summary of the clinically useful enzymes.

## ISOENZYMES

Isoenzymes are different forms of the same enzyme, which catalyze the same reaction but differ in physical properties such as optimum pH, isoelectric point, substrate affinity, electrophoretic mobility, etc., and often present in different tissues of the body and are usually derived from different genes. Examples of isoenzymes areas listed below.

## Table 3.7: Enzyme profiles in diseases.

| | | |
|---|---|---|
| I. Hepatic diseases | 1. Alanine aminotransferase (ALT)<br>2. Aspartate aminotransferase (AST)<br>3. Alkaline phosphatase (ALP)<br>4. Gamma glutamyl transferase (GGT) | Marked increase in parenchymal liver diseases<br>Elevated in parenchymal liver disease<br>Marked increase in obstructive liver disease<br>Increase in obstructive and alcoholic liver |
| II. Myocardial infarction | 1. Creatine kinase (CK-MB isoenzyme)<br>2. Cardiac troponin | CK-MB isoenzyme is specific<br>It is a better marker for myocardial infarction |
| III. Muscle diseases | 1. Creatine kinase (CK-MM isoenzyme)<br>2. Aspartate aminotransferase (AST) | CK-MM isoenzyme is elevated<br>Increase in muscle disease, not specific |
| IV. Bone diseases | 1. Alkaline phosphatase (ALP) | Marked elevation in rickets and Paget's disease<br>Heat-labile bone isoenzyme is elevated (BAP) |
| V. Prostate cancer | 1. Prostate-specific antigen (PSA)<br>2. Acid phosphatase (ACP) | Marker for prostate cancer. Metastatic bone disease especially from a primary prostate. Prostate fraction is inhibited by L tartrate |

## Creatine Kinase

This enzyme catalyzes the breakdown of creatine phosphate to creatine and phosphate. The enzyme is a dimer, with two subunits, namely, M and B. The two units can combine to give three isoenzymes, namely, CK-MM (muscle origin), CK-BB (brain origin), and CK-MB (cardiac origin). Thus, CK-MB is extensively used for the diagnosis of **myocardial infarction**, as it is the first enzyme elevated in such cases.

## Lactate Dehydrogenase

The action of the enzyme is depicted in **Figure 3.1**.

The LDH is a tetramer formed from two subunits, namely, H and M. These two subunits can unite randomly to give five isoenzymes.
1. HHHH ($H_4$) - LDH1
2. HHHM ($H_3M$) - LDH2
3. HHMM ($H_2M_2$) - LDH3
4. HMMM ($HM_3$) - LDH4
5. MMMM ($M_4$) - LDH5

The M subunits predominate in skeletal muscles and liver, and the H subunits predominate in the heart. Both LDH1 and LDH2 isoenzymes are found predominantly in heart and RBC, LDH3 in brain and kidney, and the last two are found predominantly in liver and skeletal muscles. These isoenzymes are of use in the diagnosis of myocardial infarction, but it is rarely used nowadays, because of its nonspecificity.

## SUMMARY OF THE CHAPTER

- Enzymes are biocatalysts that are essential for biochemical reactions to proceed in the human body.
- Biological activity of enzymes is dependent on the structural conformation of the enzyme protein.
- Enzymes can be classified into (i) oxidoreductases (e.g., alcohol dehydrogenase), (ii) transferases (e.g., hexokinase), (iii) hydrolases (e.g., acetylcholinesterase), (iv) lyases (e.g., aldolase), (v) isomerases (e.g., triosephosphate isomerase), and (vi) ligases (e.g., acetyl-CoA carboxylase).
- Enzymes requiring the presence of a certain metal ion for their activity are called metalloenzymes. Examples are zinc in carbonic anhydrase, iron in catalase and peroxidase, calcium in lipase, etc.
- Apoenzyme (protein part) combines with coenzyme (prosthetic group) to form the functional holoenzyme.
- Coenzymes may take part in reactions as cosubstrates but are regenerated.
- Some vitamin coenzymes are derivatives of nucleotide phosphates, e.g., $NAD^+$, flavin

adenine dinucleotide (FAD). Deficiency of coenzymes can affect the rate of enzymatic reactions.
- Area of an enzyme where the catalysis occurs is called the "active site."
- Enzymes catalyze reactions by lowering the activation energy but do not change the equilibrium constant.
- Michaelis–Menten theory states that an enzyme (E) combines with a substrate (S) to form an enzyme-substrate (ES) complex, which breakdown to give the product (P).
- Enzyme activity is influenced by enzyme concentration, substrate concentration, pH, temperature, and presence of inhibitors. Velocity is directly proportional to the concentration of enzymes.
- Velocity at saturating the concentration of substrate is called maximum velocity or $V_{max}$.
- $K_m$ value (Michaelis-Menten Constant), the substrate concentration at half maximum velocity is a constant for each enzyme-substrate pair.
- $K_m$ value is characteristic of a given enzyme. No two enzymes can have the same $K_m$ value. It denotes the affinity of the enzyme to its substrate. Lesser the $K_m$, greater the affinity and vice versa.
- $K_m$ value indicates the affinity of enzyme for substrate; higher the affinity, lower the $K_m$.
- Velocity is maximum for each enzyme at an optimum pH and temperature.
- Inhibition may be reversible or irreversible.
- Enzyme inhibition can be competitive or noncompetitive or uncompetitive. Competitive inhibition is usually reversible.
- The competitive inhibitor increases the $K_m$ and its effect can be reversed by increasing the substrate concentration.
- Many drugs are competitive inhibitors of specific enzymes, e.g., folic acid synthesis is inhibited by sulfonamides since they are structurally similar to PABA.
- Actions of drugs, such as sulfonamides, methotrexate, dicoumarol, and isoniazid are based on the principle of competitive inhibition.
- Noncompetitive inhibition is irreversible and can be caused by toxins or poisons.
- Allosteric enzymes can be regulated by the binding of positive or negative modifiers to the allosteric site, thus affecting substrate binding to active catalytic site.
- Suicide inhibition is an irreversible inhibition. The inhibitor makes use of the natural reaction of the enzyme for inhibition, e.g., ornithine decarboxylase.
- Isoenzymes are physically distinct forms of the same enzyme activity. They may be the products of the same gene or different genes.

## QUESTIONS FOR SELF-ASSESSMENT

1. Define enzymes and give its classification.
2. Define coenzymes and cofactors with examples.
3. What are isoenzymes?
4. Mention the clinically important isoenzymes and their functions.
5. Mention the factors that affect enzyme activity.
6. Mention the importance of Michelis-Menten equation and $K_m$ value.
7. What is the significance of $V_{max}$.
8. Mention the different types of enzyme inhibition with examples.
9. Mention the isoenzymes important for assessment in myocardial infarction.
10. Schematically draw a table of the important isoenzymes and their clinical role.

## MULTIPLE CHOICE QUESTIONS (MCQs)

3-1. **All enzymes are proteins, *except*:**
   a. Hydratase
   b. Metalloenzyme
   c. Ribozymes
   d. Topoisomerase
3-2. **The following factor that does affect enzyme activity is:**
   a. pH
   b. Concentration of various isoenzymes
   c. Substrate concentration
   d. Temperature

3-3. The following two enzymes are important for diagnosis of pancreatitis:
   a. ALP
   b. Lipase
   c. GGT
   d. AST

3-4. The enzyme which is used as marker for prostate cancer is:
   a. Lipase
   b. GGT
   c. Acid phosphatase
   d. Alkaline phosphatase

3-5. What is an apoenzyme?
   a. It is a protein portion of an enzyme
   b. It is a nonprotein group
   c. It is a complete, biologically active conjugated enzyme
   d. It is a prosthetic group

3-6. Name the enzyme secreted by pancreas?
   a. Pepsin
   b. GGT
   c. Trypsin
   d. Alcohol dehydrogenase

3-7. Name the enzyme that catalyzes the oxidation–reduction reaction?
   a. Transaminase
   b. Glutamine synthetase
   c. Phosphofructokinase
   d. Oxidoreductase

3-8. Which of the following reaction is catalyzed by lyase?
   a. Breaking of bonds
   b. Formation of bonds
   c. Intramolecular rearrangement of bonds
   d. Transfer of group from one molecule to another

3-9. Removal of phosphoryl groups is catalyzed by:
   a. Isomerases
   b. Dinitrogen reductase
   c. Protein phosphatases
   d. Protein kinases

3-10. In the enzyme-catalyzed reaction shown below, what will be the effect on substances A, B, C, and D, if the enzyme E2 is inactivated?
   A---(E1)--->B---(E2)--->C---(E3)--->
   a. A, B, C, and D will all still be produced
   b. A, B, and C will still be produced, but not D
   c. A and B will still be produced, but not C or D
   d. A will still be produced, but not B, C, or D

## ANSWER KEYS TO MCQs

3-1. (c)   3-2. (b)   3-3. (b)   3-4. (c)   3-5. (a)   3-6. (c)   3-7. (d)   3.8. (a)
3-9. (c)   3-10. (c)

# Chapter 4

# Carbohydrate Chemistry

At the completion of this chapter, the reader will be able to answer questions on the following topics:

- Classification of carbohydrates
- Monosaccharides
- Stereoisomerism
- Optical activity
- Epimerism
- Mutarotation and anomerism
- Reactions of carbohydrates
- Glucose, galactose, and fructose
- Deoxy sugars and amino sugars
- Sucrose, maltose, and lactose
- Starch and glycogen
- Heteroglycans
- Glycoproteins and mucoproteins

Carbohydrates are the most abundant biomolecules in nature. They can be defined as polyhydroxy alcohols with an aldehyde or keto group. Primarily they contain carbon, hydrogen, and oxygen. But they may also contain phosphorus, sulfur, or nitrogen. They have a general molecular formula of $C_n(H_2O_n)$. The ratio of hydrogen to oxygen is the same as in water and hence the name carbohydrates or hydrated carbon.

## FUNCTIONS OF CARBOHYDRATES

- Carbohydrates are the most abundant dietary sources of energy.
- They are precursors for many other organic compounds, such as fats, proteins, and so on.
- Along with proteins and lipids, carbohydrates form part of the structure of a cell membrane.
- Carbohydrates serve as short-term energy resources.

## CLASSIFICATION OF CARBOHYDRATES

Main classification of carbohydrates is shown in **Figure 4.1**.

## Simple Carbohydrate

*Monosaccharides*

Common monosaccharides are shown in **Table 4.1**.

**Simple sugars**

They contain only one sugar molecule and cannot be hydrolyzed into a simpler form.
1. Divided into aldoses and ketoses based on the functional group present.
   - Example: Glucose—aldose
   - Fructose—ketose
2. Depending on the number of carbon atoms present, they are divided into triose (3C), tetrose (4C), pentose (5C), and hexose (6C).

*Example:* Hexose (6C)—glucose (aldose), fructose (ketose) (**Table 4.2** shows hexoses of physiological importance)

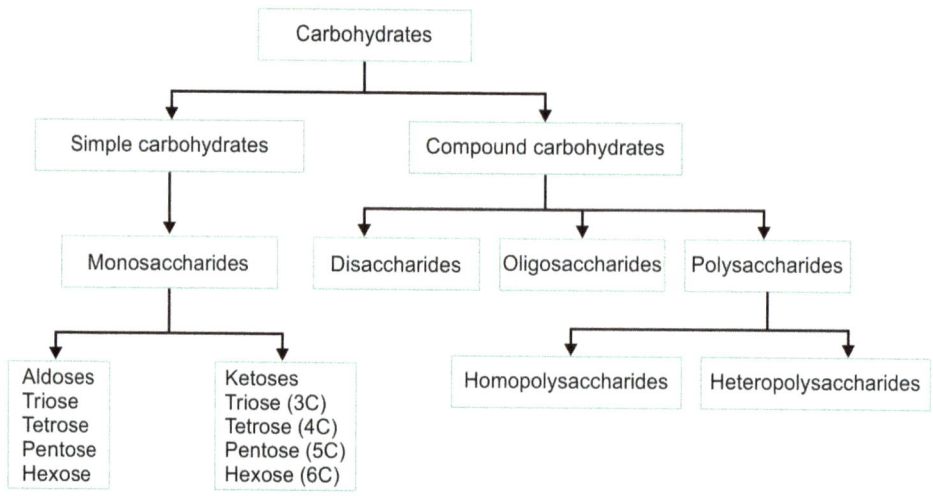

**Fig. 4.1:** Classification of carbohydrates.

**Table 4.1:** Common monosaccharides.

| No. of carbon atoms | Generic name | Aldoses (with aldehyde group) | Ketoses (with keto group) |
|---|---|---|---|
| 3 | Triose | Glyceraldehyde | Dihydroxyacetone |
| 4 | Tetrose | Erythrose | Erythrulose |
| 5 | Pentose | Xylose, ribose | Xylulose, ribulose |
| 6 | Hexose | Glucose, galactose, mannose | Fructose |

**Table 4.2:** Hexoses of physiological importance.

| Sugar | Importance |
|---|---|
| D-glucose | Blood sugar. Main source of energy in the body |
| D-fructose | Constituent of sucrose, the common sugar |
| D-galactose | Constituent of lactose, glycolipids, and glycoproteins |
| D-mannose | Constituent of globulins, mucoproteins, and glycoproteins |

## Compound Carbohydrates

They are formed by the linkage of two or more monosaccharide molecules with glycoside bonds and can be hydrolyzed into monosaccharides. They are classified as follows:

### Disaccharides

They contain two monosaccharide molecules.
*For example:*
Sucrose = glucose + fructose
Maltose = glucose + glucose
Lactose = glucose + galactose

### Oligosaccharides

They contain 3–10 monosaccharide units, joined by glycosidic bonds. They are divided into trisaccharide, tetrasaccharide, and so forth, based on the number of sugar molecules present. Disaccharides are predominant in nature.

Trisaccharides contain three monosaccharide molecules, for example, maltotriose.

### Polysaccharides (Glycans)

They contain more than 10 sugar molecules. They are divided into:

- Homopolysaccharides (homoglycan)—polymers of same monosaccharide units, for example, starch, polymer of glucose.
- Heteropolysaccharides (heteroglycan)—polymers of two or more types of monosaccharide units, for example, heparin.

## PROPERTIES OF MONOSACCHARIDES

### Stereoisomerism

Sugar molecules exhibit isomerism. Stereoisomers are compounds that have same molecular formula but differ in spatial configuration.

*Asymmetric Carbon Atom*

Asymmetric carbon atom is the carbon atom having four different atoms or groups attached to it. The simplest monosaccharide with an asymmetric carbon atom is glyceraldehyde. This is taken as reference for naming higher carbohydrates. A compound having one asymmetric carbon may exist in two forms, which are mirror images of each other.

The number of possible stereoisomers depends on the number of asymmetric carbon atoms. This is given by the formula $2^n$, where $n$ is the number of asymmetric carbon atoms present in that compound, for example, glucose has four asymmetric "C" atoms (atom numbers 2, 3, 4, and 5 in **Fig. 4.2**), and hence the number of possible stereoisomers for glucose are $2^4 = 16$ isomers.

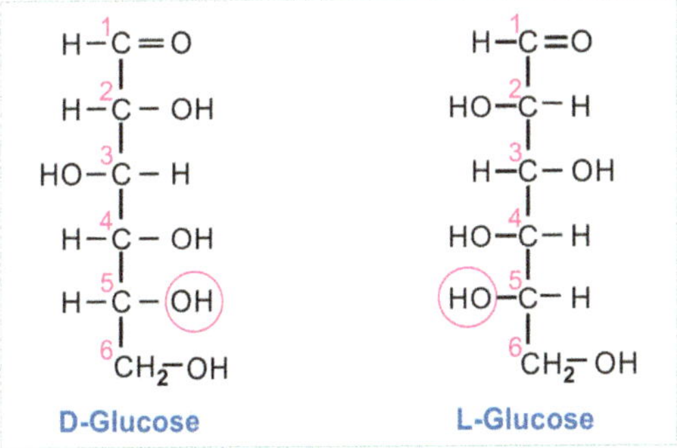

*D and L Forms of Stereoisomers*

Stereoisomers occur either as D or L form. This is based on the position of the –OH group on the reference carbon atom. In a glyceraldehyde molecule if the –OH group is on the right-hand side of the asymmetric carbon atom, then it is D-glyceraldehyde. If the –OH group is on the left-hand side of the carbon atom, then it is L-glyceraldehyde (**Fig. 4.3**). The penultimate carbon atom is used as the reference carbon while naming other higher monosaccharides. The reference carbon atom of glucose is number 5 (**Fig. 4.2**). In nature the D form is predominant, and our body utilizes only the D forms, because enzyme systems are specific to metabolize the D isomers.

### Optical Activity

Optical activity is exhibited by compounds having asymmetric carbon atoms. When a beam of plane-polarized light is passed through a sugar solution, then the light will be rotated either to the right or to the left. If the light is rotated to the right (clockwise direction), then the molecule is said to be dextrorotatory, represented as "d" or "+." If the light is rotated to the left (anticlockwise direction), then it is said to be levorotatory and is represented as "l" or "." Students should not confuse the "d" form described in this paragraph with the "D" form described in the previous paragraph. In nature, D-glucose is dextrorotatory and D-fructose is levorotatory.

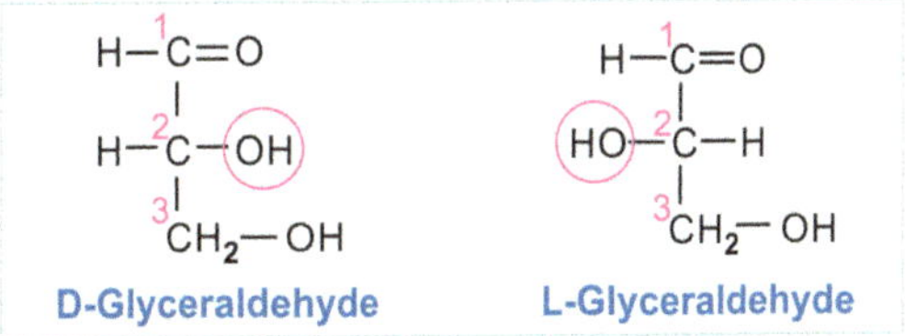

**Fig. 4.2:** D and L forms of glucose. Carbon atom numbers 2, 3, 4, and 5 are asymmetric. Carbon atom number 5 (penultimate carbon atom) is the reference carbon atom.

**Fig. 4.3:** D and L forms of glyceraldehyde.

## Racemic Mixtures

Equimolecular mixture of "d'" and "l" forms is called racemic mixture. It does not rotate the polarized light, as the "d" and "l" forms cancel each other.

## Epimerism

When two monosaccharides differ from each other in their configuration with respect to a single carbon atom (other than the reference carbon atom), then they are referred to as epimers, and they are said to be epimeric pairs. For example, D-glucose and D-galactose are C4 epimers. They differ only at carbon atom number 4 **(Fig. 4.4)**. Similarly, D-glucose and D-mannose are C2 epimers.

```
    H–C=O              H–C=O              H–C=O
     |                  |                  |
    H–²C–(OH)         (HO)–²C–H           H–C–OH
     |                  |                  |
    HO–C–H             HO–C–H             HO–C–H
     |                  |                  |
    H–⁴C–(OH)          H–C–OH           (HO)–⁴C–H
     |                  |                  |
    H–C–OH             H–C–OH             H–C–OH
     |                  |                  |
    CH₂OH              CH₂OH              CH₂OH

   D-Glucose          D-Mannose         D-Galactose
```

**Fig. 4.4:** Epimerism. D-glucose and D-galactose differ only with regard to carbon atom number 4.

## Anomerism

Sugars in solution exist in the form of rings. The ring is formed because of the high reactivity of the carbonyl group (--C=O) of both the aldehyde and ketone. In aldohexoses (e.g., glucose), a hemiacetal ring is formed between the first carbon and the fifth carbon. This results in a ring form containing five carbon atoms and one oxygen atom. This structure is similar to that of the compound pyran. Hence this six-membered ring structure of glucose is called **pyranose**. The formation of ring structure makes carbon atom number 1 also to be asymmetric. Therefore, carbon atom 1 shows α or β form, depending on the position of the –OH group of $C_1$. If the –OH group is present above the plane of the ring, then it is β form, or if the –OH group is below, it is α form **(Fig. 4.5)**.

## Fructose is in Furanose Form

In case of fructose (ketohexoses), the hemiketal ring is formed between the carbon atoms number 2 and 5. This results in a five-membered ring, having four carbon atoms and one oxygen atom. This is called the **furanose** ring. The ring formation makes $C_2$ asymmetric and leads to the α and β forms depending on the position of the –OH

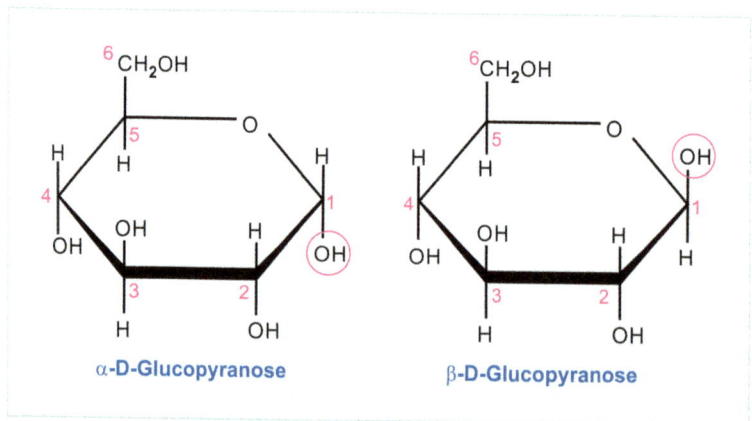

**Fig. 4.5:** Anomerism. Alpha and beta forms of glucose. Carbon atom number 1 is called anomeric carbon atom.

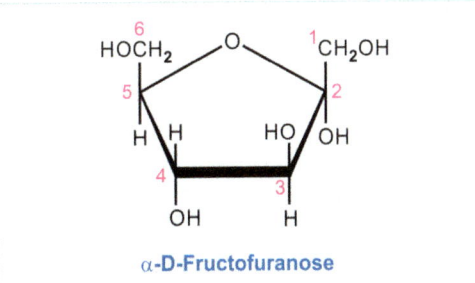

**Fig. 4.6:** Structure of fructose.

group on $C_2$ atom **(Fig. 4.6)**. The aldehyde or ketone carbon in pyranoses and furanoses becomes a new asymmetric carbon, and this is called **anomeric carbon**. This type of stereoisomerism is called **anomerism**.

## Mutarotation

Glucose is dextrorotatory in nature. A fresh solution of α-glucose rotates the plane polarized light to +112°. This goes on decreasing and reaches a constant value of +52.5° after a few hours. Rotation of a fresh solution of β-glucose is +19°. This goes on increasing and reaches a constant value of +52.5° after a few hours. This change in rotation is called mutarotation. This is due to the conversion of some α-glucose molecules into β-glucose and vice versa.

## REACTIONS OF CARBOHYDRATES

## Glycoside Formation

Two carbohydrate molecules are joined by a glycoside bond. The –OH group of anomeric carbon of one carbohydrate is joined by a covalent bond to the –OH group of another molecule; this is the **glycoside bond**. It involves the elimination of a water molecule. All oligosaccharides and polysaccharides are formed by glycosidic linkages. Only molecules containing an anomeric –OH group can form a glycoside bond.

Glucose + glucose ⟶ maltose + $H_2O$

## Action of Strong Acids

Sugar loses water on being heated with strong acids and forms furfural derivatives. These furfurals form colored complexes with phenolic compounds like a—naphthol or resorcinol.

### Molisch Test

In this test, 5 mL of sugar solution and a few drops of Molisch reagent are mixed. Concentrated sulfuric acid is added through the sides of the test tube. A violet ring is produced. Molisch reagent consists of alpha naphthol in alcohol. All carbohydrates will give a positive Molisch test.

### Seliwanoff's Test

This test is answered only by keto sugars (e.g., fructose).

Keto sugar + Seliwanoff's reagent $\xrightarrow{\text{Boil for 30 seconds}}$ Cherry red color

Seliwanoff's reagent contains resorcinol and dilute hydrochloric acid (HCl).

## Action of Alkali

On treating with alkali, sugars become good reducing agents. Only sugars with free aldehyde or ketone group can act as reducing agents as only they can form enediols (reducing agents).

### Benedict's Test

Benedict's reagent is very commonly employed to detect the presence of glucose in urine (**glucosuria**). It is a standard laboratory test employed to diagnose **diabetes mellitus**. Benedict's reagent contains sodium carbonate, copper sulfate, and sodium citrate. In alkaline medium, sugar forms enediol, which reduces cupric ions to cuprous oxide. **Glucose is a reducing sugar**. Any sugar with free aldehyde/keto group will reduce the Benedict's reagent. Therefore, this is not specific for glucose. In this test, 5 mL of sugar solution and 0.5 mL of Benedict's reagent are taken in a test tube and boiled for 2 minutes. A precipitate of green, orange, or red color will indicate a positive test.

## Action of Weak Acid

*Barfoed's Test*

Sugar solution with Barfoed's reagent, when heated, reduces cupric ions to cuprous ions. Barfoed's solution contains copper acetate and dilute acetic acid.

## Osazone Formation

All reducing sugars in alkaline medium will form osazones with an excess of phenylhydrazine when kept at boiling temperature. Osazones are insoluble. Each sugar will have the characteristic crystal form of osazones. The differences in glucose, fructose, and mannose are dependent on the first and second carbon atoms, and when the osazone is formed, these differences are masked. Hence these three sugars will produce the same **needle-shaped crystals** arranged like sheaves of corn or a broom (**Fig. 4.7**). Osazones may be used to differentiate sugars in biological fluids like urine.

*Example:*

| | | |
|---|---|---|
| Glucose and fructose | → | glucosazone (needle-shaped) |
| Maltose | → | maltosazone (sunflower-shaped) |
| Lactose | → | lactosazone (powder puff-shaped) |

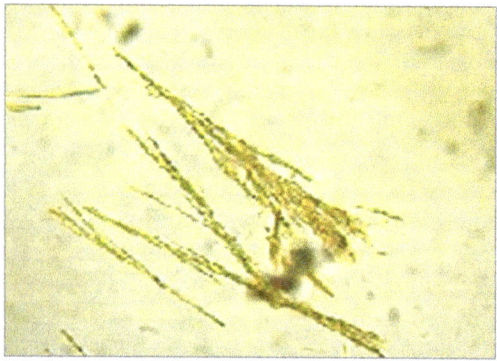

**Fig. 4.7:** Glucosazone. Needle-shaped crystals arranged like a broom.

## Oxidation Reactions

Aldoses yield different products according to the nature of oxidizing agent.
- Mild oxidizing agent (e.g., bromine water). Sugars will produce the respective **aldonic acids**. The aldehyde (-CHO) group of aldoses is oxidized to form acids.
  *Example:* Glucose → **gluconic acid**.
- Strong oxidizing agent (like nitric acid). Sugars produce respective saccharic acids. The first and last alcohol groups are oxidized to acid.
  *Example:* Glucose → **glucosaccharic acid**
- When the aldehyde group is masked, the terminal alcoholic group ($C_6$) is oxidized; this produces respective uronic acid.
  *Example:* Glucose produces **glucuronic acid**, and galactose produces galacturonic acid. Inside the body, glucuronic acid helps in conjugating water-insoluble molecules like bilirubin to make them more water-soluble. It also forms part of heteropolysaccharides.

## MODIFIED MONOSACCHARIDES

### Deoxy Sugars

In this group, the alcoholic –OH group of one carbon is deprived of its oxygen (usually it is $C_2$), for example, DNA has 2-deoxyribose as sugar component. In the ribose molecule, the $C_2$ is deprived of oxygen (**Fig. 4.8**).

### Amino Sugars

An alcoholic group in sugar is replaced by an amino ($-NH_2$) group, for example, glucosamine. The amino group is added to the second carbon atom of glucose. The $-NH_2$ (amino) group may further be acetylated (-CO-$CH_3$) or sulfated ($-SO_3^{--}$). They are components of heteropolysaccharides.

## DISACCHARIDES

### Sucrose

It is present in sugarcane. It is made of one molecule of glucose and one molecule of

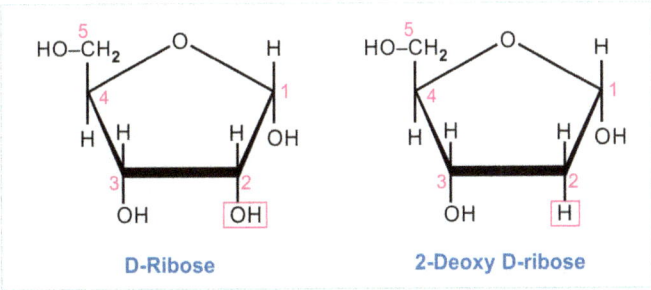

**Fig. 4.8:** Deoxy sugar.

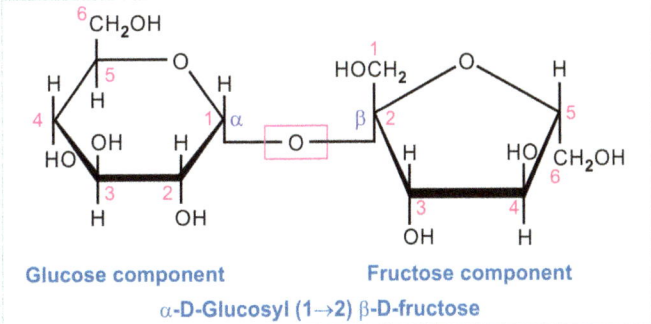

**Fig. 4.9:** Sucrose contains one glucose and one fructose residues. Sucrose is a nonreducing disaccharide.

fructose joined by an α,β-1,2 glycosidic bond **(Fig. 4.9)**. Sucrose does not have a free anomeric -OH group, and so it is nonreducing. It cannot reduce Benedict's solution and shows no mutarotation and no osazone formation. Sucrase enzyme splits sucrose into glucose and fructose. Sucrose shows the property of inversion.

Sucrose shows the phenomenon of inversion. Sucrose is dextrorotatory in nature. After hydrolysis, it gives rise to a mixture of glucose and fructose. This mixture has **levorotation**. Thus, the direction of rotation changes after the compound is hydrolyzed. This is called **inversion**. The mixture is levorotatory because fructose is more levorotatory (−92°) than glucose is dextrorotatory (+52.5°). Because of this property, hydrolyzed product of sucrose is called invert sugar (i.e., mixture of glucose and fructose), and enzyme sucrase is otherwise called **invertase**.

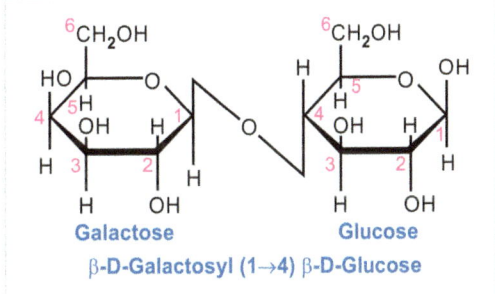

**Fig. 4.10:** Lactose contains one galactose and one glucose residues. Lactose is a reducing disaccharide.

## Lactose

It is also called milk sugar. It is made of one molecule of galactose and one molecule of glucose joined by β-1,4 glycosidic linkage **(Fig. 4.10)**. Based on the position of the -OH group of anomeric carbon of reducing end of molecule, it can be either a or b lactose. The anomeric ($C_1$) -OH group of the glucose part remains free, and hence it is a reducing sugar. It reduces Benedict's solution. Lactose shows mutarotation. It forms lactosazone crystals, which have the characteristic shape of a powder puff **(Fig. 4.11)**.

## Maltose

It is found in germinating seeds and malt. It is made of two molecules of α-glucose joined by α-1,4 glycosidic bond **(Fig. 4.12)**.

**Fig. 4.13:** Sunflower-shaped or petal-shaped crystals of maltosazone.

The anomeric –OH group of the second glucose unit is free, and so maltose is a reducing sugar. It shows mutarotation. It forms maltosazone crystals, which are sunflower-shaped **(Fig. 4.13)**. The enzyme maltase acts on maltose to give two molecules of glucose. Maltose can be either α or β maltose depending on the position of the –OH group on $C_1$ of the reducing end of the molecule.

## Glycosides

When the hemiacetal group (hydroxyl group of the anomeric carbon) of a monosaccharide is condensed with an alcohol or phenol group, it is called a glycoside **(Fig. 4.14)**. The noncarbohydrate group is called **aglycone**. Glycosides **do not reduce** Benedict's reagent, because the sugar group is masked. They may be hydrolyzed by boiling with dilute acid, so that sugar is free and can then reduce copper. Some glycosides of medical importance are given in **Table 4.3**. Digitonin is a cardiac

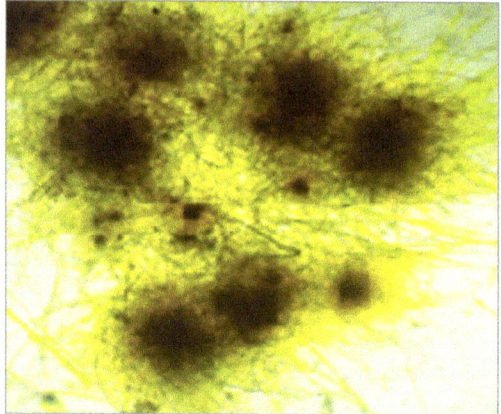

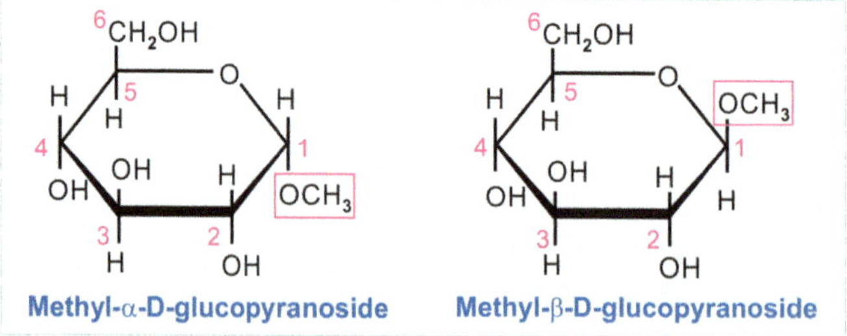

**Fig. 4.11:** Lactosazone. Hedgehog or "pincushion with pins" or flower of "touch-me-not plant" or "powder puff"-shaped crystals.

**Fig. 4.12:** Maltose contains two glucose residues. Maltose is a reducing disaccharide.

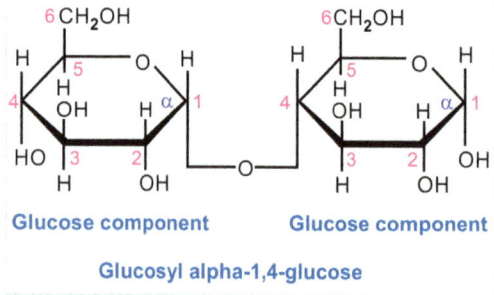

**Fig. 4.14:** Glycosides.

## Chapter 4 Carbohydrate Chemistry

**Table 4.3:** Glycosides.

| Sugar | + Aglycon | = Glycoside | Source | Importance |
|---|---|---|---|---|
| Glucose | Phloretin | Phlorizin | Rose bark | Renal damage |
| Galactose, xylose | Digitogenin | Digitonin | Leaves of foxglove | Cardiac stimulant |
| Glucose | Indoxyl | Plant indican | Leaves of Indigofera | Stain |

stimulant. Phlorizin is used to produce renal damage in experimental animals.

## POLYSACCHARIDES

Polysaccharides are polymerized products of many monosaccharide units. They are divided into:

**Homoglycans**: Single kind of monosaccharide units repeated again and again, for example, starch, glycogen.

**Heteroglycans** or mucopolysaccharides: Two or more different kinds of monosaccharide units repeated, for example, heparin.

### Homoglycans

*Starch*

It is the storage form of carbohydrates in plants. Starch is made of unbranched (**amylose**) and branched (**amylopectin**) molecules (**Fig. 4.15**). On hydrolysis both amylose and amylopectin will yield many molecules of glucose. The α-amylose molecule is an unbranched chain of about 60–600 units of α-glucose joined by α-1,4 glycosidic bond. The last glucose at one end of each chain carries a free anomeric hydroxyl group, and this is called the reducing end. The other end of each chain is called the nonreducing end, as the last glucose at that end is devoid of any free anomeric hydroxyl group. The amylopectin molecule consists of many unbranched linear chains, each made of 25–30 glucose units linked to each other by α-1,6 glycosidic bond. Such cross-linkage will give rise to a branched amylopectin molecule. An amylopectin molecule may have 5,000–6,000 glucose residues.

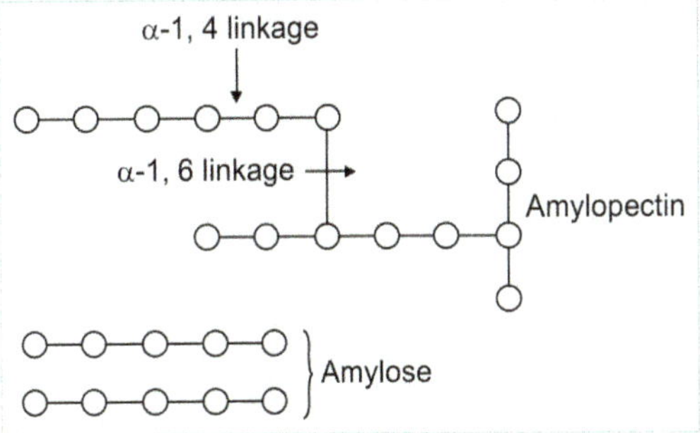

Amylose will give blue color with iodine, and amylopectin will give violet color with iodine. Starch is nonreducing, and it forms a colloidal solution in water. Starch on hydrolysis produces amylodextrins (nonreducing), which on hydrolysis produce erythrodextrin (mildly reducing), and further hydrolysis will give achrodextrin (reducing), which finally forms maltose. Salivary and pancreatic amylase will act upon starch to split it into smaller units of maltose.

*Glycogen*

It is the storage form of carbohydrate in animals. It is stored in liver and muscles. It is made of many unbranched linear chains, each

**Fig. 4.15:** Amylopectin is a highly branched molecule, while amylose is a straight-chain compound.

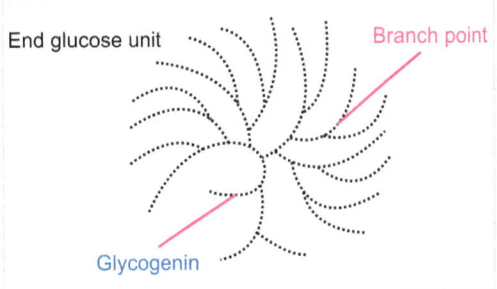

**Fig. 4.16:** Branched glycogen molecule.

having 10–15 glucose units linked by α-1,6 glycosidic linkages to form a highly branched glycogen molecule **(Fig. 4.16)**. It forms a colloidal solution in water and is nonreducing. It gives a red–violet color with iodine. The difference between starch and glycogen is that glycogen is more highly branched than starch.

*Cellulose*

It is the insoluble structural homoglycan of plant fiber. It is the main constituent of the cell wall of a plant cell. Cellulose is made of long, unbranched linear chains of 300–3,000 β-glucose units joined by β-1,4 glycosidic bond. Cellulose is nonreducing and is not digested by enzymes of the human digestive system.

*Dextrin*

It is a water-soluble homoglycan. It is made of many glucose units bound by α-1,4 glycosidic linkages. Branching is by α-1,6 linkages. Dextrins are products of the partial hydrolysis of starch.

## Heteroglycans

*Hyaluronic Acid*

It is a structural heteroglycan present in synovial fluid, vitreous humor, skin, and connective tissues. It consists of unbranched linear chain of alternating units of β-glucuronic acid and N-acetyl β-glucosamine. It is viscous in nature and thus gives the slippery nature to body fluids.

*Heparin*

It is a natural anticoagulant. It is found in the cells of liver, lungs, arterial walls, and spleen. Heparin is made of alternating units of sulfated or acetylated α-glucosamine and β-iduronic acid. Because of the presence of iduronic acid, heparin is slightly acidic in nature.

*Chondroitin Sulfate*

It is the structural heteroglycan found in bones, cartilages, cornea, skin, lungs, heart valves, and arterial wall. It is made of alternating units of β-glucuronic acid and sulfated acetyl β-galactosamine.

*Keratan Sulfate*

It forms the structural component of skin, cornea, arterial wall, and connective tissues. It is made of alternating units of β-galactose and N-acetyl β-glucosamine. It does not contain any uronic acid.

*Dermatan Sulfate*

It is widely distributed in animal tissues. It is mainly present in the cornea and sclera of eyes, skin, and blood vessels. It consists of repeating units of iduronic acid and N-acetyl galactosamine. All the heteropolysaccharides along with proteins form proteoglycan aggregates, which form the structural components of a cell. Repeating units of carbohydrates in various polysaccharides are summarized in **Table 4.4**.

*Glycoproteins and Mucoproteins*

Carbohydrates attached to polypeptide chain are proteoglycans. When the carbohydrate content is less than 10% in a proteoglycan, it is called glycoprotein, and if the carbohydrate content is greater than 10%, then it is a mucoprotein.

**Table 4.4:** Repeating units in polysaccharides.

| Polysaccharide | Repeating units |
| --- | --- |
| Hyaluronic acid | N-acetyl glucosamine, glucuronic acid |
| Heparin | Sulfated glucosamine, L-iduronic acid |
| Chondroitin sulfate | Glucuronic acid, N-acetyl galactosamine |
| Keratan sulfate | Galactose, N-acetyl glucosamine |
| Dermatan sulfate | L-iduronic acid, N-acetyl galactosamine |

# Chapter 4 Carbohydrate Chemistry

## SUMMARY OF THE CHAPTER

- Carbohydrates are polyhydroxy alcohols with an aldehyde or keto group. Simplest carbohydrates are monosaccharides, which may be trioses (3C), tetroses (4C), pentoses (5C), and hexoses (6C).
- Carbohydrates are classified into monosaccharides, disaccharides, and polysaccharides based on the number of sugar/saccharide units they possess. Disaccharides have 2 monosaccharide units, oligosaccharides around 10, and polysaccharides more than 10. They could also be classified as aldoses and ketoses based on the functional group they possess.
- Common examples of monosaccharides include glucose, fructose, galactose, and mannose.
- Common examples of disaccharides are sucrose, lactose, and maltose.
- Monosaccharides exhibit stereoisomerism, optical isomerism, anomerism, and pyranose–furanose isomerism.
- All carbohydrates are considered to be derived from glyceraldehyde by the successive addition of carbons.
- A carbon atom bound by four different groups on all its valencies is referred to as an *asymmetric carbon*. When two sugars differ from each other in the configuration around one carbon atom (other than the reference carbon), they are diastereoisomers.
- A pair of monosaccharides that differ from each other in the configuration around a single carbon atom are called epimers. Anomers of monosaccharides are produced by the spatial configuration with reference to the first carbon atom in aldoses and the second carbon atom in ketoses.
- Two anomers of glucose are alpha-D glucose and beta-D glucose. Mutarotation is the result of anomerism.
- All reducing sugars form characteristic osazone crystals. Glucose and fructose form needle-shaped crystals, maltose forms sunflower-shaped crystals, and lactose forms hedgehog-shaped crystals.
- Amino sugars form important components of mucopolysaccharides, for example, galactosamine and glucosamine.
- Action of amylase on starch yields limits dextrins.
- Mucopolysaccharides or glycosaminoglycans (GAGs) such as hyaluronic acid, chondroitin sulfate, keratan sulfate, and dermatan sulfate are associated with connective tissues.
- Keratan sulfate is the only GAG that does not contain uronic acid.
- When the carbohydrate chains are attached to a polypeptide chain, it is called a proteoglycan.

## QUESTIONS FOR SELF-ASSESSMENT

1. Define carbohydrates with a suitable example.
2. Classify carbohydrates.
3. How monosaccharides are classified further?
4. Explain the term *mutarotation*.
5. Mention the difference between anomers and epimers with examples.
6. Briefly explain the importance of mucopolysaccharide with its biomedical importance.
7. Discuss the structural difference between starch and glycogen. Mention their functions.
8. Name one nonreducing disaccharide.
9. Explain the term *glycoside*. Give an example.
10. Mention the glycosaminoglycan that serves as a lubricant and shock absorbent for joints.

## MULTIPLE CHOICE QUESTIONS (MCQs)

4-1. **Mention the nonaldose sugar:**
   a. Mannose
   b. Glucose
   c. Galactose
   d. Fructose

4-2. **The general orientation of the carbon atoms for reducing action are:**
   a. 1 and 2
   b. 2 and 3

c. 3 and 4
d. 4 and 5

**4-3.** Which among the following is an anticoagulant?
a. Hyaluronic acid
b. Heparin
c. Chondroitin sulfate
d. Keratan sulfate

**4-4.** Which one is an invert sugar?
a. Glucose
b. Galactose
c. Mannose
d. Sucrose

**4-5.** The monosaccharide present in DNA is:
a. Glucose
b. Mannose
c. Fructose
d. Ribose

**4-6.** Find out the incorrect statement regarding glycogen:
a. Stored in liver
b. A polymer of glucose
c. Stored in muscle
d. Present in plants

**4-7.** Which one among the following is not a heteropolysaccharide?
a. Dextrin
b. Heparin
c. Chondroitin sulfate
d. Hyaluronic acid

**4-8.** The fourth epimer of glucose is:
a. Galactose
b. Maltose
c. Sucrose
d. Fructose

**4-9.** Which one among the following gives negative Benedict's test?
a. Glucose
b. Sucrose
c. Fructose
d. Ascorbate

**4-10.** Seliwanoff's test is positive for:
a. Glucose
b. Galactose
c. Fructose
d. Xylulose

## ANSWER KEYS TO MCQs

4-1. (d)   4-2. (a)   4-3. (b)   4-4. (d)   4-5. (d)   4-6. (d)   4-7. (a)   4-8. (a)
4-9. (b)   4-10. (c)

# CHAPTER 5

# Carbohydrate Metabolism

At the completion of this chapter, the reader will be able to answer questions on the following topics:

- Digestion and absorption of carbohydrates
- Glycolysis pathway
- Energetics of glycolysis
- Gluconeogenesis
- Hexose monophosphate shunt pathway
- Fructose metabolism
- Galactose metabolism
- Glycogenolysis
- Glycogenesis (glycogen synthesis)
- Glycogen storage diseases

## DIGESTION

In the diet, carbohydrates are available mainly as polysaccharides (starch) and to a minor extent as disaccharides (sucrose and lactose). The digestion starts in the mouth with salivary alpha amylase. In the pancreatic juice, another alpha amylase will hydrolyze alpha-1,4-glycosidic linkages, so as to produce smaller units, such as dextrins, oligosaccharides, and disaccharides. The enzymes present in the intestinal juice such as **sucrase, maltase, isomaltase**, and **lactase** then convert the disaccharides to monosaccharides. These monosaccharides are then absorbed.

## ABSORPTION

Glucose cannot diffuse freely through the lipid bilayer of the cell membrane. Glucose is absorbed by a cotransport mechanism called **sodium-dependent glucose transporter (SGluT)** (Fig. 5.1). Glucose is transported across the membrane against the concentration gradient, coupled with the movement of sodium from a higher concentration to a lower concentration. The sodium is later expelled, and this process needs energy. The same type of mechanism is also used in kidneys to reabsorb glucose in the renal tubules. The transporter in the intestine is SGluT-1 and that in the kidney is SGluT-2. Sodium-dependent glucose transporter-2 is normally present in proximal renal tubules but will be absent in renal glycosuria.

Another transport system releases glucose from intestinal cells into blood. This is called

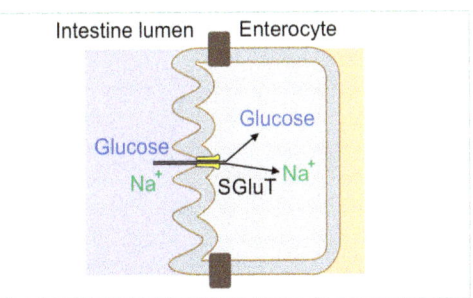

**Fig. 5.1:** SGluT. Sodium and glucose cotransport system at luminal side; sodium is then pumped out.

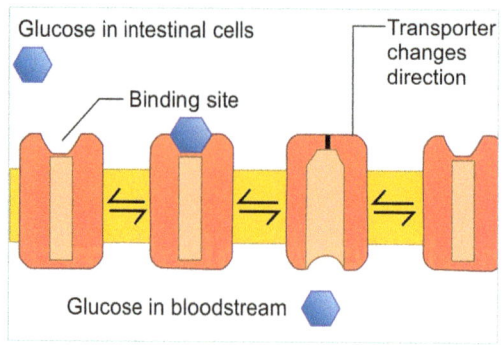

**Fig. 5.2:** Glucose absorption (GluT2).

**glucose transporter type 2 (GluT2) (Fig. 5.2).** This transporter is not dependent on sodium. It is a facilitated diffusion system. Glucose transporter type 2 is also involved in the absorption of glucose from bloodstream to cells. Glucose transporter type 2 is present in intestinal epithelial cells, liver cells, beta cells of pancreas, and kidney. This mechanism enables the pancreas to monitor the glucose level and adjust the rate of insulin secretion. Glucose transporter type 2, which does this work in the liver, is not under the influence of insulin. **Figure 5.3** shows the differences in SGluT and GluT2.

Yet another mechanism **glucose transporter type 4 (GluT4)** is the major glucose transporter in skeletal muscle and adipose tissue **(Fig. 5.4)**. Glucose transporter type 4 is under the control of insulin. But other glucose transporters are not under the control of insulin. In type 2 **diabetes mellitus**, GluT4 is reduced, leading to insulin resistance in muscle and fat cells. In diabetes, the quantity of glucose entering into muscle is only half of that of normal cells. **(Table 5.1).**

## GLUCOSE METABOLISM

The first step in the metabolism of glucose is the conversion of glucose to glucose-6-phosphate by the enzyme **glucokinase** or **hexokinase** utilizing adenosine triphosphate (ATP) in the presence of $Mg^{2+}$ ions **(Fig. 5.5)**. Hexokinase has a low $K_m$ value and high affinity for glucose and is present in all tissues, whereas glucokinase has a high $K_m$ value and low affinity for glucose and is present only in liver. The glucose-6-phosphate thus formed can then enter into many pathways **(Fig. 5.6)**. $K_m$ (Michaelis constant) is described in Chapter 3.

## GLYCOLYSIS (EMBDEN–MEYERHOF PATHWAY)

### Importance of Glycolysis

- It is the only pathway that takes place in all the tissues.
- It is the only source of energy for RBCs.
- In strenuous exercise when muscle lacks oxygen, anaerobic glycolysis is the only source of energy.

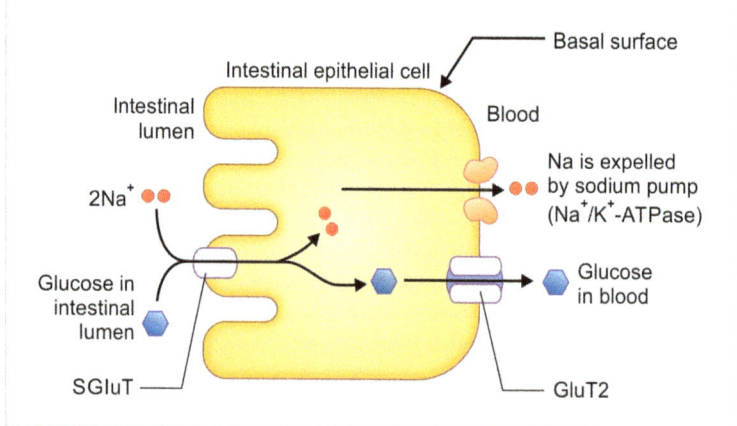

**Fig. 5.3:** Intestinal absorption of glucose. At the intestinal lumen, absorption is by SGluT, and at the blood vessel side, absorption is by GluT2.

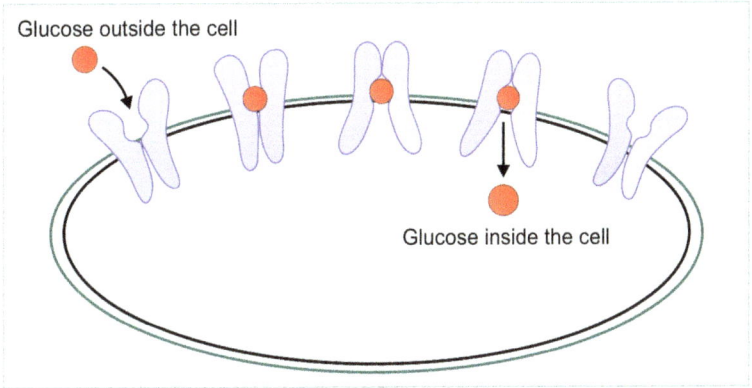

**Fig. 5.4:** Glucose transport in muscle cells by glucose transporter (GluT4).

**Table 5.1:** Glucose transporters and their clinical importance.

| Transporter | Present in | Properties |
|---|---|---|
| GluT1 | Red blood cells, brain, kidney | Glucose uptake in most of the cells |
| GluT2 | Serosal surface of intestinal cells, beta cells in pancreas | Glucose uptake in liver; glucose sensor in the beta cells of pancreas |
| GluT3 | Neurons, brain | Glucose into brain |
| GluT4 | Skeletal muscle, heart muscle, adipose tissue | Insulin-mediated glucose uptake in muscle and adipose tissue |

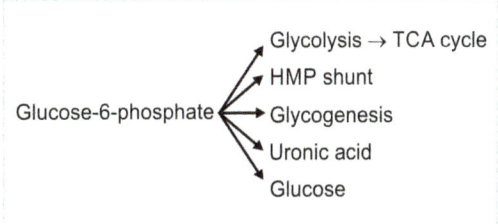

**Fig. 5.6:** Metabolic fate of glucose-6-phosphate. (TCA: citric acid cycle; HMP: hexose monophosphate)

4. It is the preliminary step before complete oxidation of glucose.
5. The pathway also provides carbon skeleton for certain nonessential amino acids as well as glycerol.

Glycolysis is the process of conversion of glucose to pyruvate in aerobic condition and to lactate in anaerobic condition with the concomitant release of energy. All the reaction steps take place in cytoplasm. Detailed steps of the glycolytic pathway is shown in **Figure 5.7**. Glycolysis is also known as the Embden–Meyerhof (EM) pathway.

Insulin favors glycolysis by activating key enzymes such as hexokinase, phosphofructokinase, and pyruvate kinase. Glucocorticoids and glucagon inhibit glycolysis and favor gluconeogenesis.

The end product of glycolysis is pyruvate. When oxygen is not available, pyruvate is converted to lactate, which enters into Cori cycle (details given later in the chapter). When oxygen is available, pyruvate is further oxidized to acetyl coenzyme A (CoA) (details given later in the chapter).

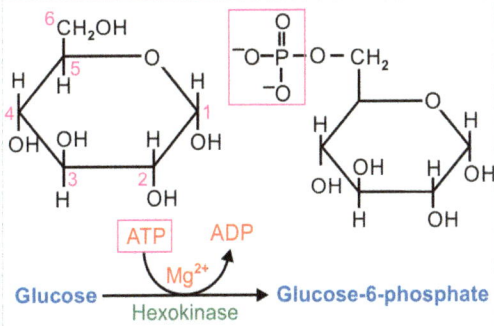

**Fig. 5.5:** Hexokinase reaction.
(ATP: adenosine triphosphate; ADP: adenosine diphosphate)

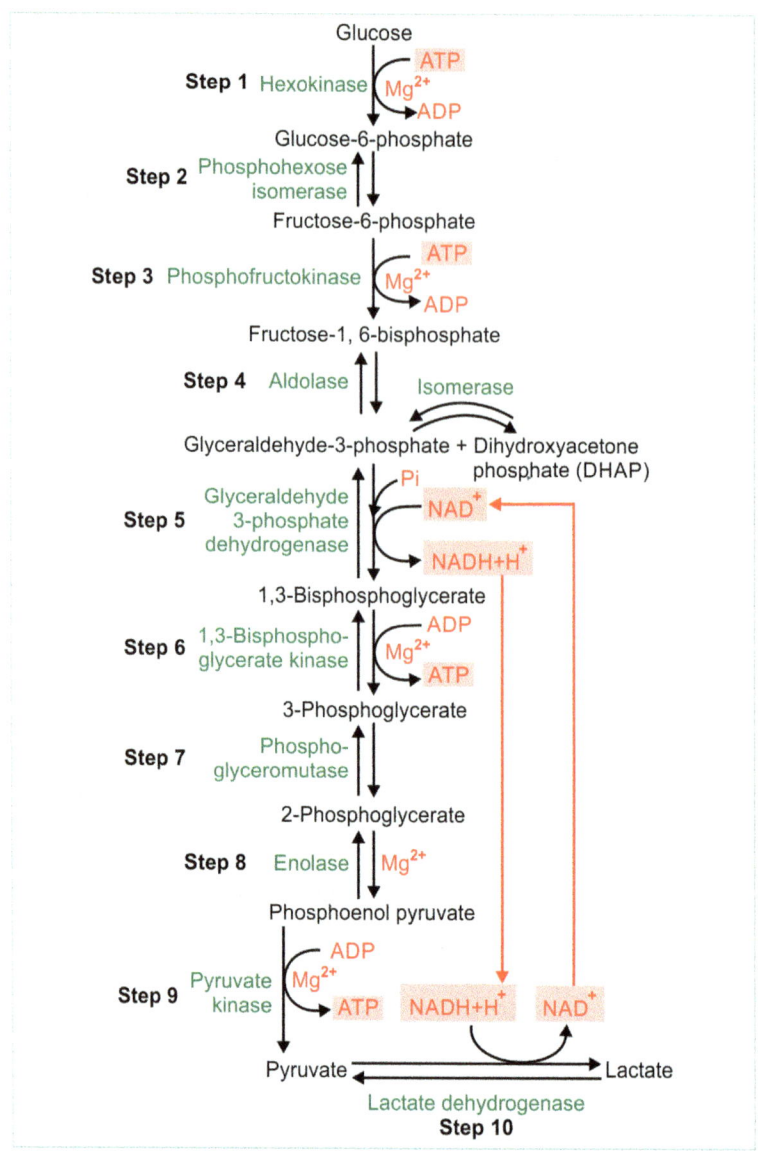

**Fig. 5.7:** Summary of glycolysis (Embden–Meyerhof pathway). Steps 1, 3, and 9 are key enzymes; these reactions are irreversible. Steps 5, 6, and 9 produce energy. Steps 5 and 10 are coupled for regeneration of nicotinamide adenine dinucleotide ($NAD^+$).
(ATP: adenosine triphosphate; ADP: adenosine diphosphate; NADH: nicotinamide adenine dinucleotide hydrogen)

### Energetics of Aerobic Glycolysis

During aerobic glycolysis (when oxygen is available), when one molecule of glucose is converted into pyruvate (glycolysis), eight ATPs are generated. The ATP generation steps are shown in **Table 5.2**.

## Chapter 5 Carbohydrate Metabolism

**Table 5.2:** Energy yield (number of ATP generated) per molecule of glucose in the glycolytic pathway, under aerobic conditions (oxygen is available).

| Step No. | Enzyme | Reaction | Source | No. of ATPs gained per glucose molecule |
|---|---|---|---|---|
| 1 | Hexokinase | Glucose to glucose-6-phosphate | | –1 |
| 3 | Phosphofructokinase | Fructose-6-phosphate to fructose-1,6-bisphosphate | | –1 |
| 5 | Glyceraldehyde-3-phosphate dehydrogenase | Glyceraldehyde-3-phosphate to 1,3-bisphosphoglycerate | 2 NADHs | 3 × 2 = 6 |
| 6 | 1,3-bisphosphoglycerate kinase | 1,3-bisphosphoglycerate to 3-phosphoglycerate | 2 ATPs | 1 × 2 = 2 |
| 9 | Pyruvate kinase | Phosphoenolpyruvate to pyruvate | 2 ATPs | 1 × 2 = 2 |
| **Total** | | | | **10 – 2 = 8** |

(ATP: adenosine triphosphate; NADH: nicotinamide adenine dinucleotide hydrogen)

*Note:*
a. From one molecule of glucose two molecules of glyceraldehyde 3-phosphate are formed; hence from step 5 onward, the molecules are doubled.
b. Calculations were made assuming that in the electron transport chain, NADH produces three ATPs. This will amount to a net generation of eight ATPs per glucose molecule in the glycolytic pathway.

## Anaerobic Glycolysis

In anaerobic condition, when oxygen is not available, pyruvate is reduced to lactate by lactate dehydrogenase (LDH). In step 5, from each molecule of glucose, 2 NADH are produced. The availability of coenzymes inside a cell is limited. Therefore, this step becomes a bottleneck in the whole reaction sequence.

For smooth operation of the pathway, the NADH is to be reconverted to $NAD^+$. This can be done by oxidative phosphorylation. However, during exercise, there is lack of oxygen. So this reconversion is not possible. Therefore the cell has to couple some other reaction in which $NAD^+$ is regenerated in the cytoplasm itself. Hence pyruvate is reduced to lactate; the $NAD^+$ thus generated is reutilized for the uninterrupted operation of the pathway. Thus steps 5 and 10 are coupled during anaerobic glycolysis.

In RBCs, there are no mitochondria. Hence RBCs derive energy only through glycolysis, where the end product is lactic acid.

In anaerobic glycolysis, the NADH produced is utilized for the conversion of pyruvate to lactate, and hence the total energy produced is 4 – 2 = 2 ATPs. Detailed steps are shown in **Table 5.3**. Please compare **Table 5.2** and **Table 5.3**.

## Significance of Lactate Production

The steps 5 and 10 of glycolysis are coupled (**Fig. 5.7**). In step 5, for each molecule of glucose entering in the pathway, two molecules of $NAD^+$ are reduced to NADH. The availability of coenzymes inside a cell is limited. Therefore, this step becomes a bottleneck in the whole reaction sequence. For smooth operation of the pathway, the NADH is to be reconverted to $NAD^+$. This can be done by oxidative phosphorylation. However, during exercise, there is lack of oxygen. So this reconversion is not possible. Therefore, the cell has to couple some other reaction in which $NAD^+$ is regenerated in the cytoplasm itself. Hence, pyruvate is reduced to lactate; the $NAD^+$ thus

**Table 5.3:** Energy yield (number of ATPs generated) per molecule of glucose in the glycolytic pathway, under anaerobic conditions (oxygen is not available).

| Step No. | Enzyme | Reaction | Source | No. of ATPs gained per glucose molecule |
|---|---|---|---|---|
| 1 | Hexokinase | Glucose to glucose-6-phosphate | | –1 |
| 3 | Phosphofructokinase | Fructose-6-phosphate to fructose-1,6-bisphosphate | | –1 |
| 6 | 1,3-bisphosphoglycerate kinase | 1,3-bisphosphoglycerate to 3-phosphoglycerate | 2 ATPs | 1 × 2 = 2 |
| 9 | Pyruvate kinase | Phosphoenolpyruvate to pyruvate | 2 ATPs | 1 × 2 = 2 |
| **Total** | | | | **4 – 2 = 2** |

generated is reutilized for the uninterrupted operation of the pathway **(Fig. 5.7)**.

## Fate of Pyruvate

When oxygen is not available, pyruvate is made to lactate, which then goes to Cori cycle, described later in the chapter. But when oxygen is available, pyruvate is oxidized further to acetyl-CoA, which then enters the citric acid cycle, when large quantity of energy is trapped as ATPs.

## Pyruvate Dehydrogenase Reaction

Inside the mitochondria, pyruvate is **oxidatively decarboxylated** to acetyl CoA by pyruvate dehydrogenase (PDH). It is a multienzyme complex with five coenzymes and three apoenzymes. The coenzymes needed are:
1. Thiamine pyrophosphate (TPP)
2. Coenzyme A (CoA)
3. FAD
4. $NAD^+$
5. Lipoamide

In this reaction, NADH is generated, which then gives rise to three ATP molecules.

**Pyruvate dehydrogenase reaction is a completely irreversible process.** There is no pathway available in the body to circumvent this step. Glucose through this step is converted to acetyl CoA from which fatty acids can be synthesized. But the backward reaction is not possible, and so **there is no net synthesis of glucose from fat.**

Pyruvate may be channeled back to glucose through gluconeogenesis. But when pyruvate becomes acetyl CoA, it cannot go back. Thus, PDH step is the **committed step** toward oxidation of glucose.

## GLUCONEOGENESIS

Gluconeogenesis is the synthesis of new glucose molecules from noncarbohydrate precursors. The precursors for gluconeogenesis are (a) lactate, (b) pyruvate, (c) glucogenic amino acids, (d) glycerol, and (e) propionyl CoA. It is to be emphasized that **fatty acids could not be converted to glucose.**

Gluconeogenesis is important for the maintenance of glucose levels during starvation or vigorous exercise. The brain and erythrocytes depend almost entirely on glucose for energy. Gluconeogenesis mainly occurs in liver. The pathway takes place partly in mitochondria and partly in cytoplasm.

In glycolysis, glucose is converted to two molecules of pyruvate. On the other hand, in gluconeogenesis, two molecules of pyruvate are converted to one molecule of glucose, utilizing six molecules of ATP for energy purposes. Part of the gluconeogenesis appears to be a reversal of glycolysis; but three steps of glycolysis are irreversible. These steps are the

**Table 5.4:** Key enzymes of glycolysis and gluconeogenesis.

| Irreversible steps in glycolysis | Corresponding key gluconeogenic enzymes |
|---|---|
| Pyruvate kinase (step 9) | Pyruvate carboxylase; phosphoenolpyruvate carboxykinase |
| Phosphofructokinase (step 3) | Fructose-1,6-bisphosphatase |
| Hexokinase (step 1) | Glucose-6-phosphatase |

reactions catalyzed by (a) **hexokinase** (step 1 of glycolysis), (b) **phosphofructokinase** (step 3 of glycolysis) and (c) **pyruvate kinase** (step 9 of glycolysis) **(Table 5.4)**. These steps are circumvented by the following reactions:
1. Pyruvate is converted to oxaloacetate by the enzyme **pyruvate carboxylase (Fig. 5.8)**. The oxaloacetate so formed is then converted to phosphoenolpyruvate (PEP) by the enzyme **phosphoenolpyruvate carboxykinase** (PEPCK) **(Fig. 5.8)**. Thus pyruvate is converted back to phosphoenolpyruvate by two steps circumventing the irreversible step catalyzed by pyruvate kinase.
2. Fructose-1,6-bisphosphate is dephosphorylated by the enzyme fructose-1,6-bisphosphatase **(Fig. 5.9)**. This reaction circumvents step 3 of glycolysis.
3. **Glucose-6-phosphatase** converts glucose-6-phosphate to glucose **(Fig. 5.10)**. This effectively circumvents step 1 of glycolysis. This enzyme is absent in muscle and brain and active in liver.

The metabolic **significance of gluconeogenesis** is the maintenance of blood glucose levels. The body's store of glycogen will be used up within 12 hours. In starvation, lipid and protein catabolism provides the

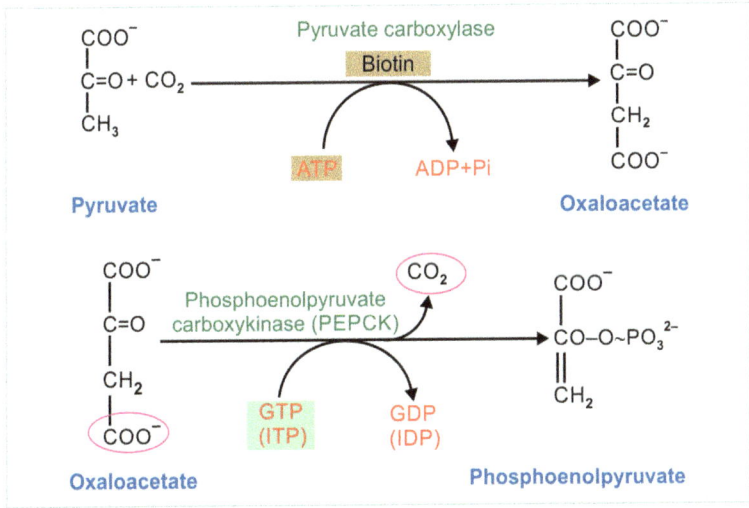

**Fig. 5.8:** Reversal of step 9 of glycolysis is achieved by two enzymes.
(GTP: guanosine triphosphate; GDP: guanosine diphosphate; ITP: immune thrombocytopenia purpura)

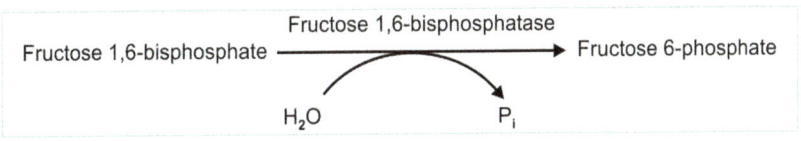

**Fig. 5.9:** Reversal of step 3 of glycolysis.

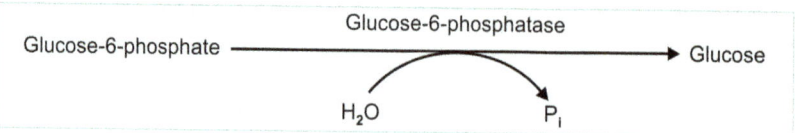

**Fig. 5.10:** Reversal of step 1 of glycolysis.

precursors for gluconeogenesis. Glucagon and glucocorticoids increase gluconeogenesis, whereas insulin has an inhibitory effect.

## CORI CYCLE

In strenuous exercise, pyruvate is converted to lactate in muscle. But lactate cannot be metabolized further without getting it converted back to pyruvate.

As oxygen is not enough in muscle, lactate could not be made to pyruvate in muscle. So, lactate from the muscle diffuses to blood stream, which is taken to the liver where it is converted back to pyruvate by the enzyme lactate dehydrogenase (LDH). This pyruvate is converted to glucose by gluconeogenesis and enters into the blood, ready to be taken up by the skeletal muscles. This cycle of reaction is called Cori cycle **(Fig. 5.11)**. By means of this cycle, the lactate generated in muscle is efficiently utilized.

Students should not confuse lactate with lactose. Lactate or lactic acid is the metabolic product of glucose. Lactose is a disaccharide, made up of galactose and glucose.

## GLYCOGENOLYSIS

The breakdown of glycogen to glucose is called glycogenolysis. Glycogen is the storage form of carbohydrate in humans. The major sites of storage are the liver and muscles. The main function of glycogen is to provide glucose for maintaining the blood glucose level. All the enzymes for this process are present in cytoplasm.

The enzyme **glycogen phosphorylase** removes glucose units one at a time from the nonreducing end of glycogen molecule by attacking the alpha-1,4-glycosidic linkages **(Fig. 5.12)**. This enzyme cannot attack the alpha-1,6 linkage at the branching point. Trisaccharide units are transferred from the branching point to another branch by the enzyme **glucan transferase**. Then **debranching enzyme** hydrolyzes the remaining glucose units as free glucose. These two enzymes together will convert the branching point to a linear one.

Phosphorylase reaction produces glucose-1-phosphate, which is converted to glucose-6-phosphate by phosphoglucomutase. Next, hepatic glucose-6-phosphatase hydrolyzes glucose-6-phosphate to glucose (see

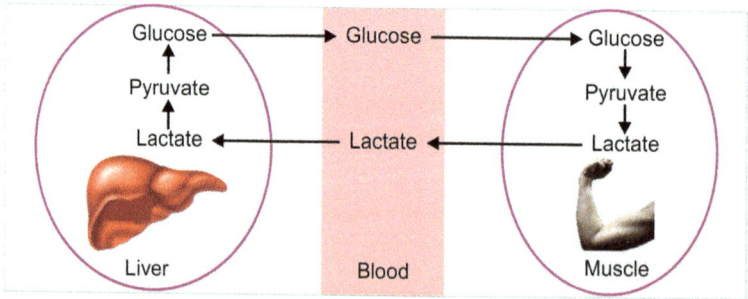

**Fig. 5.11:** Cori cycle. Contracting muscle has lack of oxygen. So pyruvate is reduced to lactate. This can be reconverted to glucose in liver by gluconeogenesis.

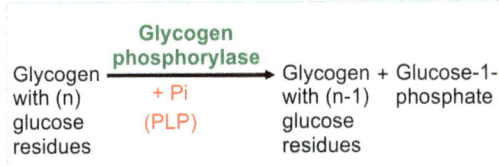

**Fig. 5.12:** Key enzyme of glycogenolysis (glycogen phosphorylase).

gluconeogenesis, **Fig. 5.10**). The free glucose is released to the bloodstream.

**Important note**: Muscle will not release glucose to the bloodstream, because muscle tissue does not contain **glucose-6-phosphatase**. Therefore, blood glucose maintenance is by liver glycogenolysis and not by muscle.

## Functions of Glycogen

- Glycogen is the storage form of carbohydrates in the human body. The major sites of storage are liver and muscle. The major function of liver glycogen is to provide glucose during starvation.
- When blood glucose level lowers, liver glycogen is broken down and helps to maintain blood glucose level.
- The function of muscle glycogen is to act as reserve fuel for muscle contraction.
- After taking food, blood sugar tends to rise, which causes glycogen deposition in liver. About 5 hours after taking food, the blood sugar tends to fall. But, glycogen is lysed to glucose so that the energy needs are met.

- After about 18 hours of fasting, most of the liver glycogen is depleted, when depot fats are hydrolyzed and energy requirement is met by fatty acid oxidation.

## GLYCOGEN SYNTHESIS (GLYCOGENESIS)

The synthesis of glycogen from glucose is known as glycogenesis. The main pathway is shown in **Figure 5.13**. Uridine diphosphate (UDP)-glucose is added to glycogen primer to lengthen the chain; the key enzyme is **glycogen synthase.**

When the glycogen primer length reaches 11 or 12 glucose units, the **branching enzyme** shifts 6 or 8 glucose residues to another site to form the branch alpha-1,6 linkage. Each branch will be formed by at least four glucose residues away from the existing branch (**Fig. 5.14**).

## Regulation of Glycogen Metabolism

The key enzyme for glycogenolysis is **glycogen phosphorylase**, which is activated by glucagon and adrenaline, under the stimulus of hypoglycemia. Epinephrine and glucagon can activate liver glycogen phosphorylase, but glucagon has no effect on the muscle.

The key enzyme for glycogen synthesis is **glycogen synthase**, the activity of which is decreased by adrenaline but is enhanced by insulin, under the stimulus of hyperglycemia. The activities of glycogen synthase and phosphorylase are reciprocally regulated.

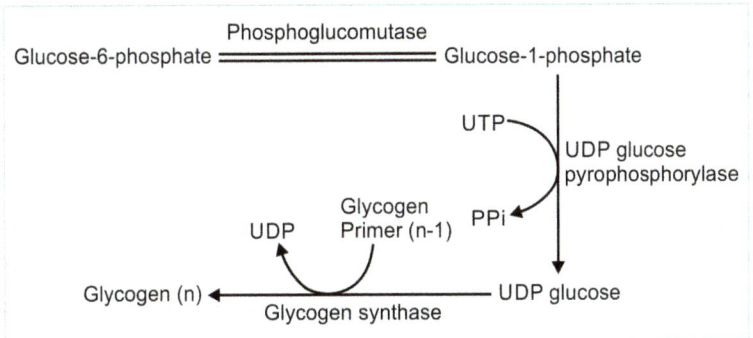

**Fig. 5.13:** Summary of glycogen synthesis pathway; glycogen synthase is the key enzyme.
(UTP: uridine triphosphate; UDP: uridine diphosphate)

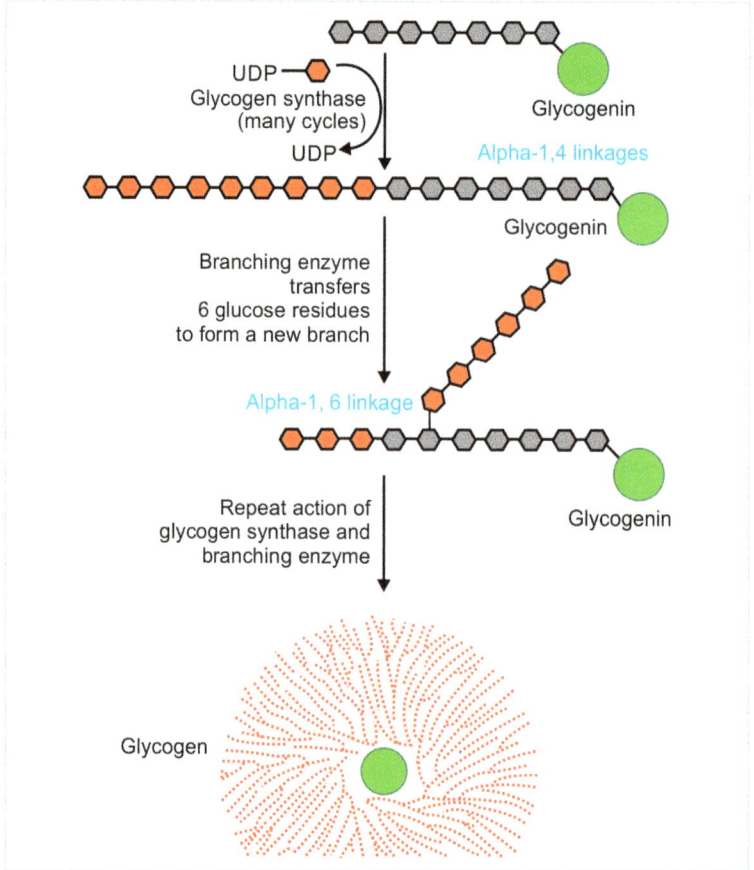

**Fig. 5.14:** Glycogen synthesis; glycogen synthase is the key enzyme.

## GLYCOGEN STORAGE DISEASES

They are a group of diseases due to inborn errors of metabolism resulting in the accumulation of abnormal quantity and/or type of glycogen in the liver and in the muscles as well as in the renal tissues. Hypoglycemia and ` are the common features. Life expectancy is very low, and the child may die at around 2–3 years.

### Glycogen Storage Disease Type I

It is also called **Von Gierke's Disease.** It is the most common type of glycogen storage diseases. The enzyme **glucose-6-phosphatase** is deficient. Salient clinical features are **fasting hypoglycemia**, **lactic acidosis**, and **ketosis**. Glycogen gets deposited in liver leading to massive liver enlargement and **cirrhosis**. Children usually die in early childhood. Treatment is to give small quantity of food at frequent intervals. Twelve different types of glycogen storage diseases are described; some important ones are shown in **Table 5.5**.

## HEXOSE MONOPHOSPHATE SHUNT PATHWAY/PENTOSE PHOSPHATE PATHWAY/ PHOSPHOGLUCONATE OXIDATIVE PATHWAY/ DICKENS–HORECKER PATHWAY

About 10% of the glucose ingested is metabolized through this pathway. The metabolism takes place in cytoplasm. This pathway is important in RBCs, adrenal cortex,

**Table 5.5:** Important glycogen storage diseases.

| Type | Name of the disease | Enzyme deficiency |
|---|---|---|
| I | Von Gierke's disease | Glucose-6-phosphatase |
| II | Pompe's disease | Lysosomal maltase |
| III | Cori's disease (limit dextrinosis) | Debranching enzyme |
| IV | Andersen's disease (amylopectinosis) | Branching enzyme |
| V | McArdle's disease | Muscle phosphorylase |

liver, testes, mammary gland, adipose tissues, ovaries, and so on.
- This pathway provides pentoses for nucleotide synthesis.
- This pathway produces NADPH.
  - NADPH is necessary for reductive biosynthetic reactions, such as fatty acids, cholesterol, and steroids synthesis.
  - NADPH helps to maintain the integrity of RBCs by maintaining the glutathione in reduced form.
  - NADPH prevents the accumulation of methemoglobin by keeping the iron in reduced form.
  - NADPH helps to maintain the transparency of the lens of the eyes.
  - NADPH also helps to produce superoxide anions inside macrophages so that microbes are killed.

The reactions of hexose monophosphate shunt (HMP) shunt pathway start with the dehydrogenation of glucose-6-phosphate by the enzyme glucose-6-phosphate dehydrogenase **(Fig. 5.15)**. This is the key enzyme of the pathway. The overall reactions of the pathway are summarized in **Figure 5.16**.

Students should note that NAD+ (nucleotide adenine dinucleotide) and NADP+ (nucleotide adenine dinucleotide phosphate) are different. NADP differs from NAD+ in having an additional phosphate group. NADH is used for reducing reactions in catabolic pathways, e.g., pyruvate to lactate. NADH generates ATP. NADPH is used for reductive biosynthetic reactions, e.g., de novo synthesis of fatty acid, synthesis of cholesterol, and so on. NADPH is generated mainly by the HMP shunt pathway. NADPH will not generate ATP.

The HMP shunt pathway is regulated by the level of NADP. The first step is catalyzed by the enzyme, **glucose-6-phosphate dehydrogenase (G6PD)**. It is the rate-limiting reaction. G6PD may be deficient in some individuals, which leads to **drug-induced hemolytic anemia**. Deficiency of this enzyme will also lead to an increased level of methemoglobin.

## GLUCURONIC ACID PATHWAY

### Importance of the Glucuronic Acid Pathway

It provides **UDP-glucuronic acid,** which is the active form of glucuronic acid. It is used for the following purposes:
- Conjugation of bilirubin
- Conjugation of steroids
- Conjugation of various drugs, which will make them more water-soluble and more easily excretable. Barbiturates, antipyrine, and aminopyrine will increase the uronic acid pathway, leading to the availability of more glucuronate for conjugation purpose.
- Synthesis of glycosaminoglycans (GAGs).
- Vitamin C synthesis in lower animals. The enzyme L-**gulonolactone oxidase is absent**

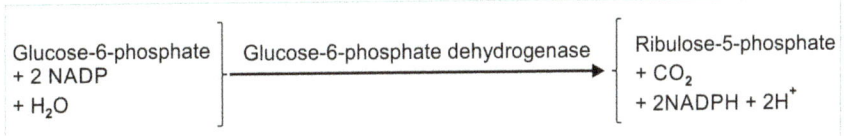

**Fig. 5.15:** First step of HMP shunt pathway.

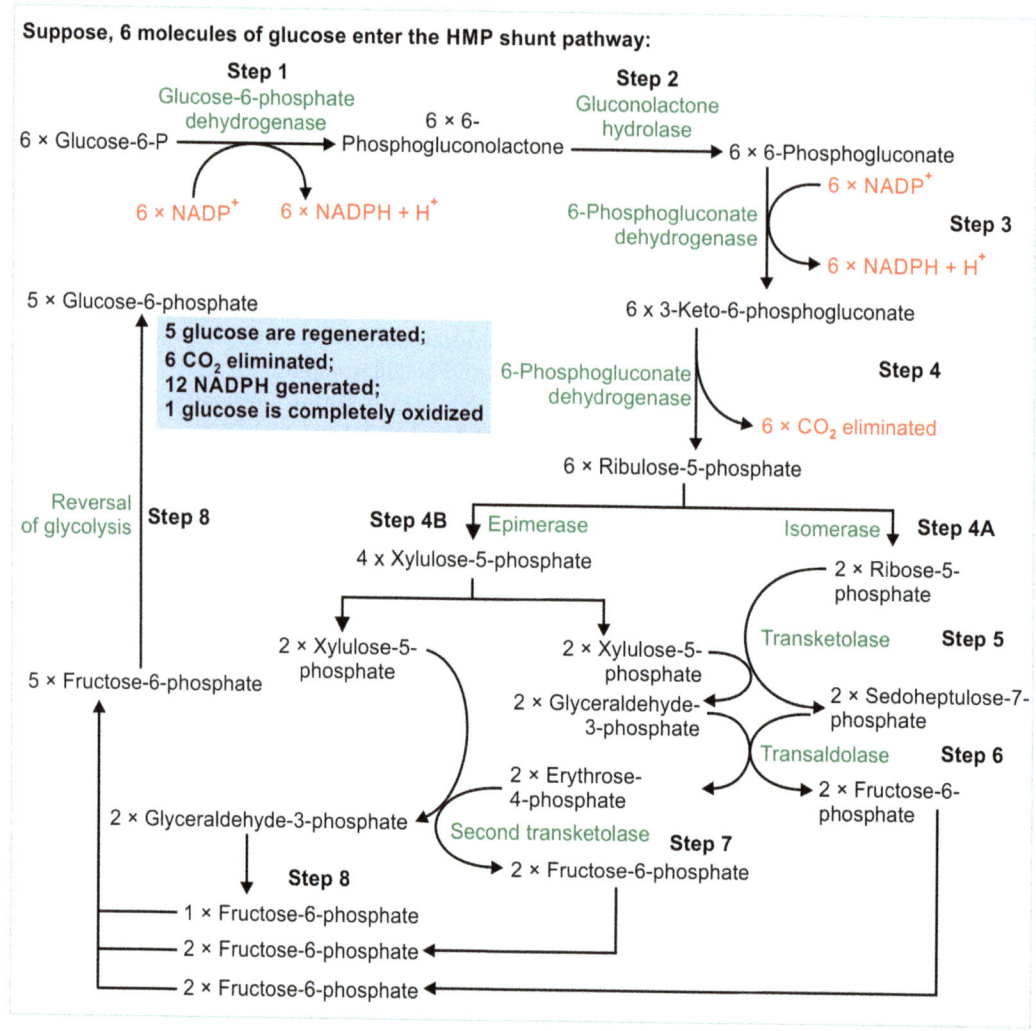

Fig. 5.16: Hexose monophosphate shunt pathway.

in human beings, primates, guinea pigs, and bats. Therefore, ascorbic acid cannot be synthesized by these organisms. Hence ascorbic acid is an essential nutrient in the diet of human beings.

**Essential pentosuria**: It is an inborn error of metabolism arising from the deficiency of the enzyme (**xylitol dehydrogenase**) of this pathway, leading to the excretion of xylitol in urine.

## FRUCTOSE METABOLISM

Fructose is present in honey and many fruits. Fructose is a constituent of sucrose. Fructose is metabolized in liver. Fructose is first phosphorylated to fructose-1-phosphate by **fructokinase (Fig. 5.17)**. Fructokinase phosphorylates the substrate at **first position**, whereas hexokinase action is on the sixth position. Fructokinase is not dependent on insulin. So fructose is more

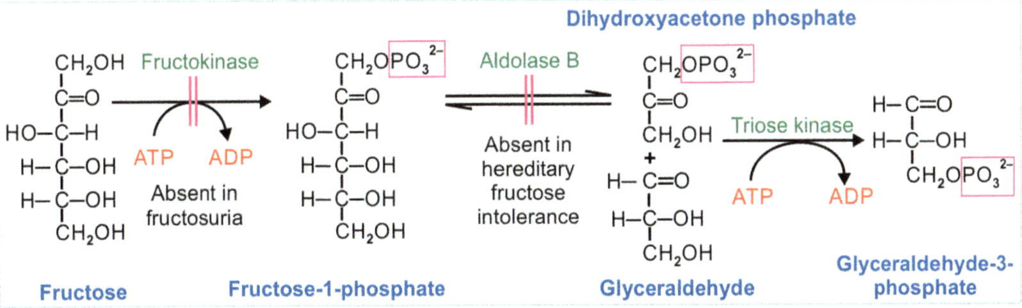

**Fig. 5.17:** Metabolism of fructose.
(ATP: adenosine triphosphate; ADP: adenosine diphosphate)

rapidly utilized in normal persons. Fructose is seen in large quantities in seminal plasma.

Deficiency of fructokinase leads to the excretion of fructose in urine (**fructosuria**). It is a benign disorder.

On the other hand, the defect of aldolase B results in the accumulation of fructose 1-phosphate in blood as it cannot be metabolized. This condition is known as **fructose intolerance**, which is a serious disorder. This is a hereditary disorder characterized by vomiting, accumulation of glycogen in liver, hypoglycemia, and often hepatomegaly and jaundice. The treatment is withdrawal of fructose from diet.

## GALACTOSE METABOLISM

Galactose is a constituent of lactose, the milk sugar. Galactose is a constituent of glycolipids and glycosaminoglycans. Inside the intestines, lactose is hydrolyzed to glucose and galactose by the enzyme **lactase**. The deficiency of lactase will result in a condition known as **lactose intolerance**. After the age of 40, people will have less lactase enzyme in their intestines, and so the condition is more prevalent in aged people. It is characterized by vomiting, diarrhea, flatulence, and abdominal distention.

Lactose is metabolized in liver. Galactose is first phosphorylated by **galactokinase** to galactose-1-phosphate (step 1, **Fig. 5.18**).

## Galactosemia

It is an inborn error of metabolism. There is deficiency of the enzyme **galactose-1-phosphate uridyltransferase (Fig. 5.18**, step; 2). Due to the block in this enzyme,

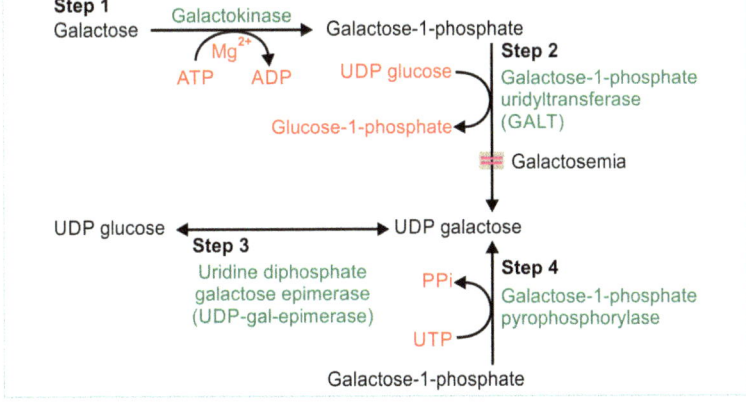

**Fig. 5.18:** Galactose metabolism.

galactose-1-phosphate will accumulate in liver. This will inhibit galactokinase as well as glycogen phosphorylase. **Hypoglycemia** is the result. Bilirubin conjugation is reduced; so **unconjugated bilirubin** level is increased in blood, leading to jaundice. There is enlargement of liver and severe **mental retardation (Fig. 5.19)**.

Free galactose accumulates, leading to **galactosemia.** It is partly excreted in urine (**galactosuria**). Accumulation of galactose in the eye lens results in **congenital cataract** and is a very characteristic feature of galactosemia. Galactose-1-phosphate may get deposited in renal tubules, producing tubular damage leading to generalized **amino aciduria**.

Diagnosis is done by detecting congenital cataract and presence of galactose in urine as well as elevated blood galactose levels. Collection of fetal cells by amniocentesis may be useful in prenatal diagnosis.

**Treatment:** If lactose is withdrawn from the diet, most of the symptoms recede. But mental retardation, when established, will not improve. Hence early detection is most important. For an affected infant, **lactose-free diet** is given.

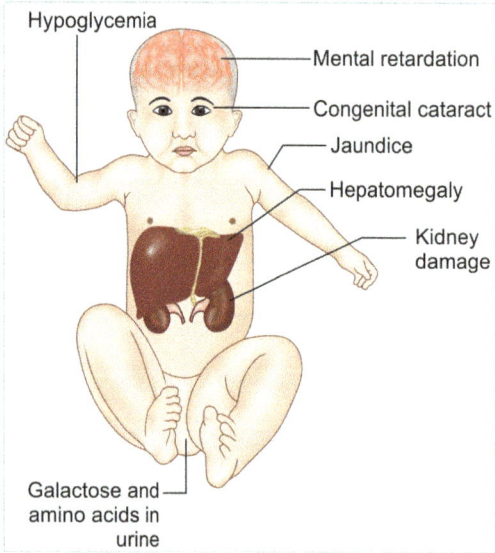

**Fig. 5.19:** Salient features of galactosemia.

## SUMMARY OF THE CHAPTER

- Glycolysis occurs both in aerobic and anaerobic conditions. Anaerobic glycolysis is the major source of energy for muscles, when the muscle tissue lacks oxygen.
- Phosphofructokinase (PFK) is the regulatory or rate-limiting enzyme of glycolysis. It is an allosteric enzyme. Adenosine monophosphate (AMP) is an allosteric activator, while citrate and adenosine triphosphate (ATP) are allosteric inhibitors.
- Energy-generating steps of glycolysis are catalyzed by the enzymes glyceraldehyde-3-phosphate dehydrogenase (NADH); 1,3-bisphosphoglycerate kinase (ATP), and pyruvate kinase (ATP).
- Cori cycle ensures efficient reutilization of lactate produced in the muscle.
- Energy yield per molecule of glucose in the glycolytic pathway under anaerobic conditions is two ATPs.
- Under anaerobic conditions, pyruvate is reduced to lactate by lactate dehydrogenase. Under aerobic conditions, it is oxidatively decarboxylated to acetyl coenzyme A (CoA) by the enzyme complex pyruvate dehydrogenase (PDH).
- Key enzymes of gluconeogenesis are pyruvate carboxylase, phosphoenolpyruvate carboxykinase, fructose-1,6-bisphosphatase, and glucose-6-phosphatase.
- The HMP shunt pathway, also known as pentose phosphate pathway (PPP) generates NADPH required for reductive biosynthesis of biomolecules such as steroids, fatty acids, and cholesterol.
- NADPH generated as a result of HMP pathway is essential to maintain the transparency of the eye lens to prevent methemoglobinemia and to maintain erythrocyte membrane integrity.
- Major substrates for gluconeogenesis are lactate and glucogenic amino acids.
- Glycogen is the storage polysaccharide of the body. It is stored mainly in the liver and muscle.
- Glycogen phosphorylase is activated by glucagon and adrenaline, while glycogen synthase is activated by insulin.
- Glycogen storage diseases (GSDs) are inborn errors of metabolism. Type 1 is called von Gierke's disease.

- GPD deficiency is a common clinical condition transmitted as X-linked recessive trait. Ingestion of fava beans (favism) and antimalarial drugs precipitate the manifestations.

## QUESTIONS FOR SELF-ASSESSMENT

1. Describe the process of digestion and absorption of carbohydrates with a schematic diagram.
2. What is the importance of glycolysis in our body? Mention the steps.
3. Write a note on gluconeogenesis.
4. Mention the role of hormones in carbohydrate metabolism.
5. Mention the importance of glycogenesis and glycogenolysis in our body.
6. Mention the steps of HMP shunt pathway and mention its importance.
7. What are the minor metabolic pathways in carbohydrate metabolism?
8. What are glycogen storage diseases?
9. Mention the various glycogen storage disease in a tabular form.
10. Write a note on galactosemia.

## MULTIPLE CHOICE QUESTIONS (MCQs)

5-1. In glycolysis pathway, glucose is aerobically converted to:
   a. Lactate
   b. Pyruvate
   c. Acetate
   d. Phosphate

5-2. All the enzymes are common in glycolysis and gluconeogenesis, *except*:
   a. Phosphofructokinase
   b. Glyceraldehyde 3-phosphate dehydrogenase
   c. Pyruvate kinase
   d. Phosphoenolpyruvate carboxykinase

5-3. The following pathway is independent of ATP:
   a. Glycogenolysis
   b. Gluconeogenesis
   c. Pentose phosphate pathway
   d. Uronic acid pathway

5-4. Cyclic adenosine monophosphate (cAMP) is involved in:
   a. Glycolysis
   b. Glycogenesis
   c. Cori cycle
   d. Gluconeogenesis

5-5. Glycogen phosphorylase is activated by:
   a. Insulin
   b. Noradrenaline
   c. Glucagon
   d. Dopamine

5-6. NADPH generated as a result of HMP pathway is essential to all, *except*:
   a. Maintain transparency of the eye lens
   b. Prevent methemoglobinemia
   c. Maintain erythrocyte membrane integrity
   d. Transport of oxygen

5-7. Galactosemia is caused due to:
   a. Galactose-1-phosphate uridyltransferase
   b. Galactokinase
   c. Epimerase
   d. Uridyl phosphate transferase

5-8. Essential pentosuria occurs due to deficiency of the enzyme:
   a. Fructokinase
   b. Aldolase B
   c. Triose kinase
   d. Xylitol dehydrogenase

5-9. Glycogen storage diseases (GSDs) are inborn errors of metabolism. Type 2 is called:
   a. Pompe's disease
   b. Cori disease
   c. McArdle's disease
   d. Von Gierke's disease

5-10. Energy yield per molecule of glucose in the glycolytic pathway under anaerobic conditions is:
   a. 10 ATPs
   b. 2 ATPs
   c. 6 ATPs
   d. 4 ATPs

## ANSWER KEYS TO MCQs

5-1. (b)   5-2. (d)   5-3. (c)   5-4. (b)   5-5. (c)   5-6. (d)   5-7. (a)   5-8. (d)
5-9. (a)   5-10. (b)

CHAPTER 6

# Regulation of Blood Glucose and Diabetes Mellitus

At the completion of this chapter, the reader will be able to answer questions on the following topics:

- Normal blood glucose level
- Regulation of blood glucose level
- Action of insulin
- Diabetes mellitus classification
- Diagnostic criteria for diabetes mellitus
- Oral glucose tolerance test
- Glycated hemoglobin (HbA1c)
- Estimation of blood glucose
- Reducing substances in urine

## BLOOD GLUCOSE LEVEL

Plasma glucose analyzed at any time of the day, without any prior preparations, is called **random blood glucose**. Sugar estimated in the early morning, before taking any breakfast is called **fasting blood glucose**. Fasting state means, glucose is estimated after an overnight fast (12 hours after the food; also known as postabsorptive state). If the test is done about 2 hours after a good meal, it is called **postprandial blood glucose**. When plasma glucose level is within normal limits, it is referred to as **normoglycemia**. Normal blood sugar values are:
- Fasting blood glucose (FBG): 70–110 mg/dL
- Random blood sugar (RBS): 80–140 mg/dL
- Postprandial blood glucose (PPBG): 80–140 mg/dL

**Hyperglycemia** is the condition when blood glucose values are above the normal level. If the blood sugar value goes above 180 mg/dL, the normal thresholds of glucose, sugar will appear in urine (**glucosuria**). If the blood sugar value is less than normal, it is called **hypoglycemia**. Hyperglycemia is harmful in the long run, but hypoglycemia is fatal.

## Factors Maintaining Blood Glucose Level

Hormones regulate the blood glucose level. Insulin reduces blood glucose (hypoglycemic hormone), while glucagon and glucocorticoids increase blood glucose (hyperglycemic hormones). Blood glucose level at an instant depends on the balance between glucose entering and leaving the blood. The major factors that cause entry of glucose into blood are:
- Absorption from intestines
- Glycogenolysis (breakdown of glycogen)
- Gluconeogenesis

On the other hand, the factors leading to depletion of glucose in blood are:
- Utilization by tissues for energy
- Glycogen synthesis
- Conversion of glucose into fat (lipogenesis)

## Regulation of Blood Glucose Level

About 2–2.5 hours after a meal, the blood glucose level falls to near fasting levels. It may go down further, but this is prevented by these two processes: (1) for another 3 hours, hepatic glycogenolysis will take care of the blood glucose level; (2) thereafter, gluconeogenesis will take charge of the situation. Liver is the major organ that supplies the glucose for maintaining blood glucose level.

The blood sugar values are controlled by many factors, which can be classified as hypoglycemic and hyperglycemic factors. Insulin is the most important hypoglycemic hormone, whereas glucagon, glucocorticoids, and catecholamines are the major hyperglycemic hormones. An overview of the regulatory mechanism is shown in **Figure 6.1**.

## Effect of Insulin

Insulin lowers blood glucose by the following mechanisms:
- Insulin favors glycogen synthesis.
- It promotes glycolysis.
- It inhibits gluconeogenesis.
- It increases lipogenesis.

Insulin increases the utilization of glucose. Insulin is secreted by the beta cells of pancreas. Its secretion is controlled by the level of glucose in blood.

## INSULIN

The word *insulin* was derived from Latin *insula*, meaning "island" (islet). In 1869, Langerhans identified the alpha and beta cells in the islets of pancreas. In 1922, Banting and Best extracted insulin from pancreas. Insulin is a protein hormone with two polypeptide chains. Insulin is synthesized and secreted by the beta cells of the islets of Langerhans of the pancreas.

Insulin acts by binding to a plasma **membrane receptor** on the target cells. In obesity, the number of receptors is decreased, and target tissue becomes less sensitive to insulin. Normally, insulin binds to its receptor which, in turn, starts many protein activation cascades. These include translocation of glucose transporter type 4 (GluT4) to the plasma membrane and influx of glucose, glycogen synthesis, glycolysis, and fatty acid synthesis. The main effect of the activation of the insulin receptor is increased glucose

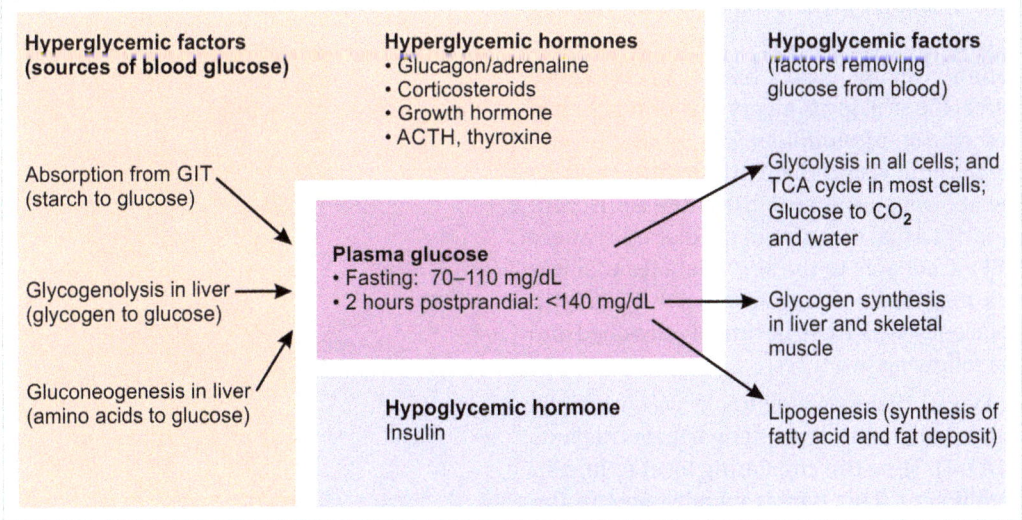

**Fig. 6.1:** Overview of regulation of blood glucose.
(GIT: gastrointestinal tract; TCA: citric acid cycle)

uptake. For this reason "insulin insensitivity," or a decrease in insulin receptor signaling, leads to diabetes mellitus type 2.

## Incretin Hormones

The incretins are hormones that work to increase insulin secretion. Nutrient intake stimulates the secretion of the gastrointestinal incretin hormones, such as glucagon-like peptide-1 (GLP-1) and glucose-dependent insulinotropic polypeptide (GIP). The predominant incretin hormone is GLP-1. Incretin stimulation of beta cells causes them to secrete *more insulin* in response to the same amount of blood glucose. Glucagon-like peptide-1 suppresses glucagon secretion. Glucagon-like peptide-1 is rapidly degraded by the enzyme dipeptidyl peptidase-4 (DPP-4). Dipeptidyl peptidase-4 inhibitors such as sitagliptin and saxagliptin increase endogenous GLP-1 concentration.

## DIABETES MELLITUS

The term *diabetes* was derived from the Greek words *dia* (= through) and *bainein* (= to go), and *diabetes* literally means "pass through." The disease causes loss of weight as the body mass is passed through the urine. The Greek word *mellitus* means "sweet," as it was known to early workers that the urine of the diabetic patient contains sugar. Charaka, in his treatise (circa 400 BC), gives a very elaborate clinical description of *madhumeha* (= sweet urine). He had the vision that carbohydrate and fat metabolisms are altered in this disease. He also described that the urine of the diabetic patient will attract ants to the site. Diabetes mellitus is a metabolic disorder caused by insulin deficiency, which can be broadly classified into the following two types:

**1. Type 1** diabetes mellitus: It was originally named as insulin-dependent diabetes mellitus (IDDM). Here the circulating level of insulin is deficient. This type is usually seen in the younger age groups, and hence it is also known as **juvenile diabetes**.

**2. Type 2** diabetes mellitus: It was originally named as non-insulin-dependent diabetes mellitus (NIDDM). It is due to the decreased biological response to insulin or insulin resistance. This is the common type usually seen after the middle age, and hence it is also known as **maturity onset diabetes.**

## Biochemical Explanation of Diabetes

Type 2 disease is commonly seen in individuals above 40 years. About 60% of patients are **obese.** These patients have insulin resistance. This is due to the defect in GluT4 receptors in muscle cells **(Fig. 6.2)**. The patients have **high or normal plasma insulin** levels. Insulin resistance develops as a consequence of excess accumulation of fat in liver and skeletal muscle. Due to the absence of insulin, glycolysis is inhibited and gluconeogenesis is favored. Blood sugar level is increased and glucose is excreted through urine. As a large quantity of glucose is to be excreted, a large quantity of water is also excreted. Fat is broken down; so acetyl coenzyme A (acetyl-CoA) is in plenty. This could

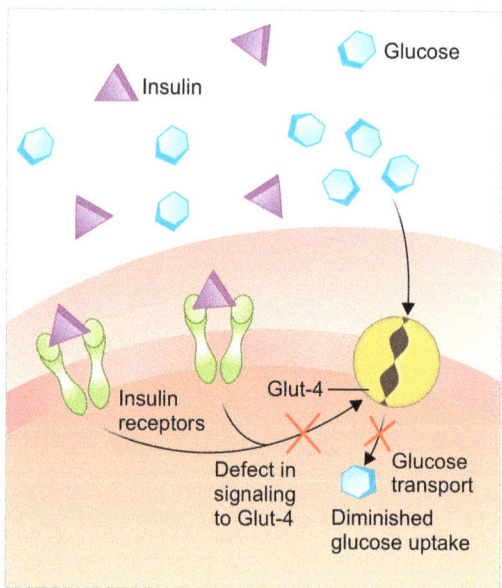

**Fig. 6.2:** Insulin resistance in diabetes mellitus type 2. The GluT4 receptors are defective in muscle cells.

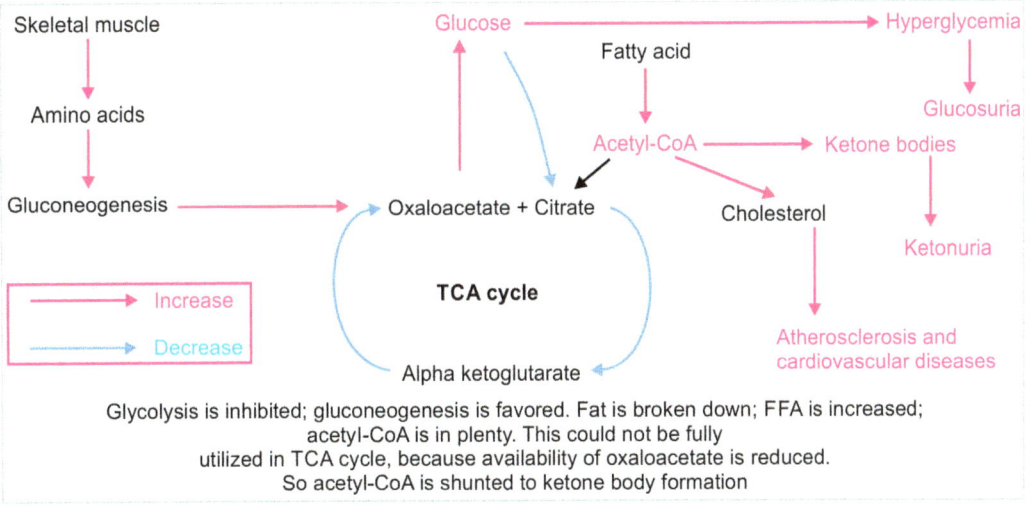

Fig. 6.3: Metabolic derangements in diabetes mellitus.

not be fully utilized in the TCA cycle, because availability of oxaloacetate is reduced. So acetyl-CoA is shunted to ketone body formation, as well as to cholesterol formation. This leads to atherosclerosis and other complications of diabetes mellitus **(Fig. 6.3)**.

## Clinical Presentations in Diabetes Mellitus

- When the blood glucose level exceeds the renal threshold, glucose is present in the urine (**glucosuria**).
- Due to the osmotic effect, more water accompanies glucose (**polyuria**).
- To compensate for this loss of water, more water is taken (**polydipsia**).
- To compensate for the loss of glucose and protein, patient will take more food (**polyphagia**). Polyuria, polydipsia, and polyphagia are called the cardinal symptoms of diabetes mellitus.
- The loss and ineffective utilization of glucose leads to the breakdown of fat and protein. This would lead to **loss of weight.**
- Often the presenting complaint of the patient may be **chronic recurrent infections** such as boils and abscesses. Any person with recurrent infections should be investigated for diabetes. In India, **tuberculosis** is commonly associated with diabetes.

## Diabetic Ketoacidosis

Diabetic ketoacidosis (DKA) is an acute complication of diabetes mellitus. Ketone body formation and ketoacidosis are explained in Chapter 8. Normally the level of ketone bodies in the blood is < 1 mg/dL, and urine contains no ketone bodies. But when the rate of synthesis exceeds, there will be accumulation of ketone bodies in blood. This leads to *ketonemia*, excretion of ketones in urine *(ketonuria)*, and smell of *acetone* in breath. All these three together constitute the condition known as *ketoacidosis*.

The presence of ketosis can be established by the detection of ketone bodies in urine by **Rothera's test**. Saturate 5 mL of urine with solid ammonium sulfate. Add a few drops of freshly prepared sodium nitroprusside followed by 2 mL of liquor ammonia along the sides of the test tube. Development of a purple ring indicates the presence of ketone bodies in urine. Strip tests based on the same principle are also available. Criteria for DKA include serum glucose above 250 mg/dL, serum anion gap above 10 mEq/L, bicarbonate below 18 mEq/L, serum pH < 7.30, and presence of ketosis.

The urine of a patient with diabetic ketoacidosis will give positive Benedict's test as well as Rothera's test. But in **starvation** ketosis, Benedict's test is negative, but Rothera's test will be positive.

*Consequences of Ketoacidosis*

- **Metabolic acidosis:** Acetoacetate and beta-hydroxybutyrate are acids. When they accumulate, metabolic acidosis results (*see* Chapter 19). There will be increased **anion gap**.
- **Reduced buffers:** The plasma bicarbonate is used up for the buffering of these acids.
- **Kussmaul's respiration:** Patients will have typical acidotic breathing due to compensatory hyperventilation.
- **Smell of acetone** in patient's breath.
- **Osmotic diuresis** induced by ketonuria may lead to dehydration.
- **Sodium loss:** The ketone bodies are excreted in urine as their sodium salt, leading to loss of cations from the body.
- **High potassium in blood:** It is due to lowered uptake of potassium by cells in the absence of insulin.
- **Dehydration:** The sodium loss further aggravates the dehydration.
- **Coma:** Hypokalemia, dehydration, and acidosis contribute to the lethal effect of ketoacidosis.

*Management of Ketoacidosis*

- Parenteral administration of insulin and glucose by intravenous route to control diabetes.
- Intravenous administration of bicarbonate to correct the acidosis.
- Correction of water imbalance by normal saline.
- Correction of electrolyte imbalance higher than insulin induces glycogen synthesis, and along with that extracellular potassium is transferred back to intracellular compartment. This may cause dangerous hypokalemia, which should be promptly corrected.
- Treatment of underlying precipitating causes, such as infection.

## Chronic Complications of Diabetes Mellitus

- **Vascular diseases:** Atherosclerosis in medium-sized vessels, **plaque formation,** and consequent intravascular thrombosis may take place. If it occurs in cerebral vessels, the result is paralysis. If it is in coronary artery, myocardial infarction results. In the case of small vessels, the process is called **microangiopathy,** which may lead to diabetic retinopathy and nephropathy.
- **Complications in the eyes:** Early development of **cataract** of lens is due to hyperglycemia. Retinal microvascular abnormalities lead to **retinopathy** and blindness.
- **Neuropathy:** Peripheral neuropathy with paresthesia is very common. Neuropathy may lead to the risk of foot ulcers and gangrene. Hence care of the feet in diabetic patients is important.
- **Nephropathy:** Nephropathy is another complication of diabetes, characterized by proteinuria and renal failure. Persistent hyperglycemia leading to the glycation of basement membrane proteins may be the cause of nephropathy.
- **Complications in pregnancy:** Diabetic mothers tend to have big babies, because insulin is an anabolic hormone. Chances of abortion, premature birth, and intrauterine death of the fetus are also more, if the diabetes is not properly controlled.

*Metabolic Syndrome*

Metabolic syndrome (MetS) is characterized by
- Impaired glucose tolerance
- Hyperinsulinemia and insulin resistance
- Obesity

- Hyperlipidemia or dyslipidemia
- Hypertension
- Excess nutrients
- Reduced physical activity

The hallmarks are abdominal obesity and insulin resistance or decreased glucose tolerance. The body cannot properly use glucose even in the presence of a normal insulin level. In other words, body cannot use insulin efficiently. Therefore, the metabolic syndrome is also called the **insulin resistance syndrome**. People with the MetS are at increased risk of coronary heart disease and type 2 diabetes. Associated conditions are polycystic ovary disease (PCOD) and nonalcoholic fatty liver disease (NAFLD). If the condition is not managed with lifestyle modifications like diet and exercise to lose weight, it may progress to type 2 diabetes mellitus.

## Laboratory Investigations in Diabetes Mellitus

- Fasting blood glucose level, random blood sugar estimation, and oral glucose tolerance tests are used for the diagnosis **(Table 6.1)**. Hyperglycemia and glucosuria will be the hallmarks. For monitoring a diabetic patient, periodic check of blood glucose level (fasting and postprandial) is to be done. Blood glucose level has to be maintained within the normal limits.
- Glycated hemoglobin or glycohemoglobin or HbA1c test needs to be done every 1-2 months. This is described later.

**Table 6.1:** The plasma glucose levels in OGTT in normal persons and in diabetic patients.

|  | Normal persons | Criteria for diagnosing diabetes mellitus |
|---|---|---|
| Fasting | <110 mg/dL | >126 mg/dL |
| 1 hour (peak) after glucose | <160 mg/dL | Not prescribed |
| 2 hours after glucose | <140 mg/dL | >200 mg/dL |

- Complete lipid profile, including total cholesterol, triglycerides, and HDL and LDL cholesterol levels may be done once in every 6 months (*see* Chapter 9).
- Kidney function tests such as blood urea and serum creatinine may be done at least once in a year (*see* Chapter 20).
- Microalbuminuria and frank albuminuria. The presence of albumin (50-300 mg/day) in urine is known as microalbuminuria (*see* Chapter 20). It is a predictor of progressive renal damage. Albumin level > 300 mg/day indicates overt diabetic nephropathy. Microalbuminuria is to be checked at least once in a year.

### Oral Glucose Tolerance Test

The diagnosis of diabetes mellitus is usually confirmed after conducting an oral glucose tolerance test. After 12-14 hours of fasting, glucose (75 g in 500 mL of water) is given orally. Blood and urine samples are collected before the administration of glucose and also at 30, 60, 90, 120, and 150 minutes after glucose administration. In normal persons, the urine samples will not contain glucose, and fasting blood glucose will be < 110 mg/dL; and the peak blood glucose will be reached within 1-1.5 hours, while the value will be < 160 mg/dL. The values may be plotted in a graph **(Fig. 6.4)**.

### Diagnostic Criteria for Diabetes Mellitus

The criteria for the diagnosis of diabetes mellitus are shown in **Table 6.1**.
- If the fasting blood glucose is > 126 mg/dL, on more than one occasion.
- Or, if 2-hour post-glucose, load value of OGTT is more than 200 mg/dL (even at one occasion).
- Or, if both fasting and 2-hour values are above these levels, on the same occasion.
- If the random blood glucose level is > 200 mg/dL on more than one occasion. Diagnosis should not be based on a single random test alone; it should be repeated.

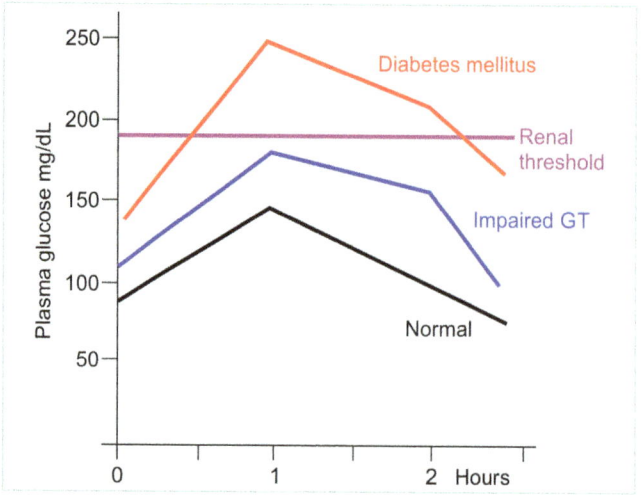

**Fig. 6.4:** Oral glucose tolerance test (OGTT).

- Glycated hemoglobin (HbA1c) level > 6.5% at any occasion. This is the preferred method for the initial diagnosis of diabetes mellitus.

*Oral Glucose Load Test*

Oral glucose load test (GLT) is performed in women who are suspected to have gestational diabetes. Here 75 g of glucose is given orally after overnight fasting, and the blood glucose value is estimated before and 1 hour after the administration of glucose. In normal condition, the fasting blood glucose will be within the normal limits, and the value 1 hour after the ingestion of glucose will be < 140 mg/dL.

**Indications for an OGTT** are: (1) patient with symptoms suggestive of diabetes mellitus, (2) excess weight gain during pregnancy, and (3) ruling out benign glucosuria.

**Contraindications** for an OGTT: (1) There is no need for doing OGTT in a confirmed case of diabetes. (2) GTT has no role in follow up of diabetes. (3) The test should not be done in acutely ill patients.

Conditions that can be assessed by OGTT are impaired glucose tolerance, impaired fasting glycemia, gestational diabetes, alimentary glucosuria, and renal glucosuria.

*Impaired Glucose Tolerance*

**Impaired glucose tolerance (IGT)** is otherwise called as **impaired glucose regulation** (IGR). Here plasma glucose values are above the normal level but below the diabetic levels **(Fig. 6.4)**. In IGT, the fasting plasma glucose level is between 110 and 126 mg/dL, and 2-hour postprandial glucose value is between 140 and 200 mg/dL **(Fig. 6.4)**. Such persons need careful follow-up because IGT progresses to frank diabetes at the rate of 2% patients per year.

## Gestational Diabetes Mellitus

The term *gestational diabetes mellitus (GDM)* is used when carbohydrate intolerance is noticed, for the first time, during a pregnancy. **A known diabetic patient, who becomes pregnant, is not included in this category.** An OGTT with a 75 g glucose load should be done to confirm or exclude GDM. Women with GDM are at increased risk for subsequent development of frank diabetes. GDM is associated with an increased incidence of **neonatal mortality**. Maternal hyperglycemia causes the fetus to secrete more insulin, causing stimulation of fetal growth and increased birth weight. After the childbirth, the women should be reassessed.

## Alimentary Glucosuria

Here the fasting and 2-hour post-glucose values are normal, but an exaggerated rise in blood glucose following the ingestion of glucose is seen. This is due to an increased rate of absorption of glucose from the intestine. This is seen in patients after a **gastrectomy** or in patients with **hyperthyroidism**.

## Renal Glucosuria

**Normal renal threshold for glucose is 175–180 mg/dL.** If blood glucose rises above this, glucose starts to appear in urine. Generally, the increased blood glucose level is reflected in urine. But even when renal **threshold is lowered,** glucose is excreted in urine. In these cases, the blood glucose levels are within normal limits. This is called renal glucosuria **(Fig. 6.4)**.

## Estimation of Blood Glucose

Estimation of glucose is the commonest analysis done in clinical laboratories. The blood is collected using an anticoagulant (potassium oxalate) and an inhibitor of glycolysis (sodium fluoride). **Fluoride** inhibits the enzyme enolase, and so glycolysis on the whole is inhibited. If fluoride is not added, blood cells will metabolize glucose, and a false low value may be obtained during the interval between blood collection in the ward and estimation in the laboratory. Capillary blood from finger tips may also be used for glucose estimation by strip method. Plasma is separated for glucose estimation. Modern techniques use serum samples also.

### Enzymatic (Glucose Oxidase Peroxidase) Method

Glucose oxidase (GOD) is very specific; it converts glucose to gluconic acid and hydrogen peroxide. Peroxidase (POD) converts the $H_2O_2$ into $H_2O$ and nascent oxygen. The oxygen oxidizes a colorless chromogenic substrate (e.g., ortho-dianisidine) to a colored one; the color intensity is directly proportional to the concentration of glucose.

As a modification, this GOD reaction mixture is immobilized on a plastic film (**dry analysis**). One drop of blood is placed over the reagent. The color is developed within 1 minute. The intensity of the dye is measured by reflectance photometry. The instrument is named as **glucometer**. It is useful for the self-monitoring of blood glucose (SMBG) by patients at home. But the instrument is less accurate.

### Glycated Hemoglobin or Glycohemoglobin or HbA1c

"Hb" stands for hemoglobin; "A1" is the normal hemoglobin; and "c" stands for the commonest form of glycated hemoglobin. HbA1c is the best index of long-term control of blood glucose level. When blood glucose level is increased, the glucose molecules are attached to all proteins in the body, including hemoglobin [seen inside the red blood cells (RBCs)]. Once attached, glucose is not removed from hemoglobin. Therefore, it remains inside the erythrocytes, throughout the life span of RBCs (120 days). Thus HbA1c level reveals the mean glucose level of the previous 10–12 weeks. It is unaffected by recent food intake or recent changes in blood glucose levels. This estimation should be done every 1–2 months in all diabetic patients, so as to understand the effectiveness of the treatment.

Normal level of HbA1c is less than 5.5% of total hemoglobin. Values between 5.6% and 6.4% are considered as impaired glucose tolerance. Value above 6.5% means that the person is diabetic. In a known diabetic patient who has well-controlled diabetes, the value should be below 7%, and levels above 8% indicate poor control of diabetes mellitus.

## REDUCING SUBSTANCES IN URINE

Normally glucose is not excreted in urine. But if blood glucose is > **180 mg/dL,** urine contains glucose. The blood level of glucose above which glucose is excreted is called **renal threshold.**

The excretion of reducing substances in urine is detected by a positive **Benedict's test**. About 0.5 mL of urine is boiled with 5 mL Benedict's reagent for 2 minutes (or kept for 2 minutes in water bath that is already boiling). The test is semiquantitative and the color of the precipitate roughly parallels the concentration of reducing sugar. Blue color indicates the absence of sugar in urine. The green precipitate means 0.5%, yellow (1%), orange (1.5%), and red indicates 2% or more of sugar (1% means 1 g per 100 mL). Nowadays strips are available, which when dipped in urine will give the color, if the urine contains sugar.

Any reducing sugar will give a positive Benedict's test. So differentiation of various sugars that may be present in urine has practical importance. Such conditions together are sometimes called as "mellituria." Differential diagnosis of a positive Benedict's test is shown in **Table 6.2**.

## Management of Diabetes Mellitus

- **Diet and exercise:** This is the first-line treatment. A diabetic patient is advised to take a balanced diet with high protein content, low calories, devoid of refined sugars and low saturated fat, adequate polyunsaturated fatty acids (PUFAs), low cholesterol, and sufficient quantities of fiber. Vegetables are the major sources of minerals, vitamins, and fiber. Alpha-glucosidase inhibitor acarbose inhibits the alpha-glucosidases present in the small intestine. So absorption of glucose is reduced. This allows the pancreas to more effectively regulate insulin secretion. It is effective at reducing the fasting plasma glucose (FPG) levels and levels of glycated hemoglobin (HbA1c).
- **Oral hypoglycemic agents:** There are several types of oral hypoglycemic agents (OHA) now in use. These are sulfonylurea and biguanides (metformin).
- **Insulin injections:** Insulin is the drug of choice in type 1 diabetes. It is also used in type 2 diabetes, where oral drugs are not sufficient. The availability of human insulin prepared by recombinant DNA technology has markedly improved the response of patients.
- **Prevention of complications:** Acute complication of diabetes is ketoacidosis. Chronic complications are atherosclerosis, cataract, neuropathy, and nephropathy. Correct treatment of diabetes will prevent most of these dreaded complications.

## Hypoglycemia

Hyperglycemia causes harm, but hypoglycemia is **fatal**. A fall in plasma glucose < 50 mg/dL is life-threatening. Most common cause is the **overdose of insulin.** The differentiation of hypoglycemic coma from hyperglycemic coma (ketoacidosis) is important, since treatment is exactly opposite. The diagnosis is mainly based on blood glucose estimation.

### SUMMARY OF THE CHAPTER

- Major factors that cause an increased level of glucose in blood are absorption in intestines, glycogenolysis, and gluconeogenesis. A continuous and adequate supply of glucose is essential for the brain, erythrocytes, and renal medulla.
- Major factors that cause depletion of glucose in blood are utilization by tissues,

**Table 6.2:** Differential diagnosis of reducing substances in urine.

1. Glucosuria
    a. Diabetes mellitus
    b. Alimentary glucosuria
    c. Renal glucosuria
2. Fructosuria
    a. Deficiency of fructokinase
    b. Fructose intolerance (aldolase B deficiency)
3. Lactosuria
4. Galactosuria (deficiency of galactose-1-phosphate uridyltransferase)
5. Pentosuria (xylulosuria)
6. Non-carbohydrate-reducing substances
    a. Glucuronides, salicylate
    b. Ascorbic acid (vitamin C)

glycogenesis, and conversion to fat. Blood glucose level varies significantly during the fasting state and in postprandial state (after food).
- The glucose levels are measured in plasma after collecting blood in a tube with oxalate and fluoride. True glucose values are given by enzymatic method (GOP-POD method). Indications for an oral glucose tolerance test (OGTT) are patients with symptoms suggestive of diabetes mellitus, excess weight gain during pregnancy, and ruling out benign glucosuria.
- Contraindications for an OGTT are known case of diabetes mellitus, following the prognosis of diabetes mellitus, and performing on acutely ill patients.
- Conditions that can be assessed by OGTT are impaired glucose tolerance, impaired fasting glycemia, gestational diabetes, alimentary glucosuria, and renal glucosuria.
- Reducing substances in urine other than glucose are fructose, lactose, galactose, pentoses, homogentisic acid, salicylates, glucuronides, and ascorbic acid.
- Insulin has the following biochemical effects: increases uptake of glucose by cells, enhances utilization of glucose, hypoglycemic, antilipolytic, antiketogenic, and favors lipogenesis.
- Insulin acts via a specific insulin receptor present in the cells of insulin-responsive tissues. This affects a signal transduction pathway, which leads to regulation of gene transcription, DNA synthesis, and activation of enzymes.
- Diabetes mellitus is of two types: type 1 and type 2. The type 1 is also known as insulin-dependent diabetes mellitus (IDDM), while the type 2 was previously known as non-insulin-dependent diabetes mellitus (NIDDM).
- Diabetic ketoacidosis (DKA), lactic acidosis, and hypoglycemia are acute metabolic complications of diabetes mellitus.
- Retinopathy, neuropathy, and vascular diseases are chronic complications of diabetes mellitus.
- Glycated hemoglobin (HbA1c) is used as an index for long-term control of blood glucose level.

- The fasting plasma glucose (FPG) denotes glucose level after overnight fasting of 8–10 hours. Postprandial glucose (2-hour PPG) is measured 2 hours after taking food.
- The term *random plasma glucose* (RPG) is used when blood is collected regardless of the time of the previous meal.
- Normal fasting blood glucose value is 110 mg/dL and 2-hour PPG value is 140 mg/dL. *Hyperglycemia* refers to elevated glucose levels.
- A plasma level of glucose below 50 mg/dL is called hyperglycemia.
- Fasting value between 111 and 125 mg/dL indicates impaired glucose tolerance.
- A diagnosis of diabetes mellitus is made when fasting plasma glucose is above 126 mg/dL and 2-hour PPG is above 200 mg/dL.
- Renal threshold for glucose is 180 mg/dL, above which glucose is excreted even at a lower level. In renal glucosuria the glucose tolerance is normal, but glucose is excreted in urine.
- Insulin is secreted in response to an increase in plasma glucose level.
- Insulin receptors are located in the plasma membrane with two alpha and two beta subunits.
- Binding of insulin to alpha subunits activates the insulin response substrate (IRS). The activated IRS in turn will activate other enzyme systems and cascades causing metabolic effects.
- Insulin recruits GluT4 in cells to the membrane, enhancing glucose uptake.

## QUESTIONS FOR SELF-ASSESSMENT

1. What is the normal serum fasting glucose level?
2. Explain with a schematic diagram how blood glucose level is maintained.
3. What are the consequences of elevated blood glucose levels?
4. What is OGTT and mention its significance.
5. Mention the role of renal threshold for glucose.
6. Define diabetes mellitus and classify it.
7. Mention the cardinal symptoms of diabetes mellitus and its long-term complications.

8. What are the tests performed to confirm diabetes?
9. What is the role of glycated hemoglobin?
10. Mention the points for the management of diabetes mellitus.

## MULTIPLE CHOICE QUESTIONS (MCQs)

6-1. **Diabetes mellitus patients are prone to develop cataracts because the elevated blood glucose concentration:**
   a. Inhibits glycogenesis
   b. Increases glycated hemoglobin
   c. Increases glycogen synthesis within the lens
   d. Allows aldose reductase to reduce glucose to sorbitol

6-2. **The normal serum fasting blood glucose level is:**
   a. 80-140 mg/dL
   b. 70-110 mg/dL
   c. 100-150 mg/dL
   d. 80-120 mg/dL

6-3. **Chronic complications of diabetes mellitus are all, *except*:**
   a. Retinopathy
   b. Neuropathy
   c. Vascular diseases
   d. Dermatitis

6-4. **Insulin is secreted in response to:**
   a. Decrease in plasma glucose level
   b. Increase in plasma glucose level
   c. Increase in fructose level
   d. Increase in glycogen level

6-5. **Insulin recruits _____ in cells to the membrane, enhancing glucose uptake:**
   a. GluT1
   b. GluT5
   c. GluT3
   d. GlutT4

6-6. **Which is not a clinical complication of type 2 diabetes mellitus?**
   a. Cardiomyopathy
   b. Retinopathy
   c. Nephropathy
   d. Neuropathy

6-7. **Which blood parameter helps to know about the 1-month past history of blood glucose?**
   a. Insulin
   b. Cholesterol
   c. C-peptide
   d. HbA1c

6-8. **All the following are the criteria for the diagnosis of diabetes, *except* one:**
   a. Persons with a family history of diabetes
   b. Persons with a BMI higher than 27 Kg/m$^2$
   c. Persons with an elevated urea level
   d. Persons with an impaired glucose tolerance

6-9. **A 72-year-old woman with a 16-year history of type 2 diabetes is brought to a family practice clinic by his elder son. The patient is unable to give a clear account of how carefully she controls her blood glucose. Which of the following laboratory parameters could be used to assess the glycemic control over the past 3–4 months?**
   a. Fasting blood glucose
   b. Blood insulin levels
   c. Glycated hemoglobin (HbA1c)
   d. Urinary glucose

6-10. **A diagnosis of diabetes mellitus is made when 2-hour PPG is above:**
   a. 200 mg/dL
   b. 126 mg/dL
   c. 140 mg/dL
   d. 150 mg/dL

## ANSWER KEYS TO MCQs

6-1. (d)   6-2. (b)   6-3. (d)   6-4. (b)   6-5. (d)   6-6. (a)   6-7. (d)   6-8. (c)
6-9. (c)`   6-10. (a)

# CHAPTER 7

# Lipid Chemistry

At the completion of this chapter, the reader will be able to answer questions on the following topics:

- Functions of lipids
- Fatty acids, saturated and unsaturated
- Essential fatty acids
- Classification of lipids
- Properties of lipids
- Waxes
- Phospholipids and glycerophospholipids
- Phosphatidic acid, lecithin, and cephalin
- Phosphatidylserine
- Sphingomyelin
- Cerebroside, ganglioside
- Lipoproteins
- Steroids

Lipids may be regarded as organic substances insoluble in water but soluble in organic solvents, such as ether, benzene, chloroform, acetone, petroleum ether, and so on. Lipids constitute a heterogeneous group of compounds of biochemical importance. They are esters of fatty acids with glycerol or alcohol.

## FUNCTIONS OF LIPIDS

- They are the concentrated fuel reserve of the body. The stored triglycerides in human body serve as a reservoir of energy during periods of restricted nutrition.
- Lipids are the constituents of membrane structures and regulate membrane permeability. Phospholipids (along with proteins) are major constituents of cellular membrane. In the myelin sheath of nerve they constitute about 60% of the dry weight. Cholesterol is also present in the cell membrane.
- They serve as a source of fat-soluble vitamins (A, D, E, and K).
- Lipids are important as cellular metabolic regulators. Steroid hormones are formed from cholesterol and prostaglandins from arachidonic acid.
- Lipids stored subcutaneously serve as an insulator against excessive heat loss to the environment.
- They also protect the internal organs and give shape and smooth appearance to the body.

## CLINICAL APPLICATIONS OF LIPIDS

- Excessive fat deposits cause obesity. Truncal obesity is a risk factor for heart attack.
- Abnormality in cholesterol and lipoprotein metabolism leads to atherosclerosis and cardiovascular diseases.
- In diabetes mellitus, the metabolisms of fatty acids and lipoproteins are deranged, leading to ketosis.

## CLASSIFICATION OF LIPIDS

Detailed classification is shown in **Table 7.1**. Based on their chemical nature, lipids are classified as:

1. **Simple lipids:** They are esters of fatty acids with glycerol or other higher alcohols.
2. **Compound lipids:** They are fatty acids esterified with alcohol, but in addition they contain other groups.
3. **Derived lipids:** They are compounds that are derived from lipids or precursors of lipids, e.g., fatty acids, steroids. Fatty acids are classified in **Table 7.2**. For details of cholesterol and steroids, see Chapter 9.
4. **Lipids complexed with other compounds:** Examples are proteolipids and lipoproteins. Plasma lipoproteins are described in Chapter 9.

**Table 7.1:** Classification of lipids.

1. Simple lipids
   - Triacylglycerol or triglycerides or neutral fat
   - Waxes
2. Compound lipids
   - Phospholipids, containing phosphoric acid
     - Nitrogen-containing glycerophosphatides
       a. Lecithin (phosphatidylcholine)
       b. Cephalin (phosphatidylethanolamine)
       c. Phosphatidylserine
     - Non-nitrogen glycerophosphatides
       a. Phosphatidylinositol
       b. Phosphatidylglycerol
     - Plasmalogens, having long-chain alcohols
       a. Choline plasmalogen
       b. Ethanolamine plasmalogen
     - Phosphosphingosides, with sphingosine
       Sphingomyelin
   - Nonphosphorylated lipids
     - Glycolipids (lipids with carbohydrate)
       a. Cerebrosides
       b. Gangliosides
     - Sulfolipids or sulfatides
3. Derived lipids
   - Fatty acids (**Table 7.2**)
   - Steroids (**Chapter 9**)
   - Prostaglandins
   - Leukotrienes
4. Lipids complexed with other compounds
   - Proteolipids
   - Lipoproteins

**Table 7.2:** Classification of fatty acids.

1. Depending on the total number of carbon atoms
   - **Even-chain fatty acids:** They have carbon atoms 2, 4, 6, and similar series. Most of the naturally occurring lipids contain even chain fatty acids
   - **Odd-chain fatty acids:** They have carbon atoms 3, 5, 7, and so forth. Odd-numbered fatty acids are seen in microbial cell walls. They are also present in milk
2. Depending on the length of the hydrocarbon chain
   - Short-chain fatty acids with 2–6 carbon atoms
   - Medium-chain fatty acids with 8–14 carbon atoms
   - Long-chain fatty acids with 16–22 carbon atoms
   - Very long-chain fatty acids with more than 24 carbon atoms
3. Depending on the nature of the hydrocarbon chain
   - Saturated fatty acids
   - Unsaturated fatty acids, which may be subclassified into:
     - Monounsaturated (monoenoic) having single double bond
     - Polyunsaturated (polyenoic) with two or more double bonds
     - Eicosanoids, derived from eicosa (20 carbon atoms) fatty acids, e.g., prostaglandins, prostacyclins, and thromboxanes

# FATTY ACIDS

Fatty acids are carboxylic acids with a hydrocarbon side chain. They are the simplest form of lipids. Fatty acids mainly occur in nature in the esterified form as constituents of lipids. They are also present as free (nonesterified) fatty acids. **Table 7.2** shows the classification of fatty acids found in fats and other lipids.

## Saturated Fatty Acids

They are the straight-chain saturated acids belonging to the acetic acid series. They have the general formula $(C_nH_{2n}O_2)$ or $(C_nH_{2n} + COOH)$. Physical properties of these fatty acids depend on their chain length. Fatty acids with 2 to 6 carbon atoms are called **short-chain** fatty acids, e.g., acetic acid (2 C), butyric acid (4 C), caproic acid (6 C). Those having 8 to 14 carbon atoms are named as **medium-chain** fatty acids, e.g., lauric acid with 12 C atoms and myristic acid with 14 C atoms. Those with 16 to 22 carbon atoms are **long-chain** fatty acids, e.g., palmitic acid (16 C) and oleic acid (18 C).

The fatty acids are named by adding the suffix "-anoic" after the hydrocarbon with the same number of carbon atoms. Thus butyric acid (4 C) is tetranoic acid, and palmitic acid (16 C) is hexadecanoic acid. The 2 C acetic acid and 4 C butyric acid are important as metabolic precursors, while palmitic acid and stearic acid are abundant in body fat.

Fatty acids are numbered starting with the carboxyl carbon as the first carbon, and so on, and the carbon adjacent to the -COO group is referred as "a" carbon, and the methyl end carbon is referred to as "w" carbon **(Fig. 7.1)**.

```
   6      5      4      3      2      1
  CH₃-CH₂-CH₂-CH₂-CH₂-COOH
  ω1    ω2    ω3    ω4    ω5
         δ     γ     β     α
```

**Fig. 7.1:** Nomenclature of carbon atoms.

## Properties of Fatty Acids

All fatty acids are weak acids. Short- and medium-chain fatty acids are liquids, whereas long-chain fatty acids are solids at room temperature. As the chain length increases, solubility in water decreases, and boiling and melting points increases with chain length.

Fatty acids form salts with alkali and esters with alcohol.

$$R\text{-}COOH + NaOH \rightarrow R\text{-}COONa + H_2O$$

Sodium and potassium salts of long-chain fatty acids are called soaps.

## Unsaturated Fatty Acids

They are characterized by the presence of one or more double bonds in their molecule. They are of three classes **(Fig. 7.2)**.

The unsaturated fatty acids are named by adding the suffix "-enoic," e.g., oleic acid (monounsaturated/monoenoic) is 18 C with one double bond. Linoleic acid (PUFA) is 18 C with two double bonds. Linolenic acid (PUFA) is 18 C with three double bonds, while arachidonic acid (PUFA) is 20 C with four double bonds. Their structures are shown in **Figure 7.3**.

### Clinical Significance of PUFA

- Linoleic and linolenic acids **(Fig. 7.3)** are called essential fatty acids because they

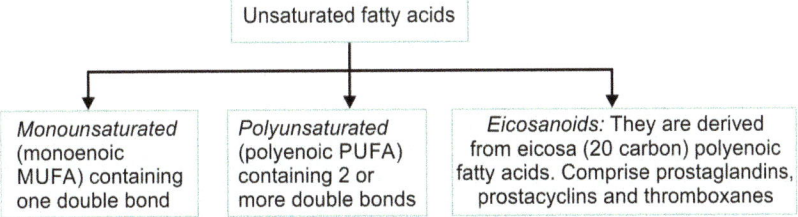

**Fig. 7.2:** Unsaturated fatty acids.

**Linoleic ($C_{18}$)** Δ9, 12 **(two double bonds)** (ω6 family)
$$CH_3-(CH_2)_4-CH=CH-CH_2-CH=CH-(CH_2)_7-COOH$$
$$\underset{18}{} \quad \underset{\omega 6}{} \quad \underset{12}{} \quad \underset{9}{} \quad \underset{1}{}$$

**Linolenic ($C_{18}$)** Δ9, 12, 15 **(three double bonds)** (ω3 family)
$$CH_3-CH_2-CH=CH-CH_2-CH=CH-CH_2-CH=CH-(CH_2)_7-COOH$$
$$\underset{18}{} \quad \underset{\omega 3}{} \quad \underset{15}{} \quad \underset{12}{} \quad \underset{9}{} \quad \underset{1}{}$$

**Arachidonic ($C_{20}$)** Δ5, 8, 11, 14 **(four double bonds)** (ω6 family)
$$CH_3-(CH_2)_4-CH=CH-CH_2-CH=CH-CH_2-CH=CH-CH_2-CH=CH-(CH_2)_3-COOH$$
$$\underset{20}{} \quad \underset{\omega 6}{} \quad \underset{14}{} \quad \underset{11}{} \quad \underset{8}{} \quad \underset{5}{} \quad \underset{1}{}$$

**Fig. 7.3:** Polyunsaturated fatty acids (PUFAs).

cannot be synthesized by the body and have to be supplied in the diet.
- Unsaturated fatty acids are also designated as
  - 3 (omega-3) family—linolenic acids **(Fig. 7.3)**
  - 6 family—linoleic and arachidonic acids **(Fig. 7.3)**
  - 9 family—oleic acid
- Arachidonic acid is the precursor of prostaglandins. Arachidonic acid can be synthesized in the body, if the essential fatty acids are supplied in the diet.
- Eicosanoids (eicosa = 20) are derived from 20 C arachidonic acid. They are polyenoic fatty acids. They are prostanoids (prostaglandins, prostacyclins, thromboxanes) and leukotrienes.

Because of the presence of the double bonds, these fatty acids exhibit geometrical isomerism. If the atoms or acyl groups are present on the same side of the double bond, it is a "*cis*" isomer and if the groups occur on opposite side it is a "*trans*" isomer **(Fig. 7.4)**. All naturally occurring fatty acids have *cis* configuration. This *cis* configuration is essential for the spatial arrangement of lipids in membranes. Trans-fatty acids are formed in the body during metabolic reactions.

*Properties of Unsaturated Fatty Acids*

The unsaturated fatty acids have lower melting point than their saturated analogue. This may be because of the loose packing of the unsaturated fatty acid molecule.

Unsaturated fatty acids form salts with alkali and esters with alcohol. Because of the presence of double bonds they undergo hydrogenation to corresponding saturated acids.

$$\text{Linolenic acid} \xrightarrow{+2H} \text{linoleic acid} \xrightarrow{+2H} \text{oleic acid}$$

*Iodination:* Unsaturated fatty acids react with halogens under mild conditions to form halogenated derivatives.

$$\text{Oleic acid} + I_2 \rightarrow \text{di-iodo oleic acid}$$

The number of halogen atoms taken up will depend on the number of double bonds and is an index of the degree of unsaturation. **Iodine number** is the number of grams of iodine absorbed by 100 grams of fat. It is a measure of the degree of unsaturation and is directly proportional to the content of unsaturated fatty acid.

$$CH_3-(CH_2)_7-CH \qquad HC-(CH_2)_7-CH_3$$
$$\| \qquad \qquad \|$$
$$HOOC-(CH_2)_7-CH \qquad HOOC-(CH_2)_7-CH$$

Cis form; Oleic acid      Trans form; Elaidic acid

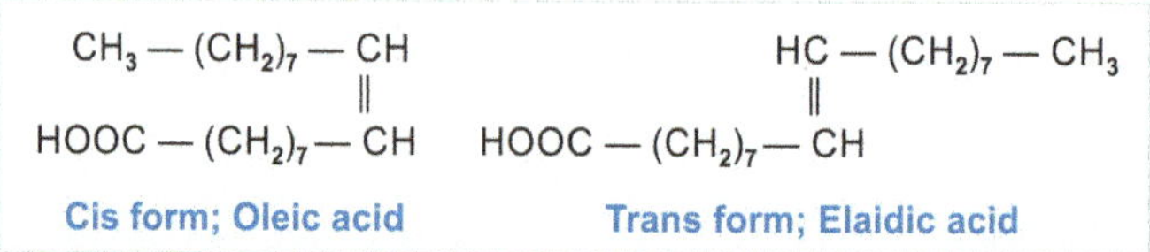

**Fig. 7.4:** Example of cis and trans forms of unsaturated fatty acid.

## Oxidation of Fatty Acids

Unsaturated fatty acids undergo autoxidation due to the presence of double bonds and a variety of products are formed, such as peroxide, epoxide, enediol, and so on. Rancidity of oils is due to the autoxidation splitting and aldehyde formation. It is prevented by naturally occurring antioxidants like vitamin E.

## ESSENTIAL FATTY ACIDS

The fatty acids that cannot be synthesized by the body and therefore should be supplied through diet are known as essential fatty acids (EFAs). There are three EFAs namely linoleic, linolenic, and arachidonic acids.

## Functions of Essential Fatty Acids

Essential fatty acids are required for the membrane structure and function. They are involved in the transport of cholesterol, formation of lipoproteins, and also in the prevention of fatty liver. They are also needed for the synthesis of prostaglandins. The deficiency of EFA leads to "phrynoderma" characterized by the presence of dermatitis, loss of hair, and poor wound healing.

## SIMPLE LIPIDS

Simple lipids are classified into (a) fats and oils and (b) waxes. Fats and oils are esters of fatty acids with glycerol. The structure of a neutral fat or triglyceride or triacylglycerols is given in **Figure 7.5**. Composition of some common oils is shown in **Table 7.3**.

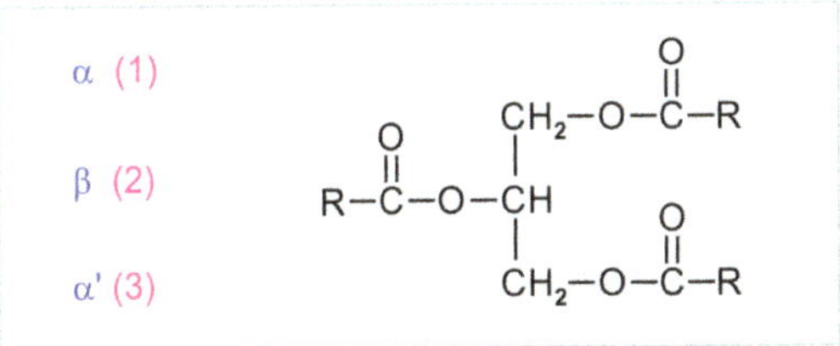

**Fig. 7.5:** Structure of a neutral fat. They are also called as triglycerides or triacylglycerols. Fatty acids are esterified with glycerol. R stands for fatty acids.

**Table 7.3:** Composition of oils and fats.

| Name | Saturated fatty acids (%) | Monounsaturated fatty acids (%) | Polyunsaturated fatty acids (%) |
|---|---|---|---|
| Coconut oil | *86 | 12 | 2 |
| Groundnut oil | 18 | 46 | 36 |
| Gingelly oil (til oil or sesame oil) | 13 | 50 | 37 |
| Palm oil | 42 | 52 | 6 |
| Cottonseed oil | 26 | 19 | 55 |
| Mustard oil | 34 | 48 | 18 |
| Safflower oil (kardi in Hindi) | 9 | 12 | 79 |
| Sunflower oil | 12 | 24 | 64 |
| Butter | 75 | 20 | 5 |
| Ox (tallow) | 53 | 42 | 5 |
| Pig (lard) | 42 | 46 | 12 |
| Fish oil | 30 | 13 | 57 |

*These saturated fatty acids are medium-chain fatty acids.

Triglycerides that contain a higher proportion of unsaturated or short- or medium-chain fatty acids are liquids at room temperature and are called **oils**. Oils are of vegetable origin. On the other hand, if they contain saturated long-chain fatty acids, they are solids at room temperature and are called **fats**. Fats are of animal origin. Fats containing medium-chain or unsaturated fatty acids are soft, e.g., butter, coconut oil. Triglycerides (TGs) are otherwise called triacylglycerols (TAGs) or neutral fat.

## Properties of Triglycerides

### Hydrolysis

Hydrolysis of TAGs (triacylglycerols) in the body is done by the enzyme lipase **(Fig. 7.6)**.

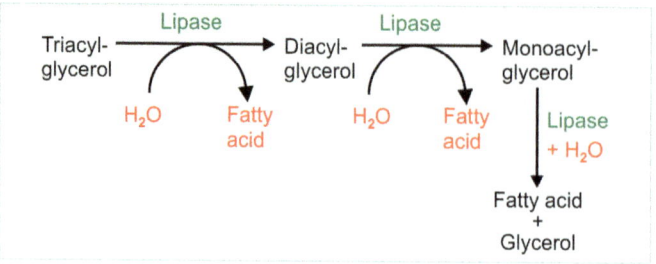

**Fig. 7.6:** Hydrolysis of triglycerides into glycerol and fatty acids.

It occurs during the digestion of dietary fat and mobilization of TG from adipose tissue.

*Saponification*

On hydrolysis with alkali, like NaOH or KOH, triglycerides yield sodium salt of fatty acid and glycerol. This process is known as saponification.

Triacylglycerol + KOH sodium salt of fatty acid + glycerol

**Saponification number** is the number of milligrams of KOH required to saponify 1 g of fat or oil. The amount of alkali required to saponify the known quantity of fat will depend upon the number of –COOH groups present. It is inversely proportional to the average molecular weight of the fatty acids in the fat (**Fig. 7.7**).

*Rancidity*

Rancidity is the deterioration of fats and oils resulting in an unpleasant smell or taste. Fats containing unsaturated fatty acids are more susceptible to rancidity. It occurs when fats or oils are exposed to air, moisture, light, bacteria, and so forth. Rancidity is of two types:

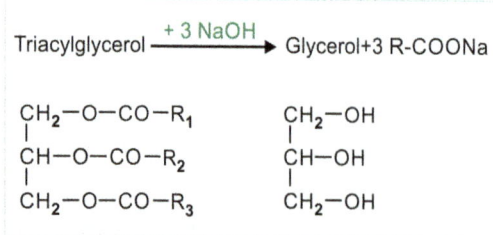

**Fig. 7.7:** Saponification.

1. **Hydrolytic rancidity:** It is due to the partial hydrolysis of triacylglycerol by traces of hydrolytic enzymes present in bacteria and also in naturally occurring fats itself. This results in the formation of unpleasant aldehydes, ketones, and so on.
2. **Oxidative rancidity:** It is the oxidation of the fatty acids with the resultant formation of peroxides. Certain antioxidants such as vitamin E or vitamin C can prevent oxidation of fats and thus the development of rancidity.

## Waxes

They are widespread in nature and form the secretions of insects, skin, and fur of animals, as well as leaves and fruits of plants. They are the esters of very long-chain fatty acids with very long-chain alcohols (other than glycerol). Examples are lanolin or wool fat and bees wax. They are used as the base for the preparation of cosmetics and ointments.

## COMPOUND LIPIDS

Compound lipids are classified as: (1) phospholipids, (2) glycolipids, and (3) sulfolipids. See the detailed classification in **Table 7.1**.

## Phospholipids

They are composed of fatty acids, glycerol, phosphoric acid, and a nitrogenous base. They are important compounds present in all plant and animal cells. They are abundantly present in heart, brain, kidneys, egg yolk, and soya

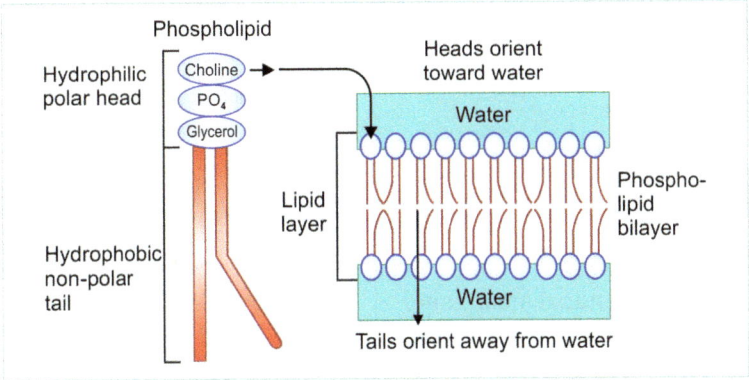

**Fig. 7.8:** Phospholipids form the bilayer.

bean. Phospholipids play an important role in the prevention of fatty liver and in the process of blood coagulation.

Phospholipids are simply divided into (1) glycerophospholipids and (2) sphingophospholipids. For detailed classification, see **Table 7.1**.

*Glycerophospholipids*

They may be regarded as derivatives of phosphatidic acid in which the phosphate is esterified with the –OH group of a suitable alcohol. An important property of phospholipids is their amphipathic nature. In general, lipids are insoluble in water since they contain predominant nonpolar (hydrocarbon) groups. However, fatty acids, phospholipids, sphingolipids, bile salts, and cholesterol contain polar groups also. Therefore, part of the molecule is hydrophobic (water-insoluble) and part is hydrophilic (water-soluble). Such molecules are known as **amphipathic (Fig. 7.8)**. They become oriented at oil–water interphases with the polar groups in the water phase and the nonpolar groups in the oil phase. A bilayer of such amphipathic lipids has been regarded as the basic structure in biomembranes **(Fig. 7.9)**.

*Phosphatidic Acid*

Phosphatidic acid **(Fig. 7.10)** is important as an intermediate in the synthesis of triacylglycerols as well as phosphoacylglycerols. It is a precursor of phosphatidylglycerol, which in turn gives rise to cardiolipin (diphosphatidylglycerol), which

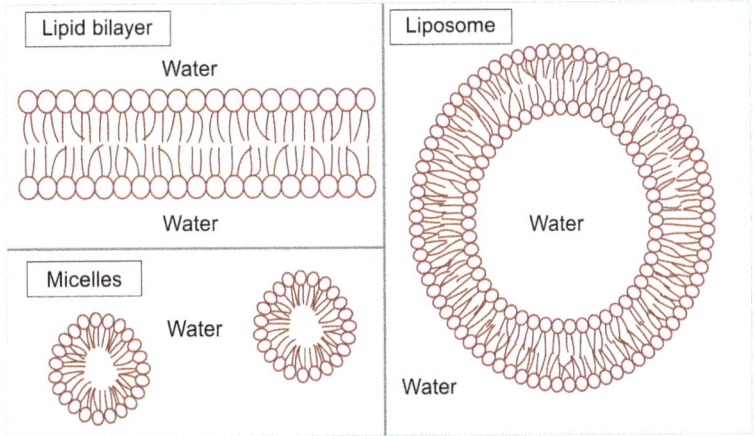

**Fig. 7.9:** Phospholipids form the bilayers, micelles, membranes, and liposomes.

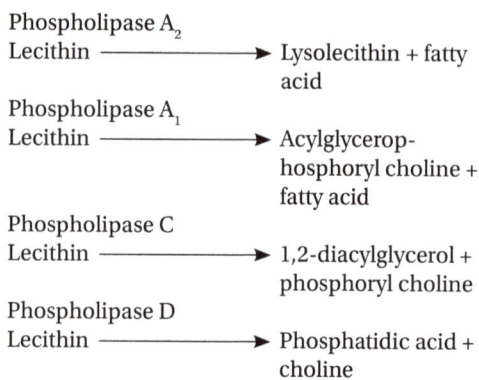

**Fig. 7.10:** Phosphatidic acid.

is a major lipid of mitochondrial membrane.

The important glycerophospholipids include lecithin, cephalin, phosphatidylserine, phosphatidylinositol, plasmalogen, and so on.

## Lecithin (Phosphatidylcholine)

Lecithin consists of one glycerol molecule esterified to two molecules of fatty acids. Structure is shown in **Figure 7.11**. One of these may be unsaturated and the other saturated, and also it contains a phosphoric acid molecule and a nitrogenous base, choline.

Lecithin is a white waxy substance, which becomes brownish when exposed to air and light. This is due to the autoxidation of unsaturated fatty acids in the molecule. They are soluble in ordinary fat solvents, such as alcohol and ether but not in acetone.

Lecithin is hydrolyzed by boiling with alkalis or dilute mineral acids and also by the enzymes lecithinases (phospholipases) to various components. Four types of lecithinases occur in nature, and each acts on lecithin at different points (*see* the numbers in **Fig. 7.11**).

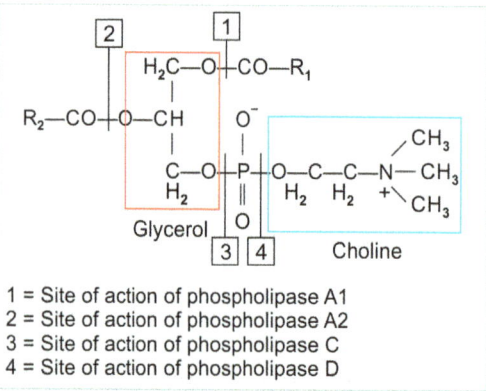

1 = Site of action of phospholipase A1
2 = Site of action of phospholipase A2
3 = Site of action of phospholipase C
4 = Site of action of phospholipase D

**Fig. 7.11:** Structure of lecithin. $R_1$ and $R_2$ are fatty acids. Choline shows polar or hydrophilic property.

Phospholipase $A_2$
Lecithin ⟶ Lysolecithin + fatty acid

Phospholipase $A_1$
Lecithin ⟶ Acylglycerophosphoryl choline + fatty acid

Phospholipase C
Lecithin ⟶ 1,2-diacylglycerol + phosphoryl choline

Phospholipase D
Lecithin ⟶ Phosphatidic acid + choline

Lecithinase $A_2$ is found in snake venom. It partially hydrolyzes lecithin to give lysolecthin. Lysolecthin is a powerful hemolytic poison, which causes hemolysis of erythrocytes.

Functions of Lecithin
- Lecithin is the most abundant lipid present in cell membrane.
- It represents a large portion of body's store of choline.
- Choline is important in nerve impulse transmission as acetylcholine.
- Choline is the main methyl group donor in biological reactions.
- Choline is important in the prevention of fatty liver. Deficiency of choline leads to fatty liver.
- Lecithin is used as an emulsifying and soothing agent in food industry.

## Cephalin (Phosphatidylethanolamine)

Cephalin contains the saturated fatty acid (stearic acid) and unsaturated fatty acids (oleic, linoleic and arachidonic acids). Cephalin differs from lecithin in having ethanolamine as nitrogenous base instead of choline (**Fig.**

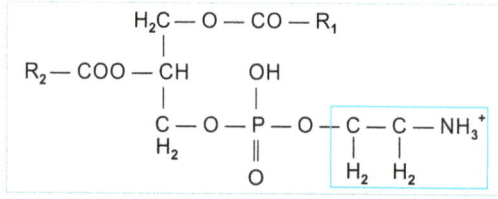

**Fig. 7.12:** Cephalin or phosphatidylethanolamine.

7.12). The properties of cephalin and lecithin are similar. Cephalin is also hydrolyzed in the same manner as lecithin. Cephalin plays an important role in the mechanism of blood coagulation. Cephalin and lecithin occur along in all animal and plant cells. Cephalin is present in brain, erythrocytes, and many other tissues.

**Phosphatidylserine**

This compound contains the amino acid serine. It is found in brain tissue.

**Phosphatidylinositol**

Phosphatidic acid is esterified to one of the hydroxyl group of inositol (**Fig. 7.13**). This compound is of importance since it plays a vital role in the mediation of hormone action in biomembranes.

**Plasmalogens**

Structurally, the plasmalogens resemble cephalins but possess an ether linkage on the first carbon of glycerol instead of the normal ester linkage found in acylglycerols (**Fig. 7.14**). The alkyl radical is an unsaturated long-chain aliphatic aldehyde. These compounds constitute as much as 10% of the phospholipids in brain and muscle.

*Sphingolipids*

The sphingosine containing lipids may be of three types: **phosphosphingosides, glycosphingolipids, and sulfatides.** All sphingolipids have the long aliphatic amino alcohol **sphingosine** having 18 carbons. Sphingosine is attached to a fatty acid to form a **ceramide.** The fatty acid has a chain length varying from 18 C to 24 C.

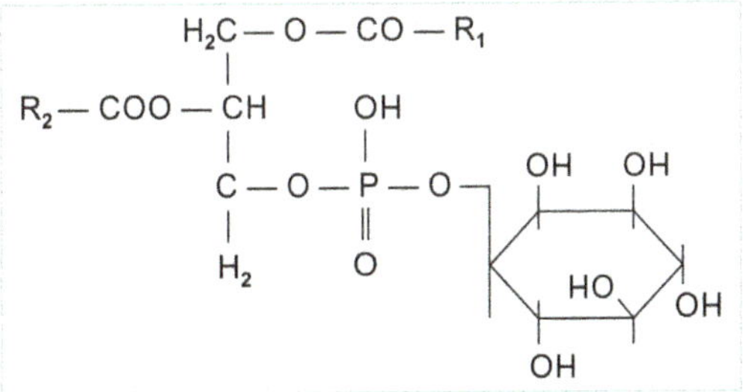

**Sphingomyelin**

It contains fatty acid, phosphoric acid, choline, and sphingosine (amino alcohol). The fatty acid is attached to the amino group of sphingosine by amide linkage. The phosphoric acid is attached to one of the hydroxyl groups of sphingosine (**Fig. 7.15**). **Sphingomyelins are the only sphingolipids that contain phosphate and have no sugar moiety.**

Sphingomyelins are found in large quantities in brain and nerve tissue. Because of their amphipathic nature, they can act as emulsifying agents and detergents. Increased concentration of sphingomyelin is observed in **Niemann-Pick disease**. This is due to the genetic deficiency of sphingomyelinase in these tissues.

**Fig. 7.13:** Phosphatidylinositol.

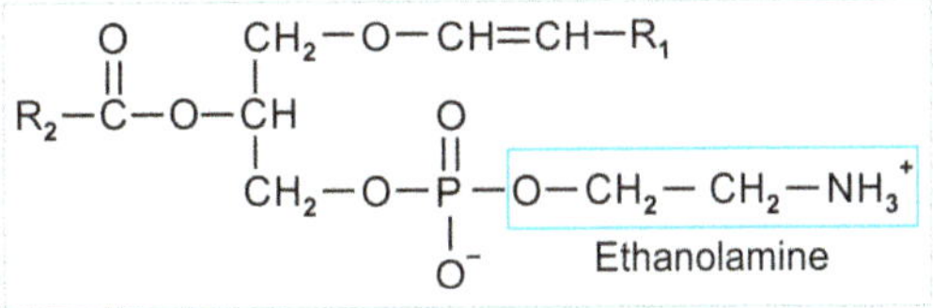

**Fig. 7.14:** Ethanolamine plasmalogen.

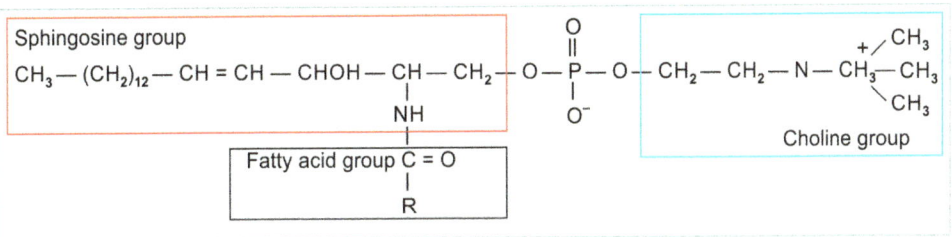

**Fig. 7.15:** Sphingomyelin.

## Glycolipids

Glycolipids are mainly classified as: (1) cerebrosides and (2) gangliosides. They are known as glycolipids because they contain a sugar molecule in addition to fatty acid and sphingosine. They do not contain phosphoric acid. They are optically active owing to the presence of asymmetric carbon atoms in the molecule of sphingosine and sugar. They are widely distributed in every tissue of the body, particularly in the white matter of brain and the myelin sheath of nerves.

### Cerebrosides

They are composed of high molecular weight fatty acid, sphingosine, and galactose. Four cerebrosides have been identified in brain and nerve, which are differentiated from each other by their fatty acid component (**Table 7.4**).

Large amounts of the cerebrosides, especially cerebron, will be accumulated in the liver and spleen of patients with **Gaucher's disease**, which is a rare hereditary disorder of lipid metabolism.

### Gangliosides

Gangliosides are glycolipids containing glucose, galactose, N-acetylgalactosamine, and neuraminic acid in addition to sphingosine and fatty acid. They occur in brain.

## Sulfolipids

These are lipids containing sulfur. They are sulfated derivatives of cerebrosides. They are most abundantly found in the white matter of the brain.

**Table 7.4:** Different cerebrosides.

| Cerebroside | Fatty acid component |
|---|---|
| Kerasin | Lignoceric acid |
| Cerebron | Cerebronic acid |
| Nervon | Nervonic acid |
| Oxynervon | Oxynervonic acid |

## Lipoproteins

They are described in detail in Chapter 9. Fats absorbed from the diet and lipids synthesized by the liver and adipose tissue must be transported between the various tissues and organs for utilization and storage. Since lipids are insoluble in water, their transportation in an aqueous environment (i.e., blood) is difficult. The nonpolar lipids (triglycerides, cholesterol) are associated with proteins to make water-soluble lipoproteins. The protein moiety of lipoprotein is called **apolipoprotein** or **apoprotein**. Classification of lipoproteins is given in Chapter 9.

### Functions of Lipoproteins

- Lipoproteins transport and deliver lipids to the tissues.
- They maintain the structural integrity of cell surface, mitochondria, and microsomes.
- They are important in clinical diagnosis, e.g., the concentration of LDL is increased in severe diabetes mellitus, atherosclerosis, and so on.

## Derived Lipids

These are substances derived from simple and compound lipids by hydrolysis, and these include fatty acids, steroids, and so on. Fatty acids are described in the early part of this chapter.

### Steroids

Steroids represent a large group of compounds that exist in nature having a common and characteristic structure, namely, cyclopentano-perhydro-phenanthrene ring (This ring is often referred to as *steroid ring*). The various substances of physiological interest having the same steroid structure are classified as:

1. *Sterols* mean solid alcohols. *Sterols*, which include cholesterol, 7-dehydrocholesterol, ergosterol, calciferol, and coprosterol, are widely present in animal and plant tissues. Cholesterol is described in Chapter 9.

2. Bile acids: These are synthesized from cholesterol. Bile acids are necessary for the absorption of fats from intestine.
3. Sex hormones: They are synthesized from cholesterol. They are produced in ovary, testes, and adrenal cortex.
4. Adrenocortical hormones are produced by adrenal cortex.
5. Vitamin D (*see* Chapter 16).
6. Sitosterols, which are of plant origin. As their structures are similar to cholesterol, they decrease the absorption of cholesterol from intestines.

## SUMMARY OF THE CHAPTER

- Lipids may be broadly classified into simple, compound, and derived lipids.
- Compound lipids are phospholipids, sphingolipids, sulfolipids, and so on. Compound lipids containing alcohol sphingosine and one or more carbohydrate residues are called glycolipids.
- Fatty acids are classified based on the: (i) number of carbon atoms, (ii) length of the hydrocarbon chain, and (iii) nature of the hydrocarbon chain. Depending on the number of carbon atoms, fatty acids may be even-chain or odd-chain compounds, which are further subdivided into short-chain (2–6 C), medium-chain (8–14 C), and long-chain (16–20 C) compounds.
- Palmitic and stearic acids are the most abundant saturated fatty acids in the body.
- Fatty acids may be saturated (no double bonds), monounsaturated (one double bond), or polyunsaturated (more than two double bonds). Polyunsaturated fatty acids (PUFA) may be essential or nonessential. Essential fatty acids are those which cannot be synthesized in the human body and have to be supplemented in the diet, e.g., linoleic acid, linolenic acid, and arachidonic acid.
- Arachidonic acid is the precursor of prostaglandins.
- Saponification number is defined as the number of milligrams of KOH required to saponify 1 gram of fat.
- The iodine number of a fat is defined as the number of grams of iodine taken up by 100 grams of fat. It is directly proportional to the degree of unsaturation.
- *Rancidity* refers to the appearance of unpleasant odor and taste to oils and fats. Rancidity can be of two types— hydrolytic and oxidative.
- Depending on the position of double bonds from the omega end, fatty acids may be omega-3, omega-6, or omega-9.
- Sodium and potassium salts of fatty acids are called soaps.
- Fatty acids can form esters with hydroxyl groups of glycerol to form mono-, di- and triacylglycerol.
- Triacylglycerols or neutral fats are the storage forms of energy in adipose tissue.
- Major fatty acids found in adipose tissue fat are oleic acid, palmitic acid, and stearic acid.
- Monounsaturated fatty acid (MUFA) and polyunsaturated fatty acids (PUFA) are commonly esterified to the second carbon (beta carbon) of glycerol.
- Oils and fats are mixtures of triacylglycerol. Oils are liquids at 25°C, and fats are solids.
- Butter contains short- and medium-chain fatty acids.
- Phospholipids may be glycerophosphatides or phosphosphingosides, depending on the alcohol present.
- The simplest glycerophosphatide is phosphatidic acid containing glycerol, two molecules of fatty acid, and one molecule of phosphoric acid.
- Phospholipids are amphipathic in nature, since they have a polar head and a nonpolar tail. Amphipathic nature is ideal for the

role of phospholipids as components of biomembranes and for micelle formation.
- Phosphatidic acid may combine with nitrogenous base to form aminophospholipids, such as phosphatidylcholine, phosphatidylethanolamine, and phosphatidylserine.
- Phosphatidylglycerol or cardiolipin is formed by the esterification of one molecule of glycerol simultaneously to two molecules of phosphatidic acid.
- Phospholipase A2 hydrolyzes the ester bond between the second hydroxyl group of glycerol and one PUFA.
- Phosphosphingosides contain sphingosine as alcohol. Sphingosine esterified to a fatty acid is called a ceramide.
- Sphingomyelin is the only phosphosphingoside that contains choline.
- Sphingolipids have ceramide attached to carbohydrate residues to form glycolipids, such as cerebrosides and lactosylceramide.
- When one molecule of N-acetylneuraminic acid (NANA) is attached to the ceramide oligosaccharide, it is called ganglioside.
- Sulfatides are formed when sulfate is esterified to ceramide oligosaccharide.
- Cholesterol is an animal sterol, which is a derived lipid. It is the precursor of all steroids in the body.

## QUESTIONS FOR SELF-ASSESSMENT

1. Define lipids and mention its biomedical importance.
2. Classify lipids according to their structure.
3. Name the essential fatty acids.
4. What are the functions of phospholipids?
5. What is PUFA? Mention its clinical importance.
6. Mention the functions of prostaglandins.
7. What are lipoproteins and mention its biological functions.
8. What are steroids and mention its functions.
9. Add a note on Niemann–Pick disease.
10. Mention the functions of omega-3 and omega-6 fatty acids.

## MULTIPLE CHOICE QUESTIONS (MCQs)

7-1. Which of the following is an amphipathic lipid?
   a. Fatty acids
   b. Phospholipids
   c. Bile acids
   d. Cephalin

7-2. The number of double bonds present in arachidonic acid is:
   a. 1
   b. 2
   c. 3
   d. 4

7-3. All the following are the examples of compound lipids, *except*:
   a. Glycolipids
   b. Plasmalogen
   c. Cholesterol
   d. Lipoprotein

7-4. One of the following is the nonessential fatty acid:
   a. Arachidonic acid
   b. Linolenic acid
   c. Palmitic acid
   d. Linoleic acid

7-5. The polyunsaturated fatty acids (PUFA) are richly present in:
   a. Sunflower oil
   b. Butter
   c. Ghee
   d. Coconut oil

7-6. All the following compounds are esterified with membrane phospholipids, *except*:
   a. Serine
   b. Inositol
   c. Ethanolamine
   d. Ribose

7-7. All the following alcohols are found in phospholipids, *except*:
   a. Sphingosine
   b. Inositol
   c. Mannitol
   d. Glycerol

7-8. A ganglioside on hydrolysis gives all the following, *except:*
   a. Fatty acid
   b. Glycerol
   c. Sphingosine
   d. N-acetylneuraminic acid

7-9. **Name of the test employed to check the adulteration of butter is:**
   a. Iodine number
   b. Saponification number
   c. Zak's method
   d. Reichert–Meissl number

7-10. **Palmitic acid is classified as:**
   a. Monounsaturated and essential
   b. Monounsaturated and nonessential
   c. Polyunsaturated and nonessential
   d. Saturated and nonessential

## ANSWER KEYS TO MCQs

7-1. (b)   7-2. (d)   7-3. (c)   7-4. (c)   7-5. (a)   7-6. (d)   7-7. (c)   7-8. (b)
7-9. (a)   7-10. (d)

# CHAPTER 8

# Lipid Metabolism

At the completion of this chapter, the reader will be able to answer questions on the following topics:

- Digestion of lipids
- Beta-oxidation of fatty acids
- Synthesis of fatty acids
- Ketogenesis
- Ketolysis
- Metabolism of triglycerides
- Fatty liver and lipotropic factors
- Eicosanoids
- Essential fatty acids
- Prostaglandins
- Prostacyclins
- Thromboxanes

## DIGESTION OF LIPIDS

The major dietary lipids are triacylglycerol (TAG), cholesterol, and phospholipids. The average normal Indian diet contains about 20–30 g of lipids per day.

### Digestion of Lipids in Intestines

**Emulsification** is a prerequisite for the digestion of lipids. The lipids are dispersed into smaller droplets, and the surface area of the droplets is increased. This process is favored by bile salts and phospholipids. The bile salts present in the bile (sodium glycocholate and sodium taurocholate) emulsify the fat droplets. The emulsification increases the surface area of the particles for enhanced activity of enzymes.

**Pancreatic lipase** is the major enzyme hydrolyzing the dietary lipids. The bile (pH 7.7) entering the duodenum serves to neutralize the acid chyme from the stomach and provides a pH favorable for the action of pancreatic lipase.

### Digestion of Triacylglycerols

**Pancreatic lipase** hydrolyzes the fatty acids esterified to the first and third carbon atoms of glycerol forming 2-monoacylglycerol and two molecules of fatty acid. Then an **isomerase** shifts the ester bond from position 2 to 1. The bond in the first position is then hydrolyzed by the **lipase** to form free glycerol and fatty acid. The major end products of the digestion of TAG are 2-monoacylglycerol (2-MAG), 1-monoacylglycerol (1-MAG), glycerol, and free fatty acids **(Fig. 8.1)**. Thus digestion of TAG is partial (incomplete).

### ABSORPTION OF LONG-CHAIN FATTY ACIDS

Long-chain fatty acids (chain length of more than 14 carbons) are absorbed into the lymph and not directly into the blood. This theory is proposed by Bergstrom (1982 Nobel Prize winner). The products of digestion, namely 2-monoacylglycerols, long-chain fatty acids, cholesterol, and phospholipids

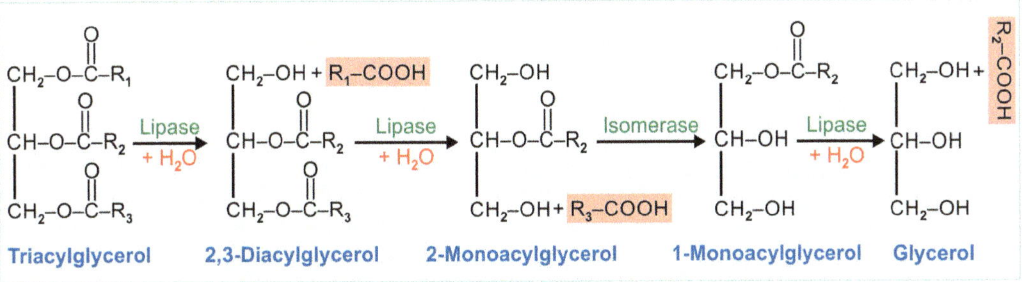

**Fig. 8.1:** Complete hydrolysis of a triglyceride. In the intestines, generally fats are only partially hydrolyzed.

are incorporated into molecular aggregates to form **mixed micelles**. The micelles are spherical particles with a hydrophilic exterior and hydrophobic interior core. The **bile salts** help to form micellar aggregates.

Once inside the intestinal mucosal cell, the long-chain fatty acids are re-esterified to form TAGs. The TAG, cholesterol ester, and phospholipid molecules are incorporated into chylomicrons. The chyle (milky fluid) from the intestinal mucosal cells, loaded with chylomicrons, is transported through the lacteals (small branches of lymph vessels) into the thoracic duct and then emptied into the systemic circulation (**Fig. 8.2**). The serum may appear milky after a high-fat meal (postprandial lipemia) due to the presence of chylomicrons in the circulation.

The bile salts thus absorbed are mostly reabsorbed from the ileum and returned to the liver to be re-excreted (enterohepatic circulation of bile salts). About 98% of dietary lipids are normally absorbed. The absorbed (exogenous) triglycerides are transported in the blood as **chylomicrons**. They are taken up by adipose tissue, skeletal muscle, and liver.

## Short-Chain Fatty Acid Absorption is Different

**Short-chain** fatty acids (SCFAs) (seen in milk, butter, and ghee) and **medium-chain** fatty acids (MCFAs) (seen in coconut oil and mother's milk) do not need re-esterification. Their absorption is rapid. They can directly enter into blood vessels, then to portal vein, and finally to liver where they are immediately utilized for energy.

## Abnormalities in the Absorption of Lipids

- **Defective digestion:** In steatorrhea, daily excretion of fat in feces is more than 6 g per day (Greek word *steat* means "fat"). It is due

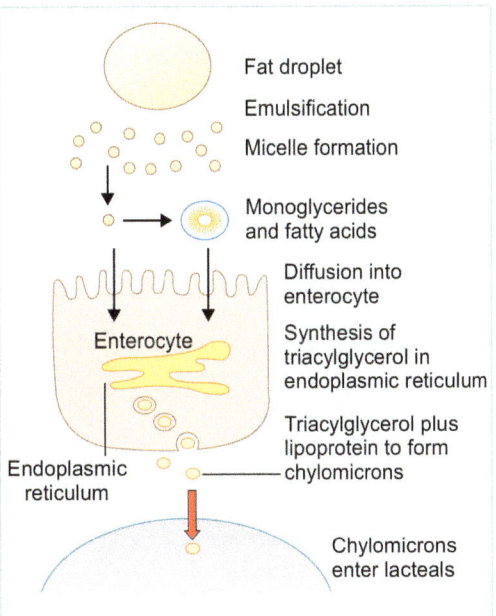

**Fig. 8.2:** Absorption of fat as chylomicrons. This needs the help of bile salts. The hydrophobic portions of bile salts intercalate into the large aggregated lipid, with the hydrophilic domains remaining at the surface. This leads to the breakdown of large aggregates into smaller and smaller droplets. Thus the surface area for the action of lipase is increased.

to the chronic diseases of the pancreas. In such cases, unsplit fat is seen in feces.
- **Defective absorption:** On the other hand, if the absorption alone is defective, most of the fat in feces may be split fat, that is, fatty acids and monoglycerides. Defective absorption may be due to celiac disease, sprue, Crohn's disease, surgical removal of intestine, obstruction of bile duct due to gallstones, tumors in the head of the pancreas, enlarged lymph glands, and so forth.
- **Chyluria**: There is an abnormal connection between the urinary tract and lymphatic drainage system of the intestine. Urine appears milky due to lipid droplets.

## Fate of Absorbed Fat

- The absorbed (exogenous) triglycerides are transported in blood as chylomicrons. They are taken up by adipose tissue and liver.
- Liver synthesizes endogenous triglycerides. These are transported as very low-density lipoproteins (VLDLs) and are deposited in adipose tissue.
- Triglycerides in adipose tissue are lysed to produce free fatty acids. In the blood, they are transported and complexed with albumin.
- Free fatty acids are taken up by the cells and are then oxidized to get energy (**Fig. 8.3**).

## BETA-OXIDATION OF FATTY ACIDS

Fatty acids are rich sources of energy. This energy is released when they undergo β-oxidation. It is a cyclic process, and in each cycle, active fatty acid [acyl coenzyme A (acyl-CoA)] is degraded to give acetyl coenzyme A (acetyl-CoA) (2 C compound) and acyl-CoA with two carbon atoms less than the original acyl-CoA. The cycle is repeated till the acyl-CoA is completely broken down to acetyl-CoA molecules. The first step of beta-oxidation is the **activation of fatty acid** (**Figs. 8.4 and 8.5**).

## Transport of Acyl-CoA

The oxidation of acyl-CoA by β-oxidation pathway occurs inside the matrix of mitochondria. But the acyl-CoA is generated outside the mitochondria and cannot cross the inner mitochondrial membrane. A carrier substance called **carnitine** helps in the transport of acyl group across the membrane. This transport also needs type I and type II translocases or **carnitine acyltransferases** (**Fig. 8.6**).

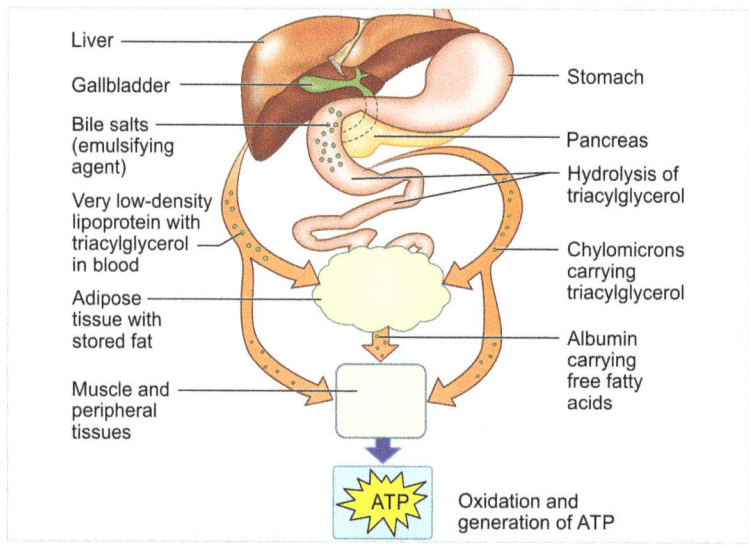

**Fig. 8.3:** Summary of utilization of fat.
(ATP: adenosine triphosphate)

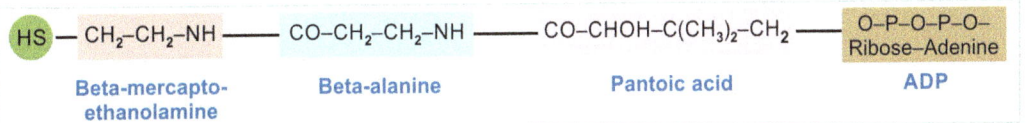

**Fig. 8.4:** Structure of coenzyme A (CoA).
(ADP: adenosine diphosphate)

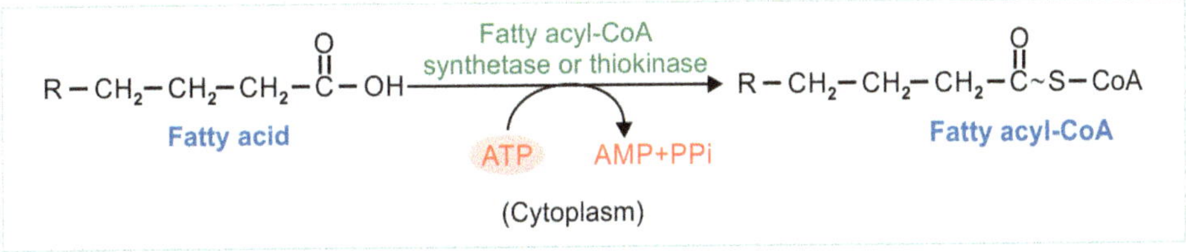

**Fig. 8.5:** First step of beta-oxidation is the activation of fatty acid.
(ATP: adenosine triphosphate; AMP: adenosine monophosphate)

The beta-oxidation takes place in the mitochondrial matrix. The next four reactions are sequentially repeated for the complete oxidation of fatty acids. After one round of the four metabolic steps, one acetyl-CoA unit is split off, and acyl-CoA with two carbon atoms less is generated. This would undergo the same series of reactions again until the fatty acid is completely oxidized.

Step 1 is catalyzed by the FAD-linked dehydrogenase (FAD = flavin adenine dinucleotide; step 1, **Fig. 8.7**). **FADH2** when oxidized in the electron transport chain will produce **1.5 ATP** (adenosine triphosphate) molecules. The enzyme for the second step is enoyl-CoA hydratase. In the third step, it is oxidized by an $NAD^+$-linked dehydrogenase (NAD = nicotinamide adenine dinucleotide). The **NADH,** when oxidized in the electron transport chain, will generate **2.5 ATPs**. In the fourth step, the enzyme is thiolase, producing one molecule of acetyl-CoA and leaving behind a fatty acid with two carbon atoms less (step 4, **Fig. 8.7**). The newly formed fatty acyl-CoA will sequentially undergo further cycles of steps 1, 2, 3, and 4 of beta-oxidation until the fatty acid is completely converted to acetyl-CoA. These steps of the pathway are shown in **Figure 8.7**.

## Energetics of Fatty Acid Oxidation

The commonly occurring fatty acid is palmitic acid (16 C) (even chain). It undergoes seven cycles before being completely broken down

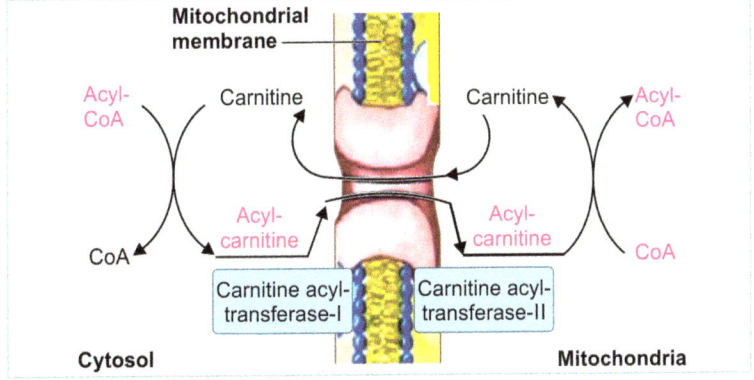

**Fig. 8.6:** Role of carnitine in the transport of acyl groups.

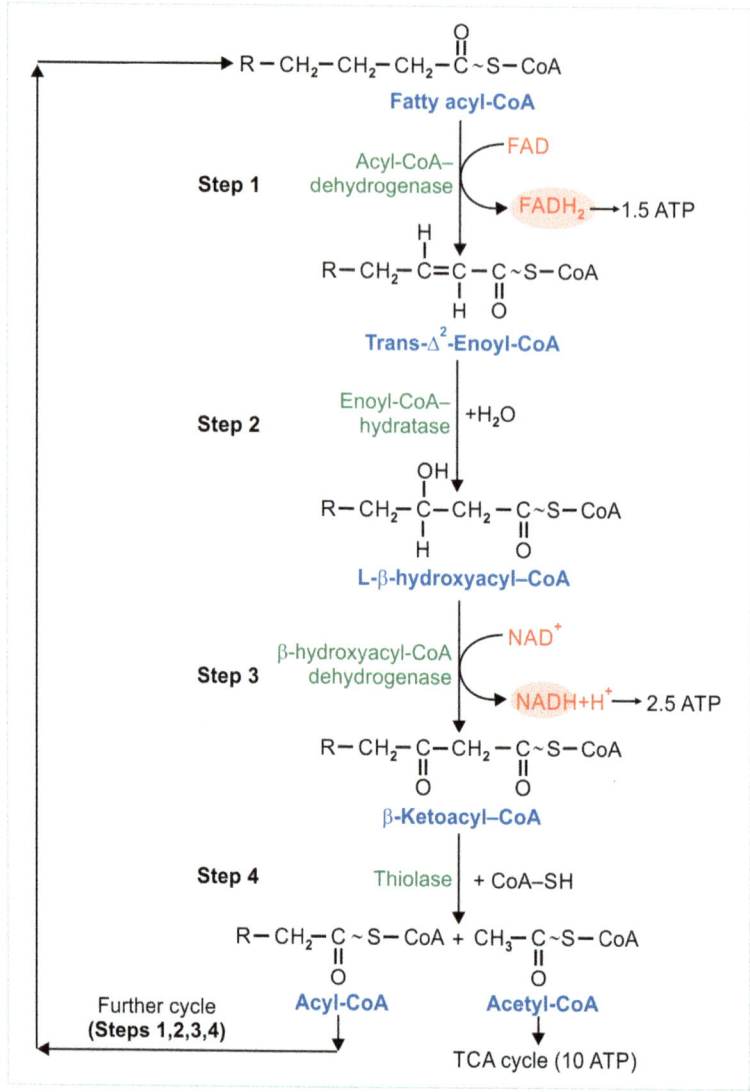

Fig. 8.7: Beta-oxidation of fatty acids. The first step is FAD dependent and the third step is NAD+ dependent.

to eight molecules of acetyl-CoA. Every molecule of acetyl-CoA when oxidized in the tricarboxylic acid (TCA) cycle gives 12 molecules of ATP. Each molecule of FADH2 produces two molecules of ATP, and each NADH generates three molecules of ATP, when oxidized in the electron transport chain. Hence the energy yield from one molecule of palmitate may be calculated as:

8 Acetyl-CoA × 10 = 80 ATPs
7 FADH2 × 1.5 = 10.5 ATPs
7 NADH × 2.5 = 17.5 ATPs
Gross total = 108 ATPs
(In the initial activation reaction, the equivalents of two high-energy bonds are utilized).

Therefore, **net yield = 108 – 2 = 106 ATPs**

Thus, from one palmitic acid molecule, 106 ATPs are formed by beta-oxidation. These calculations were made assuming that NADH produces 2.5 ATPs and FADH generates 1.5 ATPs, and one turn of citric acid cycle produces 10 ATPs. A few years back, calculations were made assuming that NADH produces 3 ATPs and FADH generates 2 ATPs. This will amount to a net generation of 129 ATPs per palmitate molecule. Later experiments show that these old values are overestimates, and net generation is only 106 ATPs.

## Summary of Beta-Oxidation

When one molecule of palmitic acid (16 C) undergoes beta-oxidation, the net reaction is:

**Palmitoyl CoA** + 7FAD + 7NAD$^+$ + 7H$_2$O + 7HSCoA $\longrightarrow$ **8 Acetyl-CoA** + 7FADH$_2$ + 7NADH + 7H$^+$

## Regulation of Beta-Oxidation

- Availability of free fatty acids (FFAs) regulates the process of beta-oxidation.
- Level of FFA is controlled by glucagon–insulin ratio. Glucagon increases while insulin decreases the level of FFA.
- Malonyl-CoA (a substrate for fatty acid synthesis) is an inhibitor of beta-oxidation.

## SYNTHESIS OF FATTY ACIDS

The pathway is also called cytosolic pathway or extra mitochondrial pathway or de novo synthesis of fatty acids or Lynen's spiral.
- It is the major pathway for the synthesis of fatty acids.
- It is present in liver and adipose tissue.
- Palmitic acid is the major fatty acid synthesized.
- All the 16 carbon atoms are derived from acetyl-CoA.

Synthesis of fatty acid (FA) takes place outside the mitochondria, but acetyl-CoA (from glucose or FA) is formed inside the mitochondria. Furthermore, acetyl-CoA cannot cross the mitochondrial membrane. Hence, it first forms citrate (which has specific transporters), which comes out into the cytoplasm. Then citrate splits to give rise to acetyl-CoA **(Fig. 8.8)**.

## Formation of Malonyl-CoA

Eight acetyl-CoA molecules are necessary to synthesize one molecule of palmitic acid. But only one acetyl-CoA takes part as acetyl-CoA. The rest seven acetyl-CoA molecules are first converted to malonyl-CoA, which is then used for the de novo synthesis (step 1, **Fig. 8.9**). **Acetyl-CoA carboxylase** is the key enzyme for the whole pathway.

## Fatty Acid Synthase System

Fatty acid synthesis involves catalyzed reactions of six enzymes. All these six enzymes are kept together as a **multienzyme complex** (MEC). This MEC is called "fatty acid synthase system." It is a dimer made of two identical monomeric units. Each monomer is made of six enzymes plus one protein called **acyl carrier protein** (ACP). The six enzymes are named as: (1) transacylase (transfers acetyl and malonyl groups), (2) keto acyl synthase, (3) keto acyl reductase, (4) dehydratase, (5) enoyl reductase, and (6) thioesterase **(Fig. 8.10)**.

The diagonal line divides the dimer into two functional units **(Fig. 8.10)**. Each functional unit has six enzymes and ACP. Besides, it contains two SH groups: one is part of cysteine (denoted as Cys-SH) of enzyme 2 (keto acyl synthase) and the other is that of phosphopantetheine (denoted as Pan-SH in **Fig. 8.10**). This Pan-SH is a part of ACP. These -SH groups are the sites of attachment of the acetyl and malonyl groups.

## Reactions of De Novo Synthesis of Fatty Acid

Enzyme 1 (**Fig. 8.10**) transfers the acetyl group of acetyl-CoA to Cys-SH. Same enzyme attaches the malonyl group of malonyl-CoA to Pan-SH,

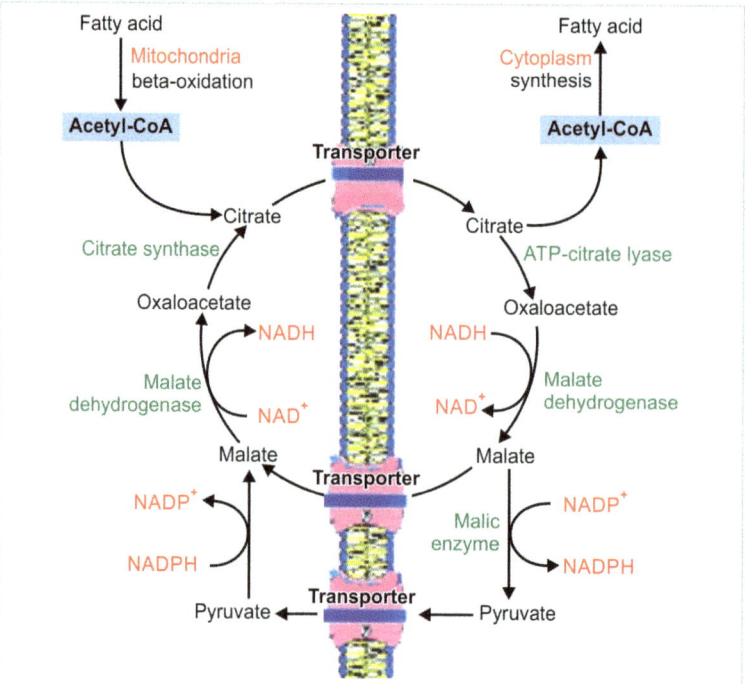

**Fig. 8.8:** Transportation of acetyl-CoA from mitochondria to cytoplasm by malate–oxaloacetate shuttle.

thus forming acetyl-malonyl enzyme. Enzyme 2 transfers the acetyl group to malonyl group. Condensation occurs with the liberation of $CO_2$ to form β-keto acyl enzyme. This undergoes reduction, dehydration, and further reduction to form acyl group **(Fig. 8.10)**. This group is shifted to cys-SH, and the next malonyl group attaches to pan-SH. The cycle continues till acyl group attains a chain length of 16 carbons. Finally, palmitic acid (16 C) is released by enzyme 6 (thioesterase) of MEC. Individual reactions of the pathway are shown in **Figure 8.9**.

The NADPH units (reducing equivalents) are required for step 4 and step 6 reactions in **Figure 8.9**. These NADPH units are provided by the HMP shunt pathway of glucose metabolism (*see* Chapter 5).

## Regulation of Fatty Acid Synthesis

Regulation of FA synthesis takes place at the conversion of acetyl-CoA to malonyl-CoA by **acetyl-CoA carboxylase** (step 1, **Fig. 8.9**). This is the rate-limiting step. Citrate and insulin are positive modifiers, which will increase the rate of the reaction. Palmitic acid is the negative modifier, which will reduce the fatty acid synthesis.

## Difference in the Beta-Oxidation and Synthesis Pathways

It is necessary to emphasize that the beta-oxidation and de novo synthesis pathways are entirely different **(Table 8.1)**.

## Adipokines (Adipose Tissue-Derived Hormones)

Adipokines are a group of active factors involved in the maintenance of energy homeostasis as well as resistance to insulin. The important adipokines are leptin, adiponectin, resistin, TNF-alpha (tumor necrosis factor), and IL-6 (interleukin-6). **Leptin** is a small peptide

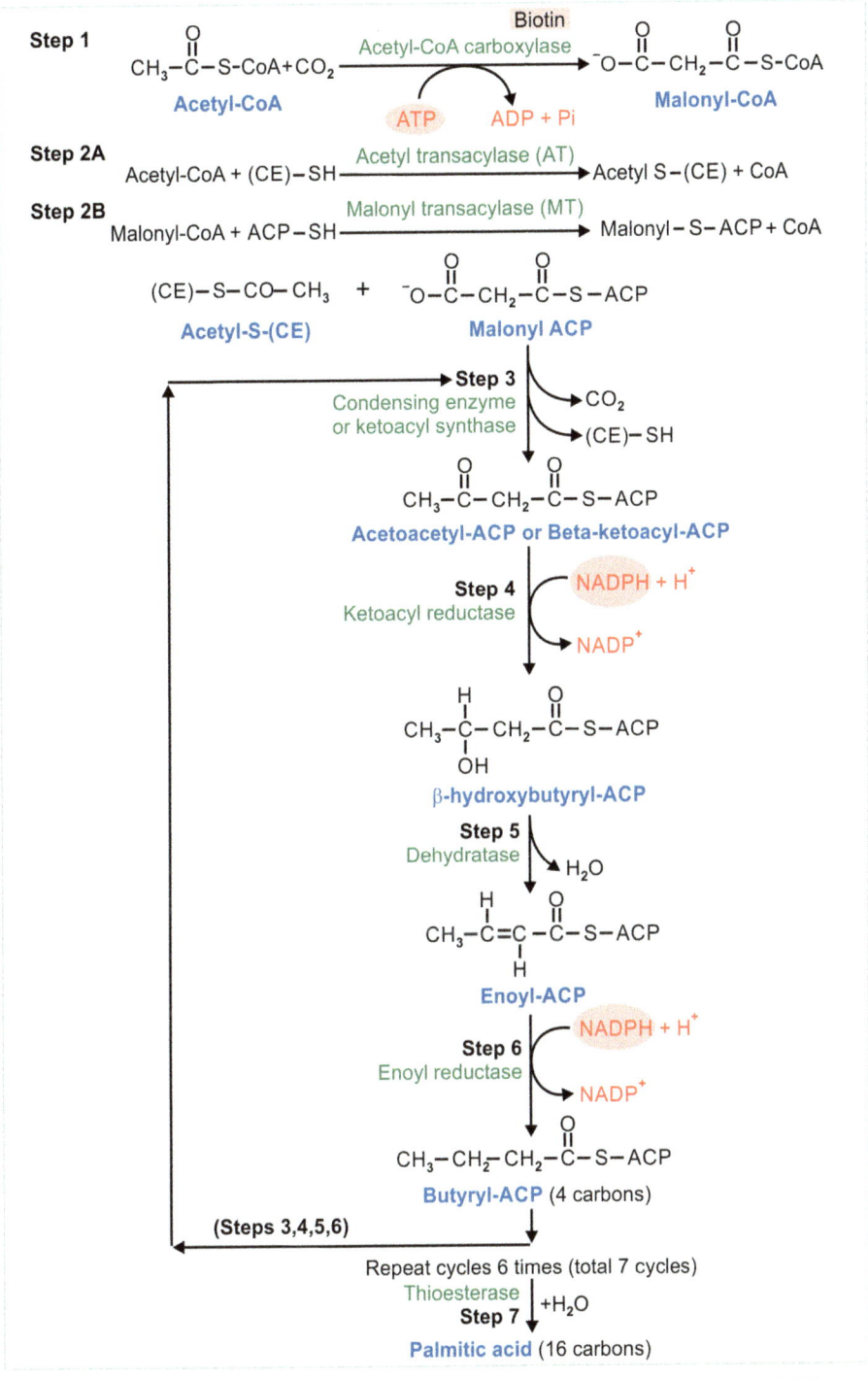

**Fig. 8.9:** De novo synthesis of fatty acid (Lynen cycle). Steps 4 and 6 utilize NADPH.

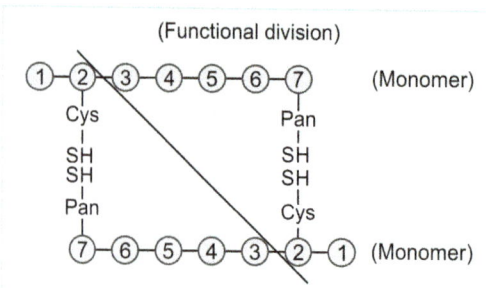

Fig. 8.10: Functional division of the fatty acid synthase system.

**Table 8.1:** Comparison between beta-oxidation and fatty acid synthesis.

|  | Beta-oxidation | Fatty acid synthesis |
|---|---|---|
| Site | Mitochondria | Cytoplasm |
| Intermediates | Present as CoA derivatives | Covalently linked to the SH group of ACP |
| Enzymes | Present as independent proteins | Multienzyme complex |
| Sequential units | Two-carbon units split off as acetyl-CoA | Two-carbon units added as three-carbon malonyl-CoA |
| Coenzymes | NAD⁺ and FAD are reduced | NADPH used as reducing power |
| Regulation | • Insulin inhibits<br>• Glucagon stimulates | • Insulin stimulates<br>• Glucagon inhibits |

produced by adipocytes. Leptin receptors are present in specific regions of the brain. The feeding behavior is regulated by leptin. A defect in leptin or its receptor can lead to obesity. **Adiponectin** is another polypeptide that increases the insulin sensitivity of muscle and liver and exerts an antiatherogenic effect. Increased secretion of TNF-alpha and IL-6 will decrease insulin action on muscle and liver.

### KETOGENESIS (FORMATION OF KETONE BODIES)

When the level of acetyl-CoA from beta-oxidation increases in excess than what is required, the acetyl-CoA gets converted to acetoacetate, beta-hydroxybutyrate, and acetone by process called ketogenesis. All these three components are collectively called as **ketone bodies**. The pathway of ketone body formation is shown in **Figure 8.11**.

In normal conditions, their concentration in blood is only about 1 mg/dL. But if the level is increased in blood, it is called **ketonemia.** The presence of acetoacetic acid will lower the pH of blood, which is called **acidosis**. Excretion of large amounts of ketone bodies in urine is called **ketonuria.** Acetone is primarily eliminated through lungs giving a fruity smell to the breath. All these three conditions are called **ketosis.**

### Causes for Ketosis

- Diabetes mellitus: Untreated diabetes mellitus is the most common cause for ketosis. Even though glucose is in plenty, the deficiency of insulin causes accelerated lipolysis, and more fatty acids are released into circulation.
- Starvation: In starvation, the dietary supply of glucose is decreased. The increased rate of lipolysis is to provide alternate source of fuel. The excess acetyl-CoA is converted to ketone bodies. The high glucagon favors ketogenesis. The brain derives 75% of energy from ketone bodies under conditions of fasting.
- Hyperemesis in early pregnancy.

### KETOLYSIS

The ketone bodies are produced in the liver, but are utilized by extrahepatic tissues to meet their energy requirement. Normally, ketone body formation and utilization are in equal quantities; but when ketogenesis exceeds the rate of its utilization, ketosis occurs. The utilization of ketone body is otherwise called ketolysis **(Fig. 8.12)**.

### Energetics

From one molecule of acetoacetate, two molecules of acetyl-CoA are formed, which will enter the TCA cycle.

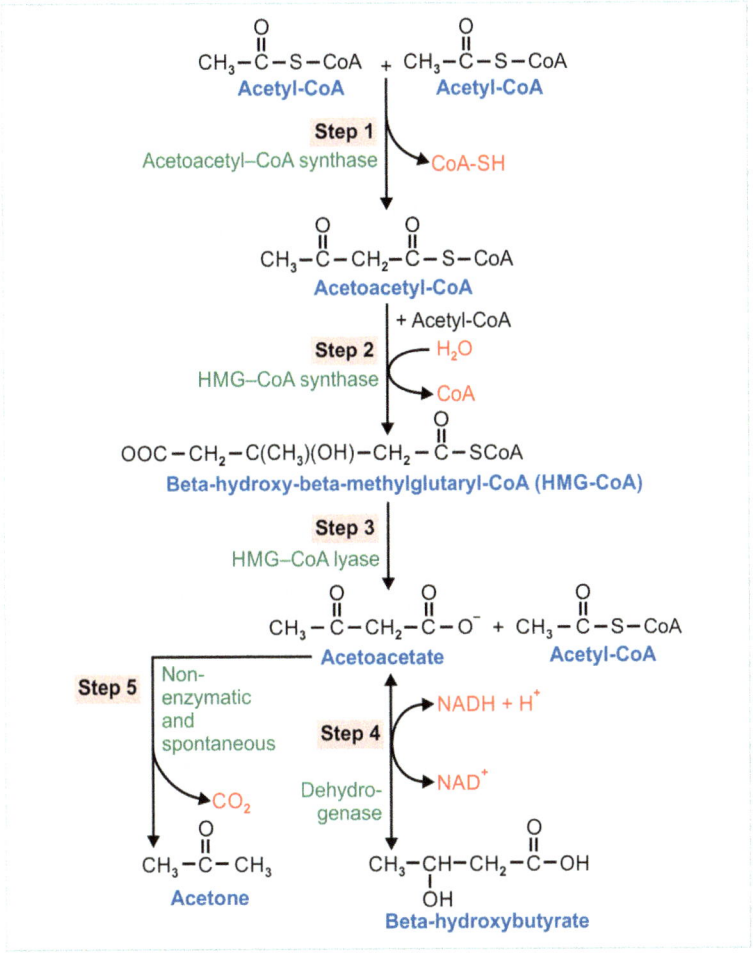

**Fig. 8.11:** Ketone body formation (ketogenesis).

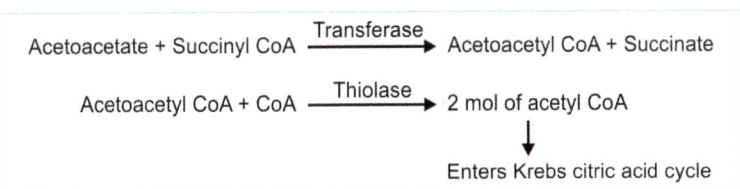

**Fig. 8.12:** Pathway of ketolysis.

1 acetyl-CoA molecule will give 10 ATPs
∴ 2 acetyl-CoA molecules will give 2 × 12 = 20 ATPs

Clinical manifestations of ketoacidosis and its treatment are given in detail in Chapter 6.

## METABOLISM OF TRIGLYCERIDES

Triglyceride (TG) or triacylglycerol consists of three fatty acid (FA) molecules esterified to a glycerol backbone. Simple TG has three identical FA molecules esterified to glycerol,

while mixed TG has two different fatty acid molecules esterified to glycerol. The TGs are stored in a specialized tissue called adipose tissue. Adipose tissue is of two types:
1. White adipose tissue, where cells are spherical with few mitochondria. Breakdown of TG leads to energy production, which is captured as ATP.
2. Brown adipose tissue, where cells are rich in mitochondria. The stored TG is broken down to produce energy, which is dissipated as heat. The main function of brown adipose tissue is thermogenesis. The TGs are circulated in the body as large lipid–protein complexes called lipoproteins.

## Synthesis of Triglycerides

TGs are synthesized from fatty acyl-CoA and glycerol 3-phosphate. Dihydroxyacetone phosphate, an intermediate in glycolysis, is reduced to glycerol 3-phosphate. This is acylated by acyltransferase enzyme. It first forms lysophosphatidic acid and then phosphatidic acid. Removal of phosphate forms diacylglycerol, which forms TG on adding one more acyl-CoA. The formation of triacylglycerol is not an ATP-dependent process.

## Breakdown of Triglycerides

Both stored fat and dietary fat are first hydrolyzed by the enzyme lipase. This enzyme releases three fatty acid molecules from the TG. These fatty acids are broken down by beta-oxidation to give energy. The glycerol backbone may be converted to dihydroxyacetone phosphate to enter in the glycolysis pathway.

Adipose tissue has **hormone-sensitive lipase**. The activation of this enzyme is regulated by certain hormones through a cascade mechanism.

Glucagon, adrenaline, and noradrenaline increase TG hydrolysis (breakdown of TG to free fatty acid) by activating the enzyme hormone-sensitive lipase. These hormones also increase free fatty acid level in blood. Glucagon and epinephrine (adrenaline) also inhibit acetyl-CoA carboxylase, thus preventing fatty acid synthesis.

Insulin has the opposite effect of glucagon. It increases the formation of TG and also activates acetyl-CoA carboxylase, thus increasing FA synthesis.

## Obesity

The fat content of the adipose tissue can increase to unlimited amounts, depending on the amount of excess calories taken in. This leads to obesity. High levels of plasma insulin are noticed. But the insulin receptors are decreased, and there is peripheral resistance against insulin action. When fat droplets are overloaded, the nucleus of adipose tissue cell is degraded, cell is destroyed, and TAG becomes extracellular. Such TAG cannot be metabolically reutilized and forms the dead bulk in obese individuals.

## Role of Liver in Fat Metabolism

- Secretion of bile salts
- Synthesis of fatty acid, triacylglycerol, and phospholipids
- Oxidation of fatty acids
- Production of lipoproteins
- Production of ketone bodies
- Synthesis and excretion of cholesterol

## FATTY LIVER

Under certain conditions, TG accumulates in liver. Extensive accumulation leads to **fatty liver**. When the accumulation becomes chronic, fibrotic changes occur in hepatic cells, which progress to **cirrhosis** and decreased liver function. Fatty liver may be produced by the following reasons:
- First type is associated with high levels of free fatty acid (FFA) in blood resulting from mobilization of fat from adipose tissue. Plasma FFA level is increased in **(a) starvation and (b) untreated diabetes**

mellitus. The large amount of FFA present in blood is taken up by liver and converted to TG. VLDL transports fat synthesized in liver to adipose tissue. But in this type of fatty liver condition, production of VLDL is not increased proportionate to the production of increased quantity of fat. This results in accumulation of fat in liver. This type of fatty liver can be relieved by treating starvation (giving a balanced diet consisting of high carbohydrate and low fat) and treating diabetes mellitus (giving insulin).
- The second type of fatty liver is due to metabolic block in the synthesis of VLDL. Plasma FFA level is normal. This type is associated with **deficiency of choline**. The lipoprotein VLDL consists of protein, fat, phospholipid, and cholesterol. Choline is necessary for the synthesis of phospholipid (lecithin). In the absence of choline, phospholipid is not synthesized, and without this, VLDL is not produced. Feeding choline will relieve this fatty liver condition. Also if the diet contains methionine, fatty liver will be relieved. Methionine could donate a methyl group for the synthesis of choline. Similarly, betaine also serves as the methyl group donor, which is used for the synthesis of choline. Choline, methionine, and betaine—the three components that relieve fatty liver—are called **lipotropic factors**.
- High cholesterol diet also decreases VLDL production, as cholesterol forms esters with essential fatty acids.
- Alcoholism: Oxidation of alcohol leads to the increase in NADH concentration. This will inhibit TCA cycle and increase the formation of TG. So, these factors also lead to the formation of fatty liver.
- Administering carbon tetrachloride, lead, or arsenic, which are inhibitors of protein synthesis, leads to decreased synthesis of VLDL.

## ESSENTIAL FATTY ACIDS

Fatty acids with one double bond are called **monounsaturated fatty acids (MUFA)**. MUFA can be synthesized in the body by introducing a double bond in a saturated fatty acid; this is called **desaturation** reaction. A number of unsaturated fatty acids can be synthesized from saturated fatty acids in the body by desaturation reactions (e.g., **palmitoleic** acid from palmitic acid). So such fatty acids are not considered essential fatty acids as they can be synthesized in the body.

Fatty acids with two or more double bonds are called **polyunsaturated fatty acids (PUFA)**. Such fatty acids are essential fatty acids as they cannot be synthesized in the body and have to be provided through diet. The essential fatty acids are **linoleic acid** and **linolenic acid**. **Arachidonic acid** is also considered as essential fatty acid; but it can be synthesized from linoleic acid, provided the diet contains enough and more linoleate.

Essential fatty acids (EFAs) occur in high concentration in vegetable oils and fish oils. Rich sources are sunflower oil, safflower oil, peanut oil, soybean oil, and corn oil. EFAs are important as their functions include:
- Formation of structural components of cell membrane
- Prevention of fat accumulation in liver
- Maintenance of proper growth
- Synthesis of eicosanoids

Deficiency of EFAs in diet leads to a clinical condition called **phrynoderma** or toad skin, characterized by scaly dermatitis on the posterior and lateral parts of limbs. Other symptoms include poor wound healing and growth impairment.

### Eicosanoids

One of the main functions of EFA is the synthesis of eicosanoids. They are so called because they are derivatives of 20 carbon eicosanoic acids. They are potent biomolecules and act as local hormones. Important eicosanoids are:

- **Prostaglandins** (PG)
- **Prostacyclins** (PGI)
- **Thromboxanes** (TX)
- **Leukotrienes** (LT)
- **Lipoxins** (LX)

**Arachidonic acid** forms the substrate for the synthesis of eicosanoids. Arachidonic acid is derived either from diet or from phospholipids by the action of enzyme phospholipase, or from linoleic acid.

**Prostaglandins** (PGs) were originally isolated from prostate tissue and hence the name. But they are present in almost all tissues. They are the **most potent biologically active substances;** as low as 1 ng/mL of PG will cause smooth muscle contraction. The diverse physiological roles of prostaglandins confer on them the status of **local hormones. PGs** exist in almost all mammalian tissues. They are synthesized from arachidonic acid (20 C) by the cyclization of carbon chain to form cyclopentane ring. Prostacyclins also have cyclopentane ring. But thromboxanes have a cyclopentane ring interrupted by an oxygen atom called oxane ring. These three components, forming the prostanoids, are synthesized by **cyclooxygenase pathway (Fig. 8.13)**. Prostaglandins are classified as A, B, D, E, F, G, H, and I, depending on the position of keto groups and hydroxyl groups. All naturally occurring PGs have an alpha-oriented **OH group at C15**.

A group of drugs known as nonsteroidal anti-inflammatory drugs (**NSAIDs**), such as **aspirin** and ibuprofen, inhibit the cyclooxygenase pathway and thus inhibit the prostanoid synthesis. So such drugs are used to treat inflammation, pain, fever, thrombosis, and so on. Aspirin inhibits **cyclooxygenase**, the enzyme necessary for the synthesis of prostaglandin. Hence there is decreased platelet aggregation. Therefore, aspirin is useful in the prevention of **heart attacks**. Other anti-inflammatory drugs (*indomethacin and ibuprofen*) also cause irreversible inhibition of the enzyme. **Paracetamol** is a reversible inhibitor of the enzyme.

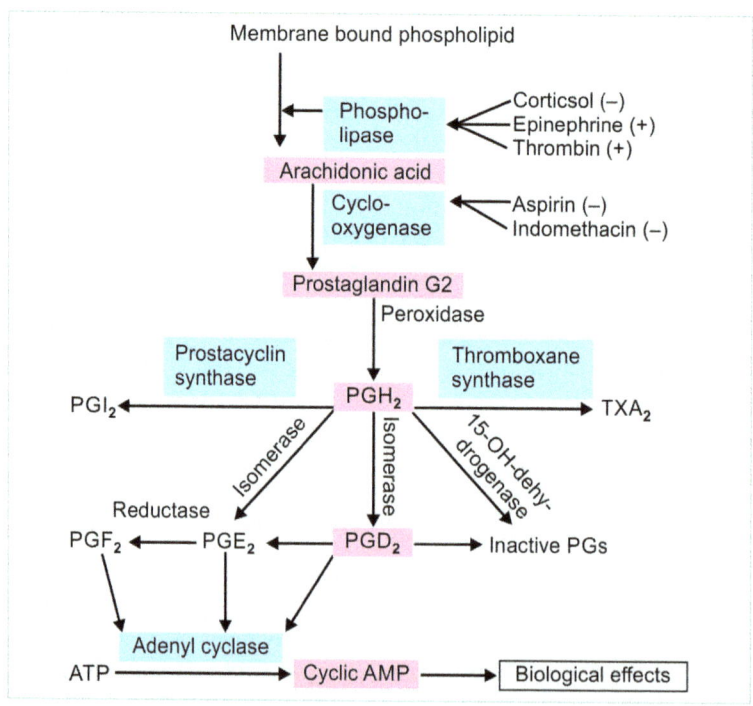

**Fig. 8.13:** Cyclooxygenase and lipoxygenase pathways.

Leukotrienes and lipoxins are formed by **lipoxygenase pathway** and are characterized by the presence of three or four double bonds. These compounds are formed in leukocytes, platelets, macrophages, and so on, in response to immunologic or sometimes nonimmunologic stimuli.

## Functions of Prostaglandins

- PGs are mediators of **inflammation.** Usually inflammation is treated by inhibiting the PG synthesis.
- "A" series of PGs are known to **induce sleep**. They act as sleep-promoting substances.
- Prostacyclin or PGI2 is synthesized by the vascular endothelium. Major effect is **vasodilatation**. It also inhibits platelet aggregation and has a protective effect on vessel wall against deposition of platelets. Thromboxane (TXA2) is the main PG produced by platelets. The major effects are vasoconstriction and **platelet aggregation**. Prostacyclin and thromboxane are opposing in activity. PGs increase the contractility and lower the blood pressure. Hence they may be used in the treatment of hypertension.
- PGs inhibit gastric acid secretion.
- PGF2 stimulates the uterine muscles. Hence PGF2 may be used for the medical termination of pregnancy. Yet another use is in inducing labor and arresting postpartum hemorrhage.
- PGF is a constrictor of bronchial smooth muscle, but PGE is a potent bronchodilator. PGE is used in aerosols for treating bronchospasm.
- Fever-inducing agents (pyrogens) activate PG synthesis thus increasing the body temperature.

### Prostacyclins

They inhibit platelet aggregation, and these are produced by blood vessel walls. PGIs are responsible for lack of adherence of platelets to healthy blood vessel walls.

### Thromboxanes

They promote platelet aggregation. They are synthesized by platelets and are responsible for spontaneous aggregation of platelets when blood vessel wall is injured. Action of TX is antagonistic to PGI action.

### Leukotrienes

They are potent bronchoconstrictors. They increase mucus secretion and act as regulators in conditions like asthma.

## SUMMARY OF THE CHAPTER

- Digestion of lipids involves the following enzymes: lingual lipase, gastric lipase, and pancreatic lipase. Pancreatic lipase is the major digestive enzyme and requires bile salts.
- Lipids are absorbed by emulsification and micelle formation with the help of bile salts.
- Short- and medium-chain fatty acids are absorbed directly without re-esterification.
- Defective absorption of lipids occurs in celiac disease and Crohn's disease.
- Mammalian tissues oxidize fatty acids primarily by the beta-oxidation pathway, which occurs in the mitochondria.
- Transport of fatty acids (long-chain acyl-CoA) through the inner mitochondrial membrane is facilitated by carnitine acyltransferase and translocase.
- Net yield of ATPs from one molecule of palmitic acid is 106 ATPs.
- Oxidation of odd-chain fatty acids produces propionyl-CoA, which may be further metabolized by the TCA cycle.
- Alpha-oxidation and omega-oxidation are two other modes of fatty acid oxidation.
- De novo synthesis of fatty acids occurs in the cytoplasm with the help of a dimeric multienzyme complex termed *fatty acid synthase*.
- Synthesis of fatty acid requires NADPH, while degradation requires NAD and FAD.
- Insulin favors fatty acid synthesis.

- The white adipose tissue is concerned with energy storage, and the brown adipose tissue is concerned with thermogenesis.
- Obesity is the result of an increase in the fat content of the adipose tissue. It is associated with insulin resistance.
- *Fatty liver* refers to the deposition of excess triglycerides in the liver cells. It is facilitated by lipotropic factors such as methionine, choline, and lecithin.
- Acetoacetate is the primary ketone body. Beta-hydroxybutyric acid and acetone are secondary ketone bodies.
- Ketosis is seen in untreated diabetes mellitus and starvation.
- Rothera's test is commonly used to detect presence of ketone bodies in urine.
- Acetyl-CoA formed from fatty acids is further oxidized in TCA cycle to generate energy, when availability of oxaloacetate is sufficient.
- Under conditions of fasting and starvation, the oxaloacetate is channeled to gluconeogenesis. Excess acetyl-CoA is then used for ketogenesis by liver.
- A similar situation is seen in uncontrolled diabetes mellitus where gluconeogenesis and lipolysis are both enhanced.
- The excess acetyl-CoA is converted to ketone bodies in hepatic mitochondria.
- The HMG-CoA formed is cleaved by liver enzyme lyase to the primary ketone body acetoacetate.
- Acetoacetate may be reduced to beta-hydroxybutyrate or spontaneously decarboxylated to acetone.
- Ketone bodies are synthesized by liver and metabolized by extrahepatic tissues, mainly cardiac muscle and skeletal muscle.
- Under conditions of starvation, brain starts metabolizing ketone bodies for energy needs.
- Since ketone bodies are acids, metabolic acidosis occurs. Excessive accumulation of ketone bodies can be dangerous, since it can result in acidosis, dehydration, and coma.

## QUESTIONS FOR SELF-ASSESSMENT

1. Mention with a flowchart the digestion and absorption of lipids.
2. Mention the role of pancreatic lipase in lipid digestion.
3. Mention the steps of beta-oxidation of fatty acids.
4. Schematically explain fatty acid synthesis.
5. What are the consequences of ketolysis?
6. Mention the steps and conditions for ketone body formation.
7. Mention the functions of prostaglandins.
8. Add a note on obesity.
9. Explain the term *fatty liver*.
10. Add a note on eicosanoids.

## MULTIPLE CHOICE QUESTIONS (MCQs)

8-1. Which among the following transfers fatty acids from cytosol to mitochondria?
   a. Carnitine
   b. Creatine
   c. Citrate
   d. GTP

8-2. Mention the correct statement regarding beta-oxidation of fatty acids in liver:
   a. Requires fatty acids with an even number of carbon atoms
   b. Produces only acetyl-CoA
   c. Occurs in the mitochondria
   d. Degrades fatty acids into $CO_2$ and $H_2O$

8-3. Mention the correct statement for fatty acid synthesis:
   a. Occurs in mitochondria
   b. Requires NADPH as a cofactor
   c. Requires NADH as a cofactor
   d. Intermediates are linked to coenzyme A

8-4. The following are the features of the fatty acid synthase complex, *except*:
   a. It is a dimer
   b. It is found within cytosol
   c. It requires pantothenic acid as a constituent
   d. Biotin is involved in the reaction as a cofactor

8-5. The concentration of the following is inversely related to the risk of cardiovascular disease:
a. HDL
b. LDL
c. VLDL
d. IDL

8-6. Mention which of the following is an abnormal form of lipoprotein:
a. LDL
b. IDL
c. Lp(a)
d. Chylomicron remnant

8-7. Which of the following is true in fatty acid elongation?
a. Occurs in mitochondria and microsome
b. Requires the provision of essential fatty acids
c. Requires biotin as a cofactor
d. Occurs only in cytosol

8-8. During electrophoretic separation, the fastest moving lipoprotein is:
a. HDL
b. LDL
c. VLDL
d. Chylomicrons

8-9. Ketosis occurs in all of the following conditions, *except:*
a. Diabetes mellitus
b. Kwashiorkor
c. Prolonged fasting
d. High-fat diet

8-10. Name the following ketone bodies that cannot be metabolized in human body:
a. Acetoacetate
b. Acetone
c. β-hydroxybutyrate
d. None of the above

## ANSWER KEYS TO MCQs

8-1. (a)   8-2. (d)   8-3. (b)   8-4. (d)   8-5. (a)   8-6. (c)   8-7. (a)   8-8. (a)
8-9. (b)   8-10. (b)

# CHAPTER 9

# Cholesterol and Lipoproteins

At the completion of this chapter, the reader will be able to answer questions on the following topics:

- Structure of cholesterol
- Biosynthesis of cholesterol
- Functions of cholesterol
- Plasma lipids
- Chylomicrons
- Very low-density lipoproteins
- Low-density lipoproteins
- High-density lipoproteins
- Atherosclerosis and coronary artery disease
- Risk factors for coronary artery disease
- Prevention of atherosclerosis

Cholesterol is a sterol widely distributed in the body. It is present in brain, muscles, liver, adipose tissue, skin, blood, intestine, and so on. It is a constituent of all cell membranes. It is necessary for the synthesis of all steroid hormones, bile acids, and also for synthesizing vitamin D. But it is not required in the diet because most tissues can synthesize it from simple precursors.

## STRUCTURE OF CHOLESTEROL

Cholesterol has cyclopentanoperhydrophenanthrene ring system. It has A, B, C, and D rings.
- There are three cyclohexane rings and one cyclopentane ring (Fig. 9.1).
- It has 27 carbon atoms.
- There is one hydroxyl group on the third carbon atom.
- There is a double bond between fifth and sixth carbon atoms.

Cholesterol is present only in foods of animal origin, such as egg, meat, and so on.

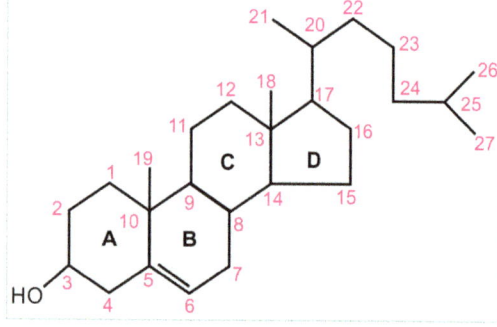

Fig. 9.1: Structure of cholesterol.

Absorption of dietary cholesterol needs micellar formation. After absorption, it is transported as chylomicrons to liver from intestine.

## BIOSYNTHESIS OF CHOLESTEROL

The major sites of synthesis of cholesterol are liver, adrenal cortex, testis, ovaries, and intestine. Liver is responsible for 80% of the endogenous cholesterol synthesis. All nucleated cells can synthesize cholesterol, including arterial walls. All the 27 carbons in cholesterol are derived from acetyl-CoA. The process may be studied in four stages.

## Summary of Biosynthesis of Cholesterol

**Stage 1:** Three acetate units condense to form six carbon mevalonate units (steps 1, 2, and 3).
**Stage 2:** Conversion of mevalonate into five carbon isoprene units (step 4).
**Stage 3:** Polymerization of six molecules of five carbon isoprene units to form linear structure of squalene (30 C) (step 5).
**Stage 4:** Cyclization of squalene to form steroid nucleus and formation of cholesterol (steps 6 and 7). These steps are elaborated in the following sections.

*Stage 1: Synthesis of Mevalonate*

Two molecules of acetyl-CoA condense to form acetoacetyl-CoA. A third molecule of acetyl-CoA condenses with acetoacetyl-CoA to form beta-hydroxy beta-methyl glutaryl CoA (HMG-CoA) **(Fig. 9.2)**. The reduction of HMG-CoA to **mevalonate** is catalyzed by **HMG-CoA reductase (Fig. 9.2)**. It uses two molecules of NADPH **(Fig. 9.2)**. NADPH required for this reaction is provided by the HMP shunt pathway.

HMG-CoA reductase is the **rate-limiting enzyme**. This regulatory enzyme is subjected to feedback regulation by cholesterol. When cholesterol level is increased, the synthesis of cholesterol is stopped at this stage, as cholesterol inhibits the enzyme reductase. Hormonal regulation includes the effect of both insulin and glucagon. **Insulin** increases reductase activity, thus increasing the synthesis of cholesterol. **Glucagon** has the inhibitory effect on reductase, thus decreasing synthesis of cholesterol. The **statin** group of drugs is competitive inhibitors of HMG-CoA reductase. So, they are used in clinical practice to reduce cholesterol level in blood.

*Stage 2: Formation of Isoprene Units*

Mevalonate is phosphorylated and then undergoes decarboxylation to give **isopentenyl pyrophosphate**, a five-carbon unit **(Fig. 9.3)**.

*Stage 3: Formation of Squalene*

Six molecules of isopentenyl pyrophosphate undergo isomerization and condensation to form squalene, which is a 30-C compound **(Fig. 9.4)**. In summary, 5C + 5C → 10C; 10C + 5C → 15C; 15C + 15C → 30C.

*Stage 4: Formation of Cholesterol*

A cyclase converts squalene to **lanosterol**. Lanosterol is the first steroid compound synthesized. It is a 30-carbon sterol. Lanosterol then undergoes a series of further reactions, such as removal of unnecessary methyl groups, migration of a double bond, and saturation of a double bond to form cholesterol **(Fig. 9.5)**.

## Regulation of Synthesis

The key enzyme of the pathway is HMG-CoA reductase. Lovastatin and other "statin" group of drugs are competitive inhibitors of HMG-CoA reductase. So, they are used in clinical practice to reduce cholesterol level in blood.

## IMPORTANCE OF CHOLESTEROL

### Significance and Functions of Cholesterol

- **Heart diseases:** The level of cholesterol in blood is related to the development of

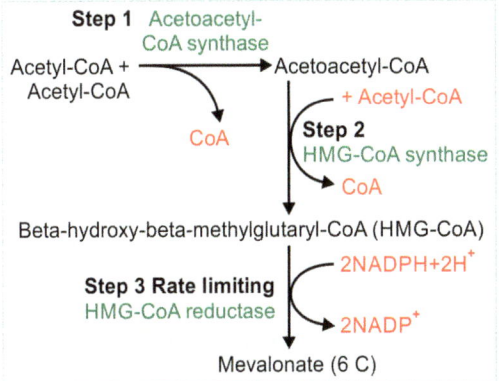

Fig. 9.2: Formation of mevalonate.

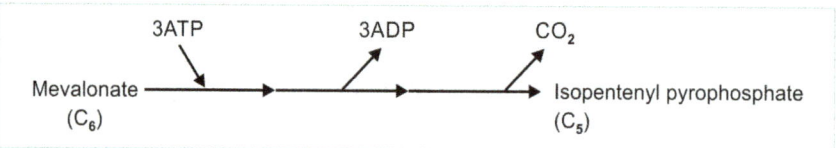

**Fig. 9.3:** Step 4: Formation of the five-carbon unit.

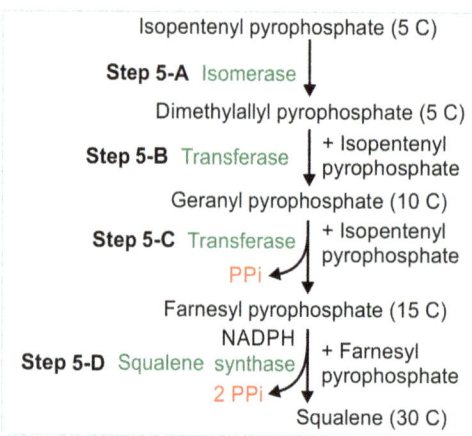

**Fig. 9.4:** Formation of squalene.

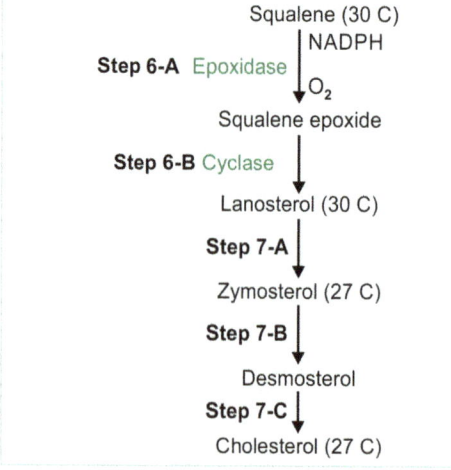

**Fig. 9.5:** Formation of cholesterol.

atherosclerosis. Abnormality of cholesterol metabolism may lead to cardiovascular accidents and heart attacks.

- **Cell membranes:** Cholesterol is a component of membranes and has a modulating effect on the fluid state of the membrane.
- **Nerve conduction:** Cholesterol is a poor conductor of electricity and is used to insulate nerve fibers.
- **Bile acids and bile salts:** The 24-carbon bile acids are derived from cholesterol. Bile salts are important for fat absorption.
- **Steroid hormones:** Glucocorticoids (21 carbon), androgens (19 carbon), and estrogens (18 carbon) are synthesized from cholesterol.
- **Vitamin D:** It is synthesized from cholesterol.

## Bile Salts

A major portion of cholesterol in liver is converted to bile salts. Bile salts are polar derivatives of cholesterol. Bile is synthesized in liver, stored and concentrated in the gallbladder, and then released into small intestine through the common bile duct. Bile salts, the major constituent of bile, help in the digestion of lipids by increasing the surface area exposed to the digestive action of lipases.

## Steroid Hormones

Cholesterol is the precursor for the synthesis of steroid hormones. The pathway is summarized in **Figure 9.6**. The major classes of steroid hormones and their functions are shown in **Table 9.1**.

Cholesterol is first converted to cholic acid (primary bile acid), which is conjugated with either glycine or taurine to form glycocholic acid or taurocholic acid. These exist as their $Na^+$ or $K^+$ salts as sodium and potassium glycocholate and sodium and potassium taurocholate. Bile salts keep cholesterol in solution.

## Chapter 9 Cholesterol and Lipoproteins

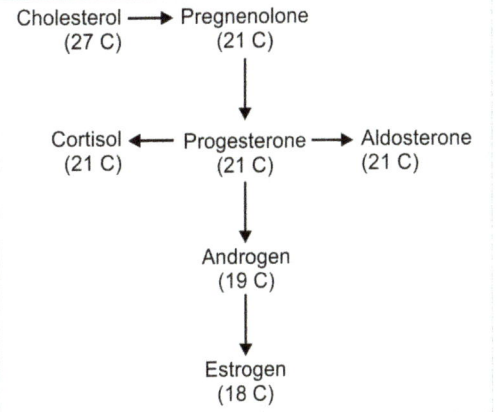

Fig. 9.6: Synthesis of steroid hormones from cholesterol.

Table 9.1: Functions of steroid hormones.

| Group of hormone | Example of hormone | Function of hormone |
|---|---|---|
| Progesteins | Progesterone | Essential for maintenance of pregnancy |
| Androgens | Testosterone | Influences the development of secondary sexual characters in males |
| Estrogens | Estradiol | Influences the development of secondary sexual characters in females |
| Glucocorticoids | Cortisol | Helps in blood glucose regulation and increases gluconeogenesis |
| Mineralocorticoids | Aldosterone | Increases $Na^+$ reabsorption by kidneys |

## Formation of Vitamin D

Vitamin D is formed by the action of sunlight (photolysis) on 7-dehydrocholesterol in skin.

## Excretion of Cholesterol

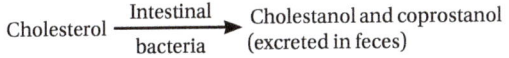

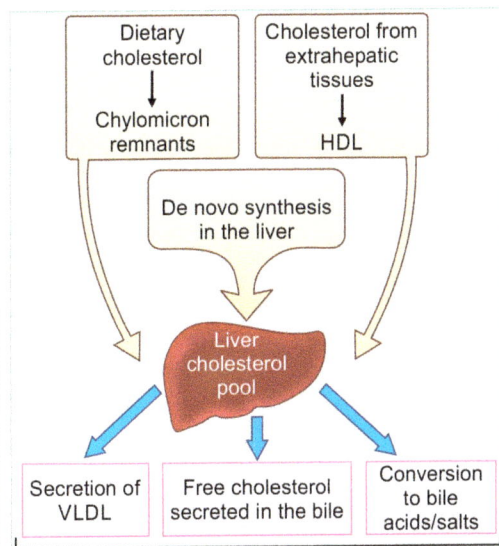

Fig. 9.7: Cholesterol pool and cholesterol excretion. (HDL: high-density lipoprotein; VLDL: very low-density lipoprotein)

Unabsorbed bile acids are excreted. This also forms a route for the excretion of cholesterol. **Figure 9.7** shows the cholesterol pool.

## PLASMA LIPIDS

Total plasma lipids amount to 400–600 mg/dL. Roughly speaking, one-third is cholesterol, one-third is triglyceride, and one-third is phospholipid. Normal values of lipid fractions are shown in **Table 9.2**.

### Transport of Lipids in Blood

Fats absorbed from the diet and lipids synthesized by the liver and adipose tissue must be transported between the various tissues and organs for utilization and storage. Since lipids are insoluble in water, their transportation in an aqueous environment (i.e., blood plasma) is difficult. This is solved by proteins carrying the lipids to make water-soluble lipoproteins. The protein part of the lipoprotein is called apolipoprotein or apoprotein. The lipoproteins are usually

**Table 9.2:** Plasma lipid profile (normal values).

| Analyte | Normal value |
| --- | --- |
| Total plasma lipids | 400–600 mg/dL |
| Total cholesterol | 150–200 mg/dL |
| HDL cholesterol, male | 30–60 mg/dL |
| HDL cholesterol, female | 35–75 mg/dL |
| LDL cholesterol, 30–39 years | 80–130 mg/dL |
| Triglycerides, male | 50–150 mg/dL |
| Triglycerides, female | 40–150 mg/dL |
| Phospholipids | 150–200 mg/dL |
| Free fatty acids (FFAs) [nonesterified fatty acids (NEFA)] | 10–20 mg/dL |

abbreviated as Lp. The lipoprotein molecules have a polar periphery made of proteins, polar heads of phospholipids, and cholesterol. The inner core consists of the hydrophobic TAGs and tails of phospholipids.

## Functions of Lipoproteins

- Lipoproteins transport and deliver lipids to the tissues.
- They maintain structural integrity of cell surface, mitochondria, and microsomes.
- They are important in clinical diagnosis, e.g., the concentration of LDL is increased in diabetes mellitus, atherosclerosis, and so on.

## Classification of Lipoproteins

1. Chylomicrons—least dense, contain mostly lipids, but protein concentration is too low
2. Very low-density lipoprotein (VLDL)
3. Intermediate-density lipoprotein (IDL)
4. Low-density lipoprotein (LDL)
5. High-density lipoprotein (HDL)—densest, protein concentration is too high

Triglyceride is predominant in chylomicrons and VLDL, whereas cholesterol and phospholipids are the predominant lipids in LDL and HDL. General characteristics of lipoproteins are shown in **Table 9.3**.

*Apolipoproteins*

The protein part of lipoprotein is called apolipoprotein (apo-Lp) or apoprotein. All apoproteins are mainly synthesized in liver. There are many apoproteins; the important ones are given in the following:

- **Apo-A-I:** It is specific for HDL. It is anti-atherogenic.
- **Apo-B-100:** It is a specific component of LDL. It is atherogenic.

**Table 9.3:** Characteristics of different classes of lipoproteins.

|  | Chylomicron | VLDL | LDL | HDL |
| --- | --- | --- | --- | --- |
| Density (g/L) | < 0.95 | 0.95–1.006 | 1.019–1.063 | 1.063–1.121 |
| Diameter (nm) | 500 | 70 | 25 | 15 |
| Electrophoretic mobility | Origin | Pre-beta | Beta | Alpha |
| Protein% | 2 | 10 | 22 | 50 |
| TAG% | 80 | 50 | 10 | 10 |
| Cholesterol% | 10 | 20 | 45 | 20 |
| FFA% | 0 | 0 | 0 | 0 |
| Apoproteins | A, B-48, C-II, E | B-100, C-II, E | B-100 | A-I, C, E |
| Transport function | TAG from gut to muscle and adipose | TAG from liver to muscle | Cholesterol from liver to heart | Cholesterol from heart to liver |

(VLDL: very low-density lipoprotein; LDL: low-density lipoprotein; HDL: high-density lipoprotein; TAG: triglycerides; FFA: free fatty acids)

- **Apo-B-48:** It is the component of chylomicrons.
- **Apo-E:** It is present in chylomicrons.

*Chylomicrons*

- Chylomicrons are formed in the **intestinal mucosal** cells and secreted into the lacteals of lymphatic system. They are **rich in triglyceride.** When the chylomicrons are synthesized by the intestinal mucosa, they contain **apo-B-48.**
- In peripheral tissues, enzyme lipoprotein lipase hydrolyzes TG to glycerol and free fatty acids (FFAs).
- FFAs are taken up by the tissue, and glycerol is returned to liver.
- Liver takes chylomicron remnants. As the TAG content is progressively decreased, the chylomicrons shrink in size. These remnants containing apo-B-48 and apo-E are taken up by hepatic cells by receptor-mediated endocytosis. Apo-E binds the hepatic receptors.

**Function of chylomicrons**

They are the transport form of dietary triglycerides from intestines to the adipose tissue for storage and to muscle or heart for their energy needs.

*Very Low-density Lipoproteins*

- **Synthesis of VLDL:** They are synthesized in the liver from glycerol and fatty acids and incorporated into VLDL. **Apo-B-100** is the major lipoprotein present in VLDL.
- **Metabolism of VLDL:** Fatty acids that are taken up by adipose tissue and muscle. The remnant is now designated as IDL (intermediate-density lipoprotein) and contains less of TAG and more of cholesterol. The IDL further loses triglyceride, so as to be converted to LDL (low-density lipoprotein). This conversion of VLDL to IDL and then to LDL is referred to as **lipoprotein cascade pathway (Fig. 9.8).**

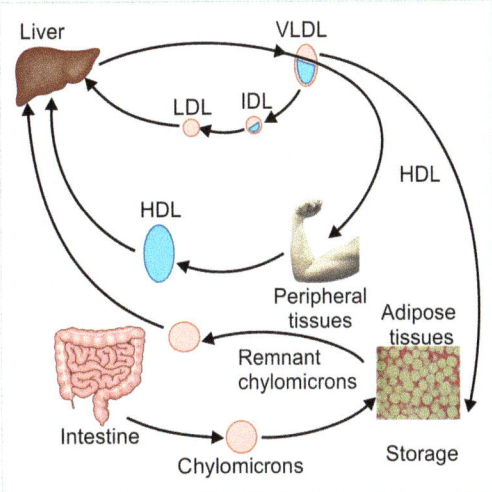

**Fig. 9.8:** Summary of lipoprotein metabolism. (VLDL: very low-density lipoprotein; LDL: low-density lipoprotein; HDL: high-density lipoprotein; IDA: intermediate-density lipoprotein)

- **Function:** VLDL carries triglycerides **(endogenous triglycerides)** from liver to peripheral tissues for energy needs.

*Low-density Lipoproteins*

- **Origin:** The LDL molecules are cholesterol-rich lipoprotein molecules containing only **apo-B-100.** Most of the LDL particles are derived from VLDL.
- **Function of LDL:** LDL transports cholesterol **from liver to the peripheral tissues.** LDL concentration in blood has positive correlation with incidence of **cardiovascular diseases.** About 75% of the plasma cholesterol is incorporated into the LDL particles.
- **LDL in clinical applications:** LDL infiltrates through arterial walls and is taken up by macrophages or scavenger cells. This is the starting event of **atherosclerosis** leading to myocardial infarction (*see* coronary artery diseases section in this chapter). Since LDL cholesterol is thus deposited in tissues, the LDL variety is called **bad cholesterol** in common parlance **(Fig. 9.9).** Normal value of LDL cholesterol in blood is shown in **Table 9.2.**

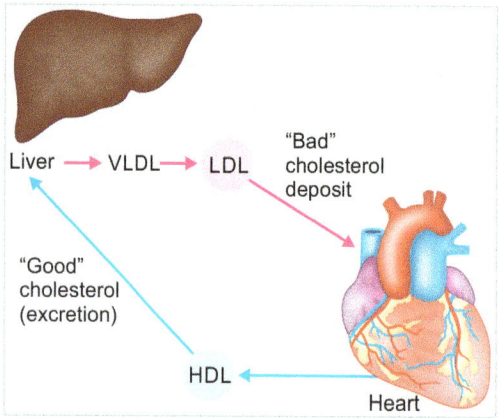

**Fig. 9.9:** Forward and reverse transport of cholesterol (HDL: high-density lipoproteins; LDL: low-density lipoproteins; VLDL: very low-density lipoproteins)

*High-density Lipoproteins*

- **Metabolism of HDL:** The intestinal cells synthesize the components of HDL and release into blood. The nascent HDL in plasma are discoid in shape (**Fig. 9.6**). It contains **apo-A-1.** Mature HDL spheres are taken up by the liver.
- **Function:** HDL is the main transport form of cholesterol from **peripheral tissue to liver,** which is later excreted through bile (**Fig. 9.9**). This is called **reverse cholesterol transport** by HDL.
- **Antiatherogenic:** Excretion of cholesterol needs prior esterification with polyunsaturated fatty acids (PUFA). Thus PUFA will help in lowering of cholesterol in the body, and so **PUFA is antiatherogenic**.
- **Clinical significance of HDL:** The level of HDL in serum is inversely related to the incidence of myocardial infarction. As it is **antiatherogenic** or "protective" in nature, HDL is known as **good cholesterol** in common parlance (**Fig. 9.9**). Normal value of HDL in blood is show in **Table 9.2**. A summary of the lipoprotein metabolism is shown in **Figure 9.6**.

## Hyperlipoproteinemia

Conditions where excessive amount of VLDL, IDL, chylomicrons, and LDL are found in serum are called hyperlipoproteinemia.

The main types of hyperlipoproteinemia are shown in **Table 9.4**.

- **Type I:** Absence of the enzyme lipoprotein lipase characterized by the presence of chylomicrons and very high TG level in serum.
- **Type IIA (primary familial hypercholesterolemia):** This is the most common type. The cause is **LDL receptor** defect. There is elevation of LDL and high cholesterol level in serum. Patients seldom survive the second decade of life due to ischemic heart disease.
- **Type IIB:** There is excessive production of apo-B. High VLDL, high LDL, high TG, and high cholesterol in serum.
- **Type IV:** It is caused by hypoinsulinemia. Overproduction of TG by liver is associated with diabetes, obesity, and impaired glucose tolerance. There will be high VLDL and high TG levels. Most hyperlipoproteinemias are associated with increased incidence of atherosclerosis because of increase in cholesterol level.

## Hypolipoproteinemia

*Abetalipoproteinemia*

Rare genetic disorder characterized by low plasma cholesterol due to absence of β-lipoprotein (LDL fraction). Main manifestation is acanthosis. Also blindness, caused by degenerative changes in retina.

*Tangier's Disease*

Genetic disorder. Deficiency of α-lipoprotein fraction (HDL). Due to accelerated catabolism of apo-A-I and HDL, plasma HDL level is low and so cholesterol ester is accumulated in tissues. Main manifestation is large orange or yellow tonsils. Atherosclerosis is common.

**Table 9.4:** Frederickson's classification of hyperlipoproteinemias.

| Type | Lipoprotein fraction elevated | Cholesterol level | TAG level | Metabolic defect | Features | Management |
|---|---|---|---|---|---|---|
| Type I | Chylomicron deficiency | ↑ | ↑↑ | Lipoprotein lipase | Eruptive xanthoma; hepatomegaly; pain in the abdomen | Restriction of fat intake; MCT supplementation |
| Type II A | LDL | ↑↑ | N | LDL receptor defect; apo-B ↑ | Atherosclerosis, coronary artery disease | Low-cholesterol diet; give PUFA |
| Type II B | LDL and VLDL | ↑↑ | ↑ | Apo-B ↑ Apo-CII ↑ | Corneal arcus | Low-cholesterol diet; give PUFA |
| Type III | Broad beta-VLDL and chylomicrons | ↑↑ | ↑ | Apo-E; apo CII ↑ | Xanthoma | Reduction of body weight, give PUFA |
| Type IV | VLDL | ↑ | ↑↑ | Apo-CII ↑ | Diabetes mellitus, heart disease | Reduction of body weight |
| Type V | VLDL and chylomicrons | N | ↑↑ | Secondary to other causes | Ischemic heart diseases | High PUFA intake |

(N: normal; ↑: increased; ; TAG: triglycerides; VLDL: very low-density lipoprotein; LDL: low-density lipoprotein; IDA: intermediate-density lipoprotein; PUFA: polyunsaturated fatty acids; MCT: medium-chain triacylglycerols)

## Clinical Significance of Cholesterol Level

### Serum Cholesterol Level

Serum cholesterol level is increased in:
- **Coronary artery disease (CAD) and atherosclerosis:** Increased cholesterol and increased LDL cause atherosclerosis. Cholesterol gets deposited in the arterial wall forming atherosclerotic plaques leading to narrowing of blood vessels. When the coronary artery gets blocked, it leads to myocardial infarction and death.
- **Familial hyperlipoproteinemias:** See type IIA discussed before. It is a genetic disorder. The blood cholesterol levels are extremely high. Affected individuals develop severe atherosclerosis in childhood. LDL receptor is defective in these persons.
- **Diabetes mellitus:** Acetyl-CoA pool is increased, and more molecules are channeled to cholesterol.
- **Obstructive jaundice:** The excretion of cholesterol through bile is blocked.
- **Hypothyroidism:** The receptors for HDL on liver cells are decreased, and so excretion is not effective.
- **Nephrotic syndrome:** Albumin is lost through urine; globulins are increased as a compensatory mechanism. So, apolipoproteins are increased, and then cholesterol is correspondingly increased.

### Risk Factors for Atherosclerosis

- **Serum cholesterol level:** Normal serum cholesterol level is given between 150–250 mg/dL; but it should be preferably below 200 mg/dL. Between 200–220 mg/dL is considered borderline and above 240 mg/dL is a definite risk of getting heart attack. Females before menopause have a lower level of cholesterol, which affords protection against atherosclerosis.

- **LDL cholesterol level:** Blood levels **under 130 mg/dL** are desirable. Levels between 130 and 159 are borderline, while above 160 mg/dL carry definite risk. Hence LDL is **bad** cholesterol.
- **HDL cholesterol level:** HDL level **above 65 mg/dL** protects against heart disease. Hence HDL is **good** cholesterol. A level below 40 mg/dl increases the risk of CAD.
- **Apoprotein levels:** Apo-A-I is a measure of HDL cholesterol (good cholesterol), and apo-B measures LDL cholesterol (bad cholesterol).
- **Cigarette smoking:** Nicotine of cigarette will increase the acetyl-CoA and cholesterol synthesis. Nicotine also causes transient constriction of coronary and carotid arteries.
- **Hypertension:** Systolic blood pressure more than 160 further increases the risk of CAD.
- **Diabetes mellitus:** In the absence of insulin, hormone-sensitive lipase is activated, more free fatty acids are formed, and these are catabolized to produce acetyl-CoA. These cannot be readily utilized, and a larger fraction may be channeled to cholesterol synthesis.
- **Serum triglyceride:** Normal level is 50–150 mg/dL. Blood level **>150 mg/dL** is injurious to health.

*Prevention of Atherosclerosis*

- **Reduce dietary cholesterol:** Cholesterol in the diet should be kept < 200 mg per day. Eggs and meat contain high cholesterol. One egg yolk contains about 500 mg of cholesterol. However, this is not very effective, as the cholesterol synthesis will be increased when the cholesterol intake is reduced.
- **Vegetable oils and PUFA:** Vegetable oils (e.g. sunflower oil) and fish oils contain PUFA. PUFA is helpful to reduce cholesterol level in blood.
- **Moderation in fat intake:** The accepted standard is that about 20% of total calories may be obtained from fat, out of which about one-third from saturated, another one-third from monounsaturated, and the rest one-third from polyunsaturated fatty acids. The recommended daily allowance will be about **20–25 g of oils** and about 2–3 g of PUFA per day for a normal adult.
- **Plenty of green leafy vegetables:** Due to their **high fiber content,** leafy vegetables will increase the motility of bowels and reduce reabsorption of bile salts. Vegetables also contain plant sterols (**sitosterol**), which decrease the absorption of cholesterol. About 400 g/day of fruit and vegetables are desired.
- **Exercise:** Regular moderate exercise will lower LDL (bad cholesterol) and raise HDL (good cholesterol) levels in blood. It will also reduce obesity.
- **Hypolipidemic drugs**: **HMG-CoA reductase inhibitors** ("statins" such as **atorvastatin** and **simvastatin**) are effective in reducing the cholesterol level and decreasing the incidence of CAD. However, such drugs should be started only when diet and exercise fail to reduce cholesterol level.

## CARDIAC BIOMARKERS

A biomarker is defined as one naturally occurring molecule, gene, or characteristic by which a particular pathological process or disease can be identified. In simple terms, a biomarker is a substance that can be used as an indicator of a particular disease.

Cardiac biomarkers may be classified as those for assessing acute cardiac injury or those used to identify long-term risks **(Table 9.5)**. Cardiac markers are tested whenever there is a doubt of a myocardial infarction. These cardiac markers include:

- Acute coronary syndrome resulting from myocardial infarction (for detecting the ischemia). The cardiac markers are used in clinical practice to detect myocardial ischemia at the earliest. Commonly used biomarkers for early detection of acute myocardial infarction are cardiac troponins

**Table 9.5:** Classification of cardiac markers.

**A. Cardiac markers of myocardial infarction**

Serial testing of the following cardiac enzymes is usually done to guide the prognosis. No single marker can successfully identify or exclude acute MI within the first 6 hours.
1. Cardiac troponins (cTnT and cTnI)
2. High-sensitivity troponin (hs cTnT and hs cTnI)
3. Creatine kinase isoenzyme (CK-MB)

**B. Markers for risk prediction**

1. Total cholesterol level in serum
2. LDL cholesterol and Apo-B-100 level
3. HDL cholesterol and Apo-A-1 level
4. Serum triglycerides

(cTnI and cTnT) and creatine kinase isoenzyme (CK-MB). Other markers are listed in **Table 9.5**.
- The risk factors (risk predictors) mainly include cholesterol, lipoproteins, and high-sensitivity C-reactive protein (hs-CRP) in plasma.

## Cardiac Troponins

Troponins are specific markers for **myocardial infarction.** Measurement of cardiac troponin has become the main test for early detection of an ischemic episode and in monitoring the patient. The troponin complex consists of three components: troponin C (calcium-binding subunit), **troponin I** (actomyosin ATPase inhibitory subunit), and **troponin T** (tropomyosin-binding subunit).

**Troponin I** is released into the blood **within 4 hours** after the onset of symptoms of myocardial ischemia, **peaks at 14–24 hours,** and remains elevated for 7–10 days postinfarction. Therefore, cTnI is very useful as a marker at any time interval after the heart attack. Serum level of **Troponin T** (cTnT or TnT) increases **within 6 hours** of myocardial infarction, peaks at **72 hours,** and then remains elevated up to 10–14 days. Serial estimations of cardiac troponins are done in any patient reporting with symptoms to the emergency clinic. An increasing blood value of troponin is the confirmation that patient has a myocardial infarction (MI). Troponin estimation could detect the MI, even before ECG changes are noticed.

## High-sensitivity Cardiac Troponins

Elevated cTn levels indicate cardiac injury, including acute coronary syndrome (ACS), acute myocardial infarction (AMI), stroke, pulmonary embolism, sepsis, acute perimyocarditis, acute heart failure, and tachycardia. Therefore, more specific tests are needed to identify myocardial infarction. To satisfy this necessity, **high-sensitivity cTnT (hs cTnT) and high-sensitive cTnI (hs cTnI)** tests have been developed. It enables the determination of very low cTnT concentrations. The higher sensitivity of this assay has allowed for improved identification of patients with AMI presenting in the first 3 hours following onset of symptoms. Even small increases are associated with a higher risk of death and other adverse outcomes. At least two measurements of high-sensitivity troponins are required for the assessment of patients with chest pain; the first measurement should be at presentation, and the second sample should be measured 3 hours after.

## Creatine Kinase

The CK enzyme catalyzes the reaction shown in **Figure 9.10**. It is further described in Chapter 11. **Normal reference serum** value for CK is 15–100 U/L for males and 10–80 U/L for females. CK value in serum is increased in **myocardial infarction.** The time course is shown in **Figure 9.11**. The CK level starts to rise **within 3–6 hours** of infarction. Therefore,

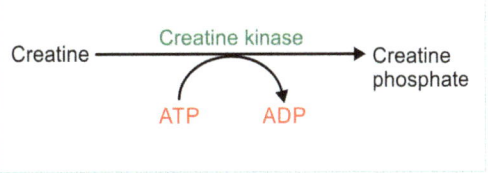

**Fig. 9.10:** Creatine kinase reaction.
(ATP: adenosine triphosphate; ADP: adenosine diphosphate)

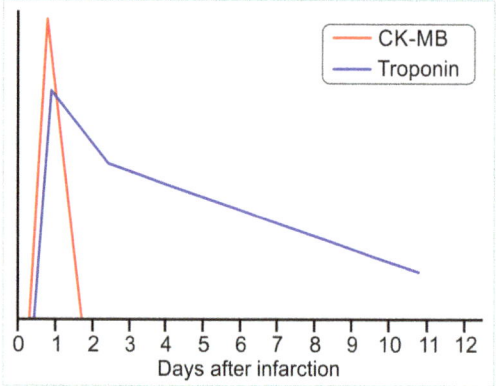

**Fig. 9.11:** Time course of the elevation of cardiac troponins and CK-MB.

CK estimation is very useful to **detect early cases,** where ECG changes may be ambiguous. Normal level of CK-MB is less than 5% of total CK. Elevation of CK can be associated with either cardiac or muscle diseases. Myocardial injury usually results in CK-MB levels higher than 5% of total CK in the blood, whereas skeletal muscle injury results in CK-MB levels lower than 5% of total CK. However, CK has now been replaced by troponins, which are more specific.

## Isoenzymes of Creatine Kinase

Creatine Kinase is a dimer; the subunits are called B for brain and M for muscle. There are, therefore, three different isoenzymes: CK-MM, CK-BB, and CK-MB. Normally CK-MB (heart isoenzyme) is only 5% of the total activity. Even doubling of the value of CK-MB isoenzyme may not be detected, if total value of CK alone is estimated. Hence the estimation of **CK-MB isoenzyme** is important for the diagnosis of myocardial infarction.

## Creatine Kinase and Muscle Diseases

The level of CK in serum is very much elevated in **muscular dystrophies**. CK level is highly elevated in crush injury, fracture, and acute cerebrovascular accidents. In short, estimation of **total CK** is employed in muscular dystrophies, and **CK-MB isoenzyme** is estimated in myocardial infarction.

## SUMMARY OF THE CHAPTER

- Cholesterol has a cyclopentanoperhydrophenanthrene ring with a total of 27 carbon atoms. Acetyl-CoA is the precursor of cholesterol. Cholesterol is a constituent of cell membranes. It is the precursor of all steroid hormones, bile acids, and vitamin $D_3$.
- Cholesterol is synthesized mainly in liver, adrenal cortex, and gonads. The rate-limiting enzyme of the cholesterol biosynthetic pathway is HMG-CoA reductase.
- Normal serum cholesterol range is 150–200 mg/dL.
- Lipoproteins are of five major types: chylomicrons, VLDL, IDL, LDL, and HDL.
- Chylomicrons contain the apo-B-48. Chylomicrons help in the transfer of triglycerides from the intestine to the muscle and adipose tissue.
- VLDL helps in the transfer of triglycerides from the liver to the peripheral tissues.
- LDL carries cholesterol from the liver to the heart, while HDL carries cholesterol from the heart to the liver. LDL contains apo-B-100, while HDL contains apo-A-I.
- LDL is "bad" cholesterol and HDL is "good" cholesterol.
- Higher concentration of lipoprotein(a) or Lp(a) increases the risk of myocardial infarction.
- Free fatty acids in plasma are transported bound to albumin and are taken up by peripheral tissues.
- Increase in TAG or cholesterol or both in plasma leads to hyperlipidemia, which can lead to early atherosclerosis and coronary artery disease.

## QUESTIONS FOR SELF-ASSESSMENT

1. Describe the pathway of cholesterol synthesis. Add a note on its clinical significance.

2. Mention the functions of cholesterol.
3. Classify lipoproteins and mention their clinical significance.
4. Mention the importance of plasma lipids.
5. Define apolipoproteins and mention their functions.
6. Mention the fate of LDL with a diagram.
7. Write a note on HDL metabolism.
8. Mention the functions of HDL.
9. What are the causes and risk factors for atherosclerosis?
10. What are lipoproteinemia? Mention its classification.

## MULTIPLE CHOICE QUESTIONS (MCQs)

9-1. **Rate-limiting step of cholesterol biosynthesis is:**
   a. Lanosterol → Cholesterol
   b. HMG-CoA → Mevalonic acid + CoA
   c. Acetoacetyl-CoA + Acetyl-CoA → HMG-CoA + CoA
   d. Squalene → Lanosterol

9-2. **The increased incidence of heart attack can be decreased by the following, *except*:**
   a. Cessation of smoking
   b. Lowered level of HDL
   c. Control of plasma cholesterol
   d. Lowered level of LDL

9-3. **The form in which dietary lipids are transported from intestinal mucosal cells is:**
   a. VLDL
   b. HDL
   c. Chylomicrons
   d. LDL

9-4. **All of the following are amphipathic lipids, *except*:**
   a. Triacylglycerol
   b. Phospholipids
   c. Cholesterol
   d. Fatty acids

9-5. **The triacylglycerol present in adipose tissue is hydrolyzed by:**
   a. Lipoprotein lipase
   b. Pancreatic lipase
   c. Hormone-sensitive lipase
   d. Phospholipase

9-6. **Mention the wrong statement for LDL:**
   a. Transports cholesterol to cells
   b. Contains apo-B-100
   c. Contains apo-C-II
   d. Is a marker for cardiovascular risk

9-7. **Free fatty acids are transported in plasma as a:**
   a. Component of VLDL
   b. Component of chylomicron remnants
   c. Part of LDL
   d. Bound to albumin

9-8. **The concentration of which of the following is inversely related to the risk of cardiovascular disease?**
   a. HDL
   b. LDL
   c. VLDL
   d. IDL

9-9. **All of the following are intermediates in the biosynthesis of cholesterol, *except*:**
   a. Acetyl-CoA
   b. Cholyl CoA
   c. Acetoacetyl-CoA
   d. HMG-CoA

9-10. **Function of LDL is the transport of:**
   a. Triglycerides from intestine to adipose tissue
   b. Cholesterol from liver to peripheral tissues
   c. Cholesterol from peripheral tissues to liver
   d. Free fatty acids [nonesterified fatty acids (NEFA)] from adipose tissue

## ANSWER KEYS TO MCQs

9-1. (b)　9-2. (b)　9-3. (c)　9-4. (a)　9-5. (c)　9-6. (c)　9-7. (d)　9-8. (a)
9-9. (b)　9-10. (b)

CHAPTER 10

# Amino Acids and Proteins: Chemistry

At the completion of this chapter, the reader will be able to answer questions on the following topics:

- Functions of proteins
- Structure of amino acids
- Isomerism of amino acids
- Classification of amino acids
- Isoelectric pH of amino acids
- Reactions of amino acids
- Functions of amino acids
- Peptide bond
- Biologically important peptides
- Classification of proteins
- Levels of organization of proteins

## GENERAL FUNCTIONS OF PROTEINS

- Proteins play a central role in cell functions and cell structure.
- Proteins are necessary for the formation of membranes, muscles, and connective tissue.
- All enzymes are proteins.
- Many hormones are proteins.
- Antibodies that confer immunity against viral and bacterial infections are proteins.
- Proteins carry (transport) compounds across cell membranes.
- Proteins function as buffers to maintain the pH of the cell.

## AMINO ACIDS

Amino acids are the simplest units of a protein molecule. They form the building blocks of protein. An amino acid has an amino group ($-NH_2$) and a carboxyl group ($-COOH$). The structure of an amino acid is shown in **Figure 10.1**.

The carbon atom to which carboxyl group is attached is called the alpha-carbon atom. In the case of amino acids, both amino and carboxyl groups are attached to the α-carbon atom (**Fig. 10.1**). Thus all amino acids are α-amino acids. Proteins contain 20 different amino acids. They differ in the "R" group. The structure of these amino acids is shown in **Figures 10.2 to 10.11**. Detailed classification is shown in **Table 10.1**.

Amino acids exhibit two types of isomerism due to the presence of asymmetric carbon atom (the carbon atom attached to four different groups).

1. Stereoisomerism
2. Optical isomerism

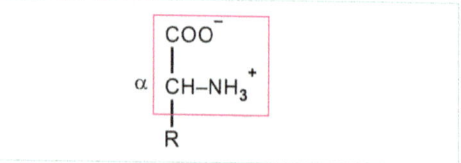

Fig. 10.1: Alpha-amino acid.

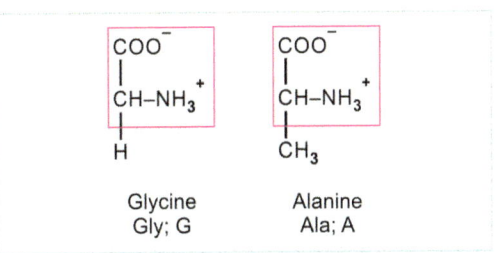

Fig. 10.2: Simple amino acids.

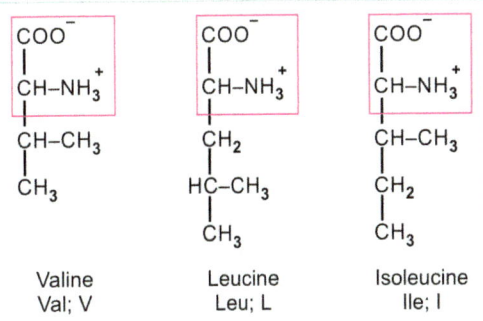

Fig. 10.3: Branched-chain amino acids.

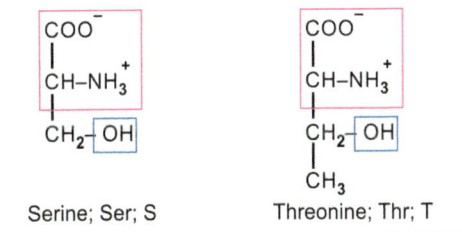

Fig. 10.4: Hydroxy amino acids.

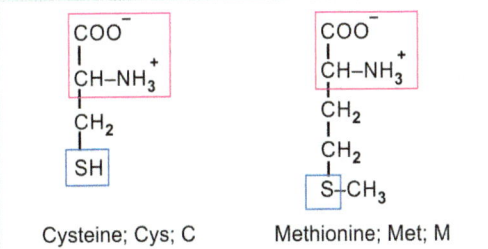

Fig. 10.5: Sulfur-containing amino acids.

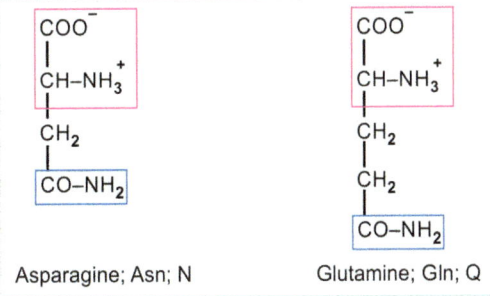

Fig. 10.6: Amino acids with amide groups.

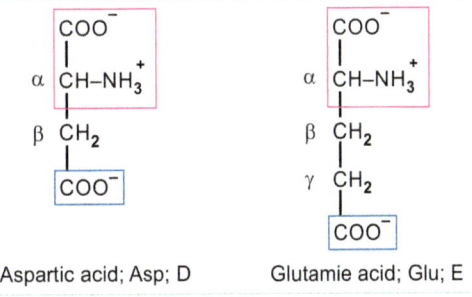

Fig. 10.7: Dicarboxylic amino acids.

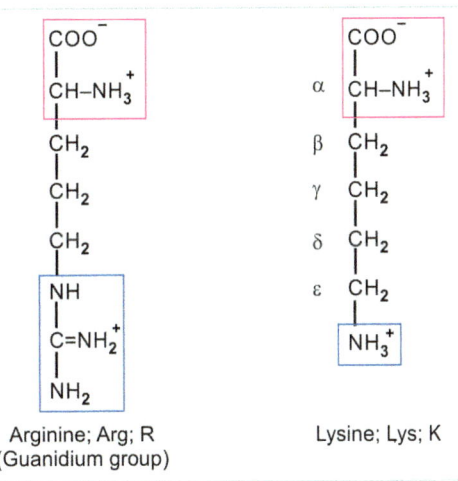

Fig. 10.8: Dibasic amino acids.

**Stereoisomerism:** All amino acids except glycine exist in D and L forms. In D-amino acids the $-NH_2$ group is on the right-hand side, whereas in L-amino acid it is oriented to the left. They are mirror images of each other **(Fig. 10.12)**.

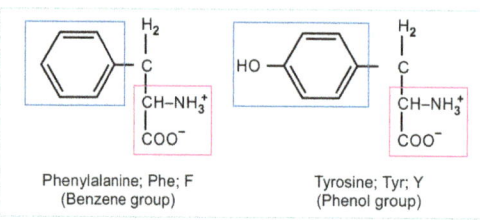

Fig. 10.9: Aromatic amino acids

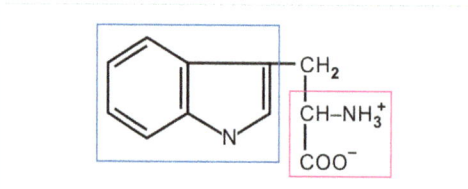

Fig. 10.10: Tryptophan (Trp) (W) with indole group.

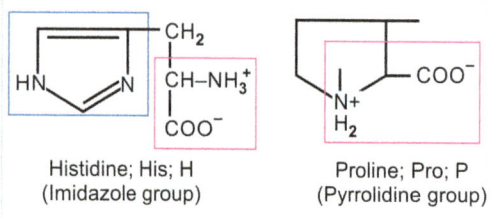

Fig. 10.11: Histidine and proline.

```
      CHO              COOH              COOH
       |                |                  |
HO — C — H         H₂N — C — H       H — C — NH₂
       |                |                  |
      CH₂OH             R                  R

L-glyceraldehyde    L-amino acid       D-amino acid
```

Fig. 10.12: Mirror images of amino acids.

Only L-amino acids are utilized in our body. Body proteins and food proteins contain L-amino acids. D-amino acids are seen in small amount in microorganisms and as constituents of certain antibiotics.

**Optical isomerism:** All amino acids except glycine exhibit optical activity as they rotate the plane of polarized light. So they exist as dextrorotatory or levorotatory isomers. Optical activity depends on the pH and side chains.

## Classification of Amino Acids

The 20 different amino acids can be classified in the following ways:

A. According to the chemical structure **(Table 10.1)**
B. According to the charge they carry **(Table 10.2)**
C. According to the nutritional importance **(Table 10.3)**
D. According to the metabolic fate **(Table 10.4)**.

*Note:* The 21st amino acid is selenocysteine and the 22nd is pyrrolysine, which are rarely seen in human body.

### According to the Charge they Carry

Those that carry a net negative charge at pH 6 are called acidic amino acids. They are also called monoamino dicarboxylic amino acids.

Those that carry a net positive charge at pH 6 are basic amino acids. They are also called diamino monocarboxylic amino acids.

Those that carry no net charge (equal number of +ve and −ve charges) at pH 6 are called neutral amino acids **(Table 10.2)**.

### According to the Nutritional Importance

Our body cannot synthesize eight amino acids. We have to obtain them through our food proteins. Hence, they are called **essential amino acids** (or indispensable amino acids). Essential amino acids are methionine (Met), threonine (Thr), tryptophan (Trp), valine (Val), isoleucine (Ile), leucine (Leu), phenyl alanine (Phe), and lysine (Lys). So, a code word to remember the essential amino acid is MeTTVILPhLy.

Histidine and arginine are called **semi-essential amino acids** because they are synthesized by the body only in small quantities. So, the synthesis is not sufficient during the period of growth and they need to be supplied through diet. They become essential for growing children, during pregnancy, and during lactation. However, for adults they are not essential.

# Chapter 10 Amino Acids and Proteins: Chemistry

**Table 10.1:** Classification based on structure.

| Group | Subgroup | Structure | Name of the amino acid | Abbreviation | Special group present | Figure/s |
|---|---|---|---|---|---|---|
| 1. Aliphatic amino acids | 1-a. Monoamino monocarboxylic acids | Simple amino acids | 1. Glycine<br>2. Alanine | 1. Gly<br>2. Ala | —<br>— | 10.2 |
| | 1-a. Monoamino monocarboxylic acids | Branched-chain amino acids | 3. Valine<br>4. Leucine<br>5. Isoleucine | 3. Val<br>4. Leu<br>5. Ile | —<br>—<br>— | 10.3 |
| | 1-a. Monoamino monocarboxylic acids | Hydroxy amino acids | 6. Serine<br>7. Threonine | 6. Ser<br>7. Thr | 6. Hydroxyl<br>7. Hydroxyl | 10.4 |
| | 1-a. Monoamino monocarboxylic acids | Sulfur-containing amino acids | 8. Cysteine<br>9. Methionine | 8. Cys<br>9. Met | 8. Sulfhydryl<br>9. Thioether | 10.5 |
| | 1-a. Monoamino monocarboxylic acids | Amino acids with amide group | 10. Asparagine<br>11. Glutamine | 10. Asn<br>11. Gln | 10. Amide<br>11. Amide | 10.6 |
| | 1-b. Monoamino dicarboxylic acids | | 12. Aspartic acid<br>13. Glutamic acid | 12. Asp<br>13. Glu | 12. Beta-carboxyl<br>13. Gamma-carboxyl | 10.7 |
| | 1-c. Diamino monocarboxylic acids | | 14. Lysine<br>15. Arginine | 14. Lys<br>15. Arg | 14. Extra amino group<br>15. Guanidinum | 10.8 |
| 2. Aromatic amino acids | | | 16. Phenylalanine<br>17. Tyrosine | 16. Phe<br>17. Tyr | 16. Benzene<br>17. Phenol | 10.9 |
| 3. Heterocyclic amino acids | | | 18. Tryptophan<br>19. Histidine<br>20. Proline | 18. Trp<br>19. His<br>20. Pro | 18. Indole<br>19. Imidazole<br>20. Pyrrolidine | 10.10 and 10.11 |

**Table 10.2:** Classification based on charge.

| Acidic amino acids (2 amino acids) | Basic amino acids (3 amino acids) | Neutral amino acids (15 amino acids) |
|---|---|---|
| • Aspartic acid<br>• Glutamic acid | • Lysine<br>• Arginine<br>• Histidine | • Glycine, alanine<br>• Valine, leucine, isoleucine, serine, threonine<br>• Cysteine, methionine, asparagine, glutamine, phenylalanine<br>• Tyrosine, tryptophan, proline |

Our body has the ability to synthesize the remaining 10 amino acids. Even if they are absent in our dietary proteins, our body can synthesize them. Hence they are referred to as **nonessential amino acids** (dispensable amino acids). Please note that all these nonessential amino acids are also required for protein synthesis, because they are also constituents of the body proteins **(Table 10.3)**.

*Memory Aid for Essential Amino Acids*

"Any Help In Learning These Little Molecules Proves Truly Valuable."

**Table 10.3:** Classification of amino acids according to nutritional importance.

| Essential amino acids (8 amino acids) | Semi-essential amino acids (2 amino acids) | Nonessential amino acids (10 amino acids) |
|---|---|---|
| • Methionine<br>• Threonine<br>• Tryptophan<br>• Valine<br>• Isoleucine<br>• Leucine<br>• Phenylalanine<br>• Lysine (MeTTVILPhLy) | • Arginine<br>• Histidine | • Glycine<br>• Alanine<br>• Serine<br>• Cysteine<br>• Asparagine<br>• Glutamine<br>• Aspartic acid<br>• Glutamic acid<br>• Tyrosine<br>• Proline |

This stands for "Arginine, Histidine, Isoleucine, Leucine, Threonine, Lysine, Methionine, Phenylalanine, Tryptophan, and Valine" in that order. Arginine and histidine are semi-essential amino acids, while others are essential.

*According to the Metabolic Fate of Amino Acids*

If the carbon skeleton can be converted into glucose after the removal of the amino group of an amino acid, such amino acids are called as **glucogenic amino acids (Table 10.4)**. If the carbon skeleton can be converted into ketone body (acetoacetic acid), such amino acids are called as **ketogenic amino acids (Table 10.4)**. After the removal of the amino group, if the carbon skeleton splits into two parts, one of which can be converted into glucose and other part becomes the ketone body, such amino acids are both glucogenic and ketogenic in nature.

**Table 10.4:** Classification of amino acids according to their metabolic fate.

| Ketogenic amino acids (2 amino acids) | Glucogenic amino acids (4 amino acids) | Both ketogenic and glucogenic amino acids (14 amino acids) |
|---|---|---|
| • Leucine<br>• Lysine | • Isoleucine<br>• Phenylalanine<br>• Tyrosine<br>• Tryptophan | • Glycine<br>• Alanine<br>• Valine<br>• Serine<br>• Threonine<br>• Cysteine<br>• Methionine<br>• Asparagine<br>• Glutamine<br>• Aspartic acid<br>• Glutamic acid<br>• Arginine<br>• Histidine<br>• Proline |

## Amino Acids as Zwitter Ions

In solution, amino acids will ionize. The $NH_2$ group will become $NH_3^+$, and the COOH group will lose one $H^+$ so as to become $COO^-$ (acid group). Thus, amino acid behaves as an acid and/or as a base. So, one amino acid molecule will carry both positive and negative charges. The same molecule carrying a positive and a negative charge is called a **zwitter ion** or **dipolar ion** or **amphoteric substance (Fig. 10.13)**.

Amino acids can both donate proton ($H^+$) and accept proton. Thus they are said to be amphoteric, that is, they behave as an acid when alkali is added and behave as a base when acid is added **(Fig. 10.14)**.

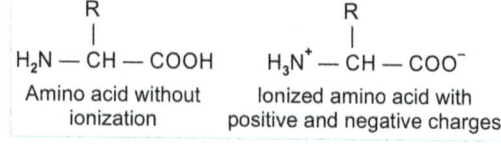

**Fig. 10.13:** Amino acid as zwitter ion.

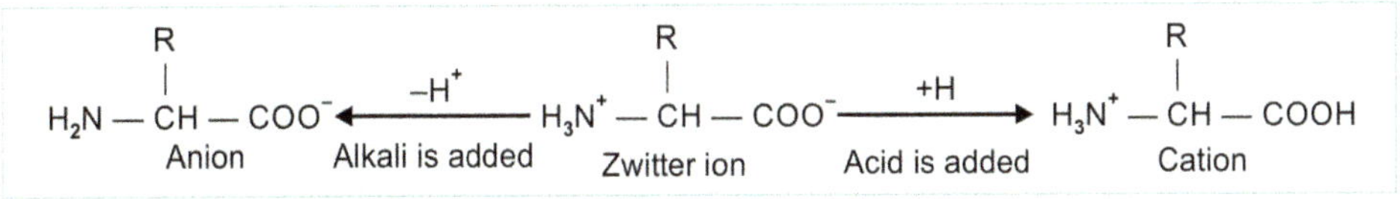

$H_2N—CH(R)—COO^- \xleftarrow{-H^+} H_3N^+—CH(R)—COO^- \xrightarrow{+H} H_3N^+—CH(R)—COOH$

Anion — Alkali is added — Zwitter ion — Acid is added — Cation

**Fig. 10.14:** Amino acid remains as anion in alkaline medium and behaves as cation in acidic medium.

Thus in acidic medium, the amino acid exists as cation. In alkaline medium, it exists as anion, and at a certain pH, it remains as zwitter ion.

## Isoelectric pH

**Isoelectric pH** or **isoelectric point** of an amino acid is the pH at which it exists as a zwitter ion **(Fig. 10.14)**. The positive and negative charges will cancel each other; so it bears no net charge. Hence it does not move in an electric field. Solubility and buffering capacity will be minimum at this pH.

It has been found that for most of the amino acids the isoelectric pH comes in physiological pH range. So, most amino acids exist as zwitter ions in physiological pH. Buffering capacity will be minimum at this pH; therefore, amino acids are not useful as physiological buffers. But the pK value of the imidazolium group of histidine is 6.1, and therefore it is effective as a buffer at a physiological pH of 7.4. The buffering action of plasma proteins and hemoglobin is mainly due to histidine residues.

## Reactions of Amino Acids

*Reactions due to —COOH Group*

**Decarboxylation**

The amino acids will undergo alpha-decarboxylation to form the corresponding amine of biological importance **(Fig. 10.15)**. For example:

Histidine ⟶ Histamine + $CO_2$
Glutamic acid ⟶ Gamma-amino butyric acid (GABA) + $CO_2$

Histamine is the mediator of allergic reactions. GABA is a neurotransmitter.

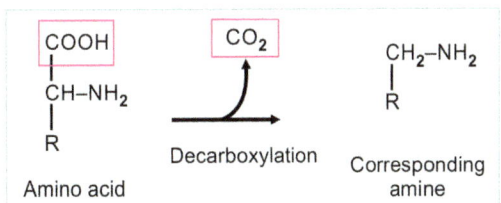

**Fig. 10.15:** Decarboxylation of amino acid to form amine.

**Amide Formation**

The –COOH group of dicarboxylic amino acids (other than α-carboxyl group) can combine with ammonia to form the corresponding amide.

For example:
Aspartic acid + $NH_3$ ⟶ Asparagine
Glutamic acid + $NH_3$ ⟶ Glutamine

Both asparagine and glutamine **(Fig. 10.6)** are constituents of proteins.

*Reactions due to Amino (–$NH_2$) Group*

**Transamination**

The α-amino group of amino acid can be transferred to α-keto acid to form the corresponding new amino acid and α-keto acid. This reaction needs pyridoxal phosphate (PLP) as a coenzyme. This reaction is of importance because (a) it is involved in the interconversion of amino acids and (b) it is also in the synthesis of nonessential amino acids. For example, alanine amino transaminase transfers the amino group from glutamate (amino acid) to pyruvate (keto acid) to make the keto acid to another amino acid (alanine) **(Fig. 10.16)**.

**Oxidative Deamination**

The α-amino group is removed from the amino acid to form corresponding keto acid.

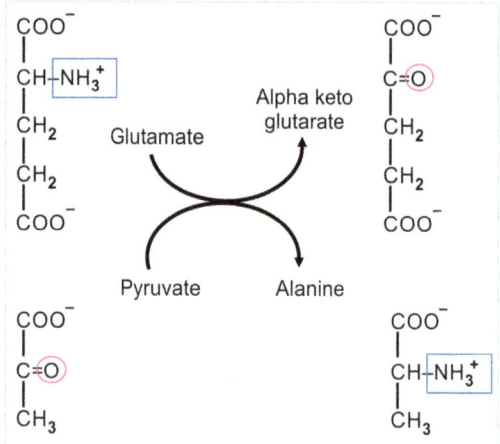

**Fig. 10.16:** Alanine amino transferase is an example of transamination reaction.

For example,

Glutamic acid ⟶ α-keto glutarate + $NH_3$

**Formation of Carbamino Compound**

Carbon dioxide adds to the α-amino group of amino acids to form carbamino compounds.

$Hb - NH_2 + CO_2$ ⟶ $Hb - NH - COOH$
(carbamino Hb)

This reaction serves as a mechanism for the transport of $CO_2$ from tissues to lungs by hemoglobin.

Color reactions of amino acids are shown in **Table 10.5**.

**Table 10.5:** Color reactions of amino acids.

| Reaction | Answered by specific group |
|---|---|
| 1. Ninhydrin | Alpha-amino group |
| 2. Biuret reaction | Peptide bonds |
| 3. Xanthoproteic test | Benzene ring (Phe, Tyr, Trp) |
| 4. Millon's test | Phenol (tyrosine) |
| 5. Aldehyde test | Indole (tryptophan) |
| 6. Sakaguchi's test | Guanidinium (arginine) |
| 7. Sulfur test | Sulfhydryl (cysteine) |
| 8. Nitroprusside test | Sulfhydryl (cysteine) |
| 9. Pauly's test | Imidazole (histidine) |

## Functions of Amino Acids

- Amino acids are the monomer constituents of proteins and peptides; amino acids serve a variety of functions.
- Some amino acids are converted to carbohydrate and are called as glucogenic amino acids.
- Specific amino acids give rise to specialized products. Examples are:
  - Tyrosine forms hormones, such as thyroid hormones ($T_3$ and $T_4$), epinephrine, and norepinephrine, and a pigment called melanin.
  - Niacin, a vitamin, is synthesized from tryptophan.
  - Glycine, arginine, and methionine are necessary for the synthesis of creatine.
  - Glycine and cysteine help in the synthesis of bile salts.
  - Glutamate, cysteine, and glycine form glutathione.
  - Histidine is decarboxylated to histamine.
  - Serotonin is formed from tryptophan.

## Peptide Bond

In proteins, successive amino acids are joined by peptide bonds. The –COOH group of one amino acid can be joined to the –$NH_2$ group of another amino acid by a covalent bond called as peptide bond (**Fig. 10.17**).

A dipeptide contains one peptide bond. If the first amino acid is alanine and the second one is glycine, this dipeptide can be written as Ala-Gly.

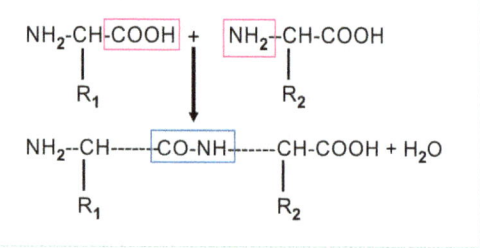

**Fig. 10.17:** Two amino acids are joined together through a peptide bond to form a dipeptide.

## Numbering of Amino Acids in Proteins

In a polypeptide chain, at one end there will be one free alpha-amino group. This end is called the **amino terminal (N-terminal) end,** and the amino acid contributing the alpha-amino group is named as the **first amino acid** (*see* dipeptide in **Fig. 10.17**). The other end of the polypeptide chain is the **carboxy terminal end (C-terminal),** where there is a free alpha-carboxyl group, which is contributed by the **last amino acid**. All other alpha-amino and alpha-carboxyl groups are involved in peptide bond formation.

## Biologically Important Peptides

Peptides varying from the simplest dipeptide to very long polypeptides are present in human body and serve specific functions.
- **Dipeptide:** A dipeptide is made up of two amino acids. Examples are **carnosine** and **anserine.** They contain two amino acids, β-alanine and histidine. Anserine is a derivative of carnosine. These dipeptides are present in the muscle.
- **Tripeptide:** A tripeptide is made up of three amino acids. An example is **glutathione** (GSH), which contains glutamic acid, cysteine, and glycine. The -SH group of cysteine is the active component. It is a reducing agent. There is a high concentration of GSH in red blood cell (RBC). In RBC, GSH protects (a) the –SH groups of various proteins, (b) decomposes toxic hydrogen peroxide, (c) maintains the integrity of cell membrane, and (d) keeps hemoglobin in the reduced form. Whenever Hb is oxidized to methemoglobin, GSH reduces it back to Hb.
- **Pentapeptides:** They are made up of five amino acids, for example, enkephalins, which are natural analgesics.
- **Nonapeptide:** They contain nine amino acids. Examples are (a) bradykinin, which relaxes smooth muscles; and (b) oxytocin and vasopressin, which are peptide hormones secreted from posterior pituitary.

Vasopressin is otherwise called antidiuretic hormone (ADH).

## Other Peptides

- Angiotensin–I has 10 amino acids and angiotensin–II has 8 amino acids; they cause hypertension.
- Gastrin, secretin, and pancreozymin are polypeptide hormones that stimulate the secretion of bile and other enzymes of digestive juices.
- Gramicidin, polymyxin, and actinomycin are polypeptides that act as antibiotics.

## CLASSIFICATION OF PROTEINS

Proteins are classified on the basis of the following characteristics:
A. Classification based on functional properties
B. Classification based on composition and solubility
C. Classification based on size and shape
D. Classification based on nutritional value

## Classification Based on the Functional Properties of Proteins

**A-1. Defense proteins (protective proteins):** Proteins that confer immunity against infections, e.g., antibodies (immunoglobulins), interferons, clotting factors.

**A-2. Catalytic proteins:** Proteins that act as enzymes, e.g., hexokinase.

**A-3. Structural proteins:** Proteins with structural function, e.g., (a) collagen, present in fibrous connective tissues, such as tendon, bone, and cartilage; (b) elastin, present in elastic connective tissues, such as ligaments; (c) keratin, present in hair, nail, and so forth.

**A-4. Transport proteins:** Proteins involved in the transport of various substances in plasma. Examples are:
a. Hemoglobin transports $O_2$ to tissues and $CO_2$ from tissues.

b. Albumin transports fatty acids and bilirubin.
c. Lipoproteins transport lipids in blood.
d. Transferrin transports iron in plasma.

**A-5. Regulatory proteins (hormones):** Proteins involved in the modulation of metabolic activity. Examples are insulin (regulates glucose metabolism), adrenocorticotropic hormone (ACTH), growth hormones.

**A-6. Contractile proteins:** Proteins of skeletal muscle that are involved in muscle contraction and relaxation, for example, actin, myosin.

**A-7. Proteins with storage function:** Ferritin is the storage form of iron in liver, bone marrow, and spleen.

**A-8. Protein as buffers:** (a) Plasma proteins are involved in buffering in plasma, while (b) hemoglobin is an important buffer inside RBCs.

**A-9. Proteins as toxins:** *Clostridium botulinum* toxin, which causes bacterial food poisoning.

**A-10. Proteins as antivitamin:** Avidin of raw egg white, which binds biotin (a vitamin) and interferes with its absorption.

## Classification Based on Composition and Solubility

Proteins may be divided into three major groups: (i) simple, (ii) conjugated, and (iii) derived.

**B-1. Simple proteins:** They are proteins that consist of only amino acids **(Table 10.6)**.

**B-2. Conjugated proteins:** They are complexes of simple protein with a non-protein part. The non-protein part is called **prosthetic** group. Examples of conjugated proteins are shown in **Table 10.7**.

*Derived Proteins*

These are formed by partial hydrolysis of high-molecular-weight proteins. Examples are: (a) denatured proteins, (b) peptone, and (c) gelatin. Gelatin is the degraded product of collagen.

## Classification Based on Shape and Size

**C-1. Fibrous proteins:** They are elongated or needle-shaped molecules. Examples are (a) keratin of hair; (b) myosin, the contractile proteins of muscle; and (c) collagen of connective tissue.

**Table 10.6:** Simple proteins.

|   | Name | Solubility | Example |
|---|---|---|---|
| 1 | Albumins | Soluble in water. Coagulated by heat | Lactalbumin, egg albumin |
| 2 | Globulins | Insoluble in water, but soluble in dilute salt solutions. Coagulated by heat | Serum globulins |
| 3 | Protamines | Soluble in water, dilute acids, and alkalies. Not coagulated by heat. They are basic proteins (contain large number of arginine and lysine) | Salmine of salmon sperm |
| 4 | Prolamines | Insoluble in water, soluble in 70–80% alcohol. They are rich in proline but lack in lysine | Zein from corn, gliadin of wheat |
| 5 | Lectins | Proteins having high affinity to sugar mostly derived from plant sources | Phytohemagglutinin |
| 6 | Scleroproteins | Insoluble in water and salt solution | Connective tissue proteins, such as collagen, elastin |

**Table 10.7:** Conjugated proteins.

| | Conjugated protein | Prosthetic group (non-protein part) | Examples |
|---|---|---|---|
| 1 | Glycoproteins | Carbohydrate | Blood group antigens, serum proteins |
| 2 | Lipoproteins | Lipids | Serum lipoproteins (LDL, HDL) |
| 3 | Nucleoproteins | Nucleic acids | Histones |
| 4 | Chromoproteins | Colored prosthetic group | • Hemoglobin (heme, red)<br>• Visual purple (vitamin A, purple)<br>• Flavoprotein (riboflavin, yellow) |
| E5 | Phosphoproteins | Phosphorus | Casein of milk, vitellin of egg yolk |
| F6 | Metalloproteins | Metal ions | • Hemoglobin (iron)<br>• Cytochrome (iron)<br>• Carbonic anhydrase (zinc) |

**C-2. Globular proteins:** These are spherical in shape. Examples are: (a) hemoglobin and (b) albumin.

## Classification Based on Nutritional Value

**D-1. Nutritionally rich proteins (complete proteins):** They contain all the essential amino acids in the required proportion, e.g., casein of milk.

**D-2. Incomplete proteins:** They are proteins that lack one essential amino acid. Examples are: (a) proteins from pulses (deficient in methionine) and (b) proteins from cereals (deficient in lysine).

**D-3. Poor proteins:** They lack many essential amino acids, e.g., zein from corn (deficient in tryptophan and lysine).

## LEVELS OF ORGANIZATION OF PROTEINS

- **Primary structure** of protein means the order of amino acids in the polypeptide chain and the location of disulfide bonds, if any.
- **Secondary structure** is the steric relationship of amino acids, close to each other.
- **Tertiary structure** denotes the overall arrangement and interrelationship of the various regions, or domains, of a single polypeptide chain.
- **Quaternary structure** results when the proteins consist of two or more polypeptide chains held together by non-covalent forces. These are schematically represented in **Figure 10.18**.

## DENATURATION OF PROTEINS

Mild heating, treating with urea, high pressure, vigorous shaking, and similar physicochemical agents produce denaturation. There will be nonspecific alterations in secondary, tertiary,

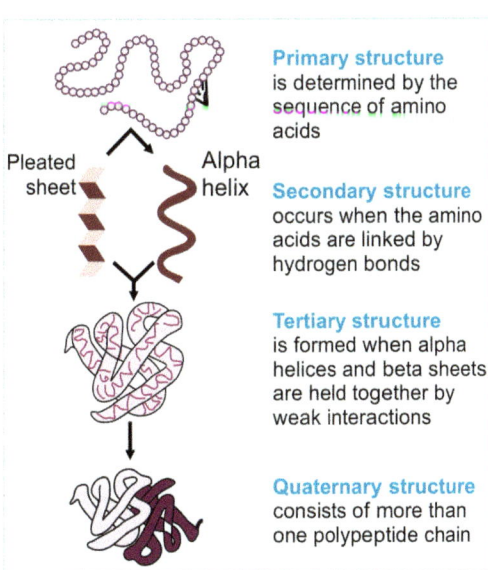

**Fig. 10.18:** Levels of organizations of proteins.

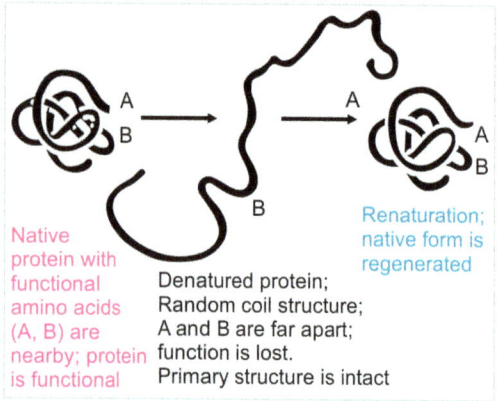

**Fig. 10.19:** Denaturation of protein.

and quaternary structures of protein molecules. Primary structure is not altered during denaturation. In general, during the process the solubility is decreased. It often causes loss of biological activity **(Fig. 10.19)**. Denatured proteins are sometimes reversible. For example, immunoglobulin chains are dissociated when treated with urea. When the urea is removed, the subunits are reassociated and biological activity of immunoglobulin is regained. But many proteins undergo irreversible denaturation. For example, albumin once heated cannot be denatured by cooling.

## HEAT COAGULATION

When heated at isoelectric point, some proteins will denature irreversibly to produce thick conglomerates called coagulum. This process is called heat coagulation. **Albumin** is easily coagulated, and globulin to a lesser extent. This is the basis of **heat and acetic acid test**, very commonly employed to detect the presence of albumin in urine.

## SUMMARY OF THE CHAPTER

- Most amino acids in the body are alpha-amino acids.
- Amino acids can be classified based on their (i) structure, (ii) side chain characters, (iii) metabolic fate, and (iv) nutritional requirements.
- All proteins are made of 20 common amino acids linked by peptide bonds.
- Amino acids may be classified based on the nature of side chains into aliphatic, aromatic, and imino acids.
- Aliphatic amino acids may be simple, branched-chain, hydroxyl, sulfur-containing amino acids and those with amide groups.
- Acidic amino acids have more than one COOH group, and basic amino acids have more than one amino group.
- Derived amino acids may be found in proteins, or they may be non-protein amino acids formed as metabolic intermediates.
- A functional classification based on the polar or nonpolar and ionized and non-ionized nature of side chains is more useful.
- Based on the metabolic fate, amino acids may be glucogenic, ketogenic, or both.
- Depending on nutritional requirements, amino acids may be essential or non-essential.
- In solution, amino acids exist as "zwitter ions" or "ampholytes" at their characteristic isoelectric pH (pI).
- Isoelectric pH is the pH at which amino acids have no net charge, no mobility in electric field, and are less soluble.
- Each amino acid has a specific isoelectric pH or pI depending on the pK value of the ionizable groups.
- Amino acids possessing an asymmetric carbon atom have optical activity, and all naturally occurring amino acids are L-amino acids.
- Glycine has no asymmetric carbon atom and therefore has no optical activity.
- Decarboxylation and amide formation are reactions involving alpha-COOH group.
- Transamination and oxidative deamination are reactions where amino group takes part.
- Peptide bonds are formed between the alpha-amino group of one amino acid and the alpha-carboxyl group of the next with elimination of a water molecule.
- Polymers of amino acids linked by peptide bonds are called polypeptides.

## Chapter 10 Amino Acids and Proteins: Chemistry

- Depending on the number of amino acids, oligopeptides (10), polypeptides (10–50), and proteins (> 50) are formed.
- Nitrogen content of ordinary proteins is on the average 16% by weight.
- Protein structure can be defined and studied at four levels, namely, primary, secondary, tertiary, and quaternary.
- Proteins have the primary level of structure, which denotes the linear sequence of amino acids linked by peptide bonds.
- The primary sequence is genetically determined and is unique and fixed for each protein produced by a particular species of organism.
- Primary structure determines the biological activity of the protein. Alterations lead to loss of functional capacity.
- The N-terminal amino acid is the first amino acid having a free alpha-NH2 group and the C-terminal amino acid (the last amino acid) has a free alpha-COOH group.
- Secondary and tertiary levels of structure are maintained by noncovalent bonds.
- The noncovalent bonds maintaining the higher levels of structure are hydrogen bonds, ionic bonds (electrostatic bonds), and hydrophobic interactions between the side chains of amino acids.
- Quaternary structure is present only in certain proteins having more than one polypeptide chain, for example, hemoglobin.
- The function of a protein is dependent on subunit interaction.
- Denaturation of protein results in the loss of biological activity but not the primary structure. Denaturation may be reversible.
- Proteins can be classified based on their (i) functions, (ii) composition, (iii) shape, and (iv) nutritional value.

## QUESTIONS FOR SELF-ASSESSMENT

1. Mention the classification of proteins with one example for each case.
2. How will you classify amino acids? Give one example for each class.
3. Mention the functions of amino acids.
4. What are peptides? Name some biologically important peptides.
5. Mention the important functions of proteins in our body.
6. Explain the different levels of organization of protein structure.
7. Mention the different color reactions of proteins.
8. What is isoelectric point? Mention its importance.
9. What are conjugated proteins? Give examples.
10. Mention the characteristic feature of the primary structure of protein with an example.

## MULTIPLE CHOICE QUESTIONS (MCQs)

10-1. Which of the following is an example of a neutral amino acid?
 a. Arginine
 b. Glutamate
 c. Aspartate
 d. Leucine

10-2. Which is an essential amino acid?
 a. Glutamine
 b. Glycine
 c. Proline
 d. Phenylalanine

10-3. Name the non-conjugated protein:
 a. Casein
 b. Rhodopsin
 c. Serum albumin
 d. Haptoglobin

10-4. Following are examples of globular proteins, *except:*
 a. Hemoglobin
 b. Albumin
 c. Myoglobin
 d. Collagen

10-5. Quaternary structure conjugated protein is:
 a. Insulin
 b. Myoglobin
 c. Hemoglobin
 d. Albumin

10-6. **Name the nonpolar amino acid:**
   a. Phe
   b. Tyr
   c. Asp
   d. His

10-7. **Name the nonessential amino acid:**
   a. Lysine
   b. Valine
   c. Glutamine
   d. Phenylalanine

10.8. **The optically inactive amino acid is:**
   a. Glutamine
   b. Alanine
   c. Glycine
   d. Tyrosine

10-9. **The force maintaining the primary structure of a protein is:**
   a. Peptide bonds
   b. Hydrophobic forces
   c. Hydrogen bonds
   d. Electrostatic (ionic) bonds

10-10. **Primary structure of a protein decides the:**
   a. Rate of synthesis of protein
   b. Biological activity of the protein
   c. Rate of degradation of the protein
   d. Effect of proteolytic enzymes on protein

## ANSWER KEYS TO MCQs

10-1. (d)   10-2. (d)   10-3. (c)   10-4. (d)   10-5. (c)   10-6. (b)   10-7. (c)   10-8. (c)
10-9. (a)   10-10. (b)

# CHAPTER 11

# Amino Acids and Proteins: Metabolism

At the completion of this chapter, the reader will be able to answer questions on the following topics:

- Digestion of proteins
- Absorption of amino acids
- Amino acid pool
- Ammonia formation
- Urea cycle
- General reactions of metabolism of amino acids
- Metabolism of glycine
- Metabolism of alanine
- Metabolism of serine
- Metabolism of methionine
- Metabolism of threonine
- Metabolism of branched-chain amino acids
- Metabolism of phenylalanine and tyrosine
- Metabolism of tryptophan

## DIGESTION OF PROTEINS

The dietary proteins are almost completely broken down to their constituent amino acids by a number of proteolytic enzymes in the gastrointestinal tract. The important proteolytic enzymes of gastrointestinal tract are **pepsin** in gastric juice; **trypsin**, chymotrypsin, carboxypeptidases, and aminopeptidases in pancreatic juice; and dipeptidases in intestinal juice.

The dietary proteins are denatured on cooking and therefore more easily digested. All these enzymes are of hydrolases nature. Proteolytic enzymes are secreted as inactive zymogens, which are converted to their active form in the intestinal lumen. This would prevent autodigestion of the secretory acini. The proteolytic enzymes include the following:

- **Endopeptidases:** They act on peptide bonds inside the protein molecule, so that the protein units become successively smaller and smaller. This group includes pepsin, trypsin, chymotrypsin, and elastase.

- **Exopeptidases:** They act at the peptide bond only at the end region of the chain. This group includes carboxypeptidase, acting on the peptide bond only at the carboxy terminal end of the chain, and aminopeptidase, which acts on the peptide bond only at the amino terminal end of the chain.

### Gastric Digestion of Proteins

In the stomach, hydrochloric acid (HCl) is secreted. It makes the pH optimum for the action of pepsin and also activates pepsin. The acid also denatures the proteins. However, at body temperature the hydrochloric acid cannot break the peptide bonds. Thus, in the stomach, to digest proteins, HCl needs enzymes.

*Rennin*

Rennin, otherwise called chymosin, is active in infants and is involved in the curdling of milk. It is absent in adults. Milk protein casein is converted to paracasein by the action of rennin. This denatured protein is easily digested further by pepsin. **Rennin** is the proteolytic enzyme present in gastric juice (**Renin** is a proteolytic enzyme secreted by kidneys. It is involved in the conversion of angiotensinogen to angiotensin, a hypertensive agent).

*Pepsin*

It is secreted by the chief cells of stomach as the inactive pepsinogen. The conversion of pepsinogen to pepsin is brought about by the HCl. The optimum pH for activity of pepsin is around 2. Pepsin is an endopeptidase. By the action of pepsin, proteins are broken into proteoses and peptones.

## Pancreatic Digestion of Proteins

The optimum pH for the activity of pancreatic enzymes (pH 8) is provided by the alkaline bile and pancreatic juice. The secretion of pancreatic juice is stimulated by the peptide hormones, cholecystokinin and pancreozymin.

Pancreatic juice contains the important endopeptidases, namely, trypsin, chymotrypsin, elastase, and carboxypeptidase.

*Trypsin*

Trypsinogen is activated by enterokinase present on the intestinal microvillus membranes. Once activated, the trypsin activates other enzyme molecules. Trypsin catalyzes the hydrolysis of the bonds formed by carboxyl groups of *arginine* (Arg) and *lysine* (Lys).

**Acute pancreatitis:** Premature activation of trypsinogen inside the pancreas itself will result in the autodigestion of the pancreatic cells. The result is acute pancreatitis. It is a life-threatening condition.

*Chymotrypsin*

Trypsin will act on chymotrypsinogen, so that the active site is formed. Thus, selective proteolysis produces the catalytic site.

*Carboxypeptidases*

Trypsin and chymotrypsin degrade the proteins into small peptides; these are further hydrolyzed into dipeptides and tripeptides by carboxypeptidases present in the pancreatic juice. They are metalloenzymes requiring zinc.

## Intestinal Digestion of Proteins

Complete digestion of the small peptides to the level of amino acids is brought about by enzymes present in intestinal juice (succus entericus). The luminal surface of the intestinal epithelial cells contains aminopeptidases, which release the N-terminal amino acids successively.

## ABSORPTION OF AMINO ACIDS

The products of digestion are rapidly absorbed. Absorption of amino acids mainly occurs in the small intestine. They are then carried to liver through portal blood and some of them enter the circulation and taken up by the tissue cells. On the other hand, tissue proteins are continually undergoing degradation to release amino acids that also enter the circulation. In addition, a continuous synthesis of amino acids is observed (nonessential amino acids).

## Clinical Applications

- The allergy to certain food proteins (milk, fish) is believed to result from the absorption of partially digested proteins.
- Partial gastrectomy, pancreatitis, carcinoma of pancreas, and cystic fibrosis may affect the digestion of proteins and absorption of amino acids.

## AMINO ACID POOL

Amino acids from all these sources get mixed up to constitute the body amino acid pool (**Fig. 11.1**). From this pool, they are drawn for the synthesis of various tissue proteins and other important compounds. So the body amino acid pool is always in a dynamic steady state.

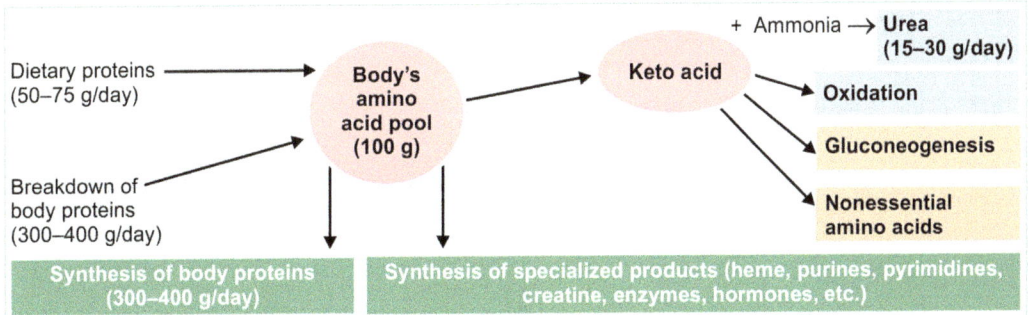

Fig. 11.1: An overview of metabolism of proteins and amino acids.

The surplus amino acids undergo catabolism to the corresponding α-keto acids and $NH_3$. The keto acids formed may have any one of the following three fates.
1. They may combine with ammonia to give rise to amino acids.
2. Majority of them can be converted to glucose (gluconeogenesis) and a few keto acids may produce acetoacetic acid (ketone body).
3. The keto acids may be oxidized in the Krebs citric acid cycle to give rise to energy (ATP).

## UREA CYCLE

Ammonia is formed by the catabolism of amino acids (see transamination reaction shown below). The ammonia, even in very small quantity, is highly toxic to brain cells. Therefore, ammonia is immediately detoxified to form urea. The ammonia from all over the body reaches liver. It is then detoxified to urea by liver cells and then excreted through kidneys. Urea is the end product of protein metabolism. Urea is formed through the urea cycle (**Fig. 11.2**). As ornithine is the first member of the reaction, it is also called as the **ornithine cycle**. When one of the enzymes of the urea cycle is deficient, certain disorders are manifested, which are summarized in **Table 11.1**.

## Hepatic Coma (Acquired Hyperammonemia)

In liver diseases, hepatic failure can finally lead to hepatic coma and death. Hyperammonemia

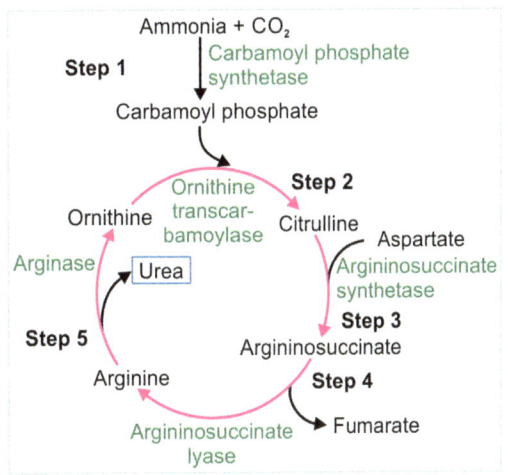

Fig. 11.2: Urea cycle.

is the characteristic feature of liver failure. Normally the ammonia and other toxic compounds produced by intestinal bacterial metabolism are transported to liver by portal circulation and detoxified by the liver. But when there is portal systemic shunting of blood, the toxins bypass the liver and their concentration in systemic circulation rises. The signs and symptoms are mainly pertaining to central nervous system (CNS) dysfunction (altered sensations and convulsions), or manifestations of failure of liver function (ascites, jaundice, hepatomegaly, edema, and hemorrhage).

## Table 11.1: Urea cycle disorders.

| Disease | Enzyme deficient | Features |
| --- | --- | --- |
| Hyperammonemia type I | Carbamoyl phosphate synthetase-I (CPS-I) | Very high NH$_3$ levels in blood. Autosomal recessive. Mental retardation. Incidence is 1 in 200,000 |
| Hyperammonemia type II | Ornithine transcarbamylase (OTC) | Ammonia level high in blood. Increased glutamine in blood, cerebrospinal fluid, and urine. Orotic aciduria due to the channeling of carbamoyl phosphate into pyrimidine synthesis—X-linked |
| Hyperornithinemia | Defective ornithine transporter protein | Elevated blood level of ammonia and ornithine. Decreased level of urea in blood. Autosomal recessive condition |
| Citrullinemia | Argininosuccinate synthetase | Autosomal recessive inheritance. High blood levels of ammonia and citrulline. Citrullinuria (1–2 g/day) |
| Arginino succinic aciduria | Argininosuccinate lyase | Argininosuccinate in blood and urine. Friable brittle tufted hair (trichorrhexis nodosa). Incidence is 3/200,000 |
| Hyperargininemia | Arginase | Arginine increased in blood and cerebrospinal fluid (CSF). Instead of arginine, cysteine and lysine are lost in the urine. Incidence is 1 in 100,000 |

## GENERAL REACTIONS OF AMINO ACID METABOLISM

While each of the 20 amino acids undergoes certain unique special enzyme reactions, there are a few reactions which are common to most of the amino acids. These reactions called the general reactions of amino acid metabolism are (a) transamination, (b) oxidative deamination, and (c) decarboxylation.

## Transamination

This is catalyzed by **transaminases** (also called aminotransferases). Here the amino group from amino acid number 1 is transferred to a keto acid number 2. This results in the formation of a keto acid number 1 and amino acid number 2 (**Fig. 11.3**). Pyridoxal phosphate (PLP) is the coenzyme for all transamination reactions. Transamination takes place in all the cells of the body.

*Biological significance:* Since these reactions are reversible, they bring about:
- Deamination of amino acid (the catabolism).

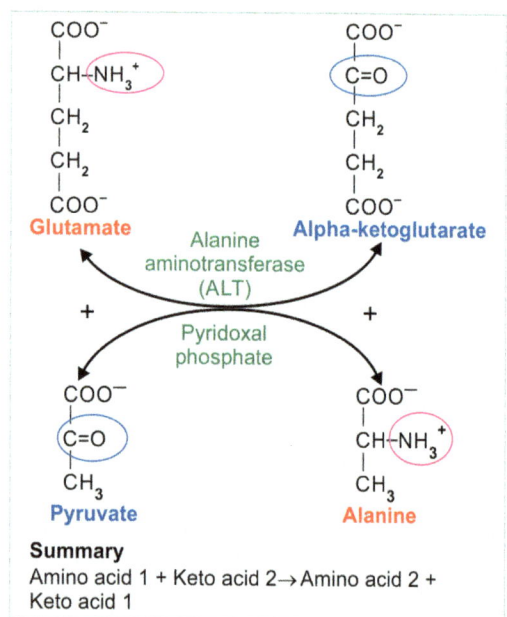

**Summary**
Amino acid 1 + Keto acid 2 → Amino acid 2 + Keto acid 1

**Fig. 11.3:** Transamination reaction. In this example, the enzyme is alanine aminotransferase (ALT) and the coenzyme is pyridoxal phosphate. The reaction is readily reversible. The ALT was previously known as glutamate pyruvate transaminase (GPT).

- Biosynthesis of nonessential amino acids. Thus, they maintain a proper concentration of various amino acids in the body.

*Clinical significance:* Transaminases are of 20 different types. The most common ones are aspartate aminotransferase (AST) and alanine aminotransferase (ALT). These are increased in cardiac and liver diseases. Serum AST is significantly elevated in myocardial infarction and is moderately increased in liver diseases.

*Exceptions:* Lysine, threonine, and proline are not transaminated.

## Oxidative Deamination

Deamination is the process by which amino group of the amino acid is removed as $NH_3$. Only liver mitochondria contain glutamate dehydrogenase which deaminates glutamate (**Fig. 11.4**). It is an allosteric enzyme which is activated by ADP and inhibited by ATP.

*Transdeamination*

The amino group of most of the amino acids is released by a coupled reaction called transdeamination, which involves **transamination followed by oxidative deamination**. Transamination takes place in the cytoplasm of all the cells of the body; the amino group is transported to liver as **glutamic acid** which is finally oxidatively deaminated in the liver cells. Thus, the two components of the reaction are physically far away, but physiologically they are coupled. Hence, the term transdeamination (**Fig. 11.5**).

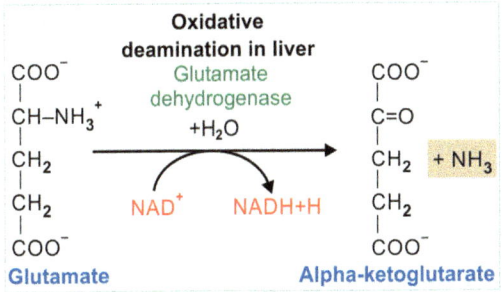

**Fig. 11.4:** Glutamate is deaminated to α-ketoglutarate.

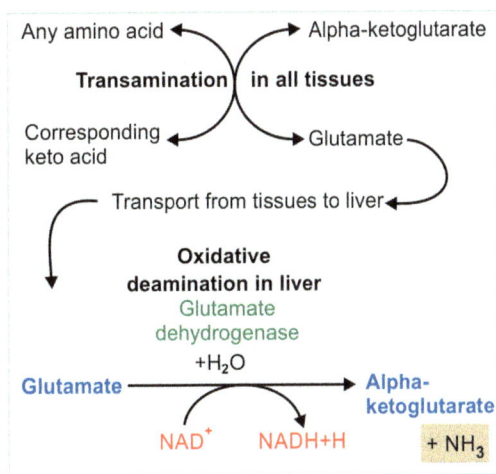

**Fig. 11.5:** Transamination + deamination = transdeamination.

## Nonoxidative Deamination

Enzymes catalyzing this reaction are dehydratase. They act on hydroxy amino acids

Serine $\xrightarrow{\text{Serine dehydratase}}$ Pyruvate + $NH_3$

Threonine $\xrightarrow{\text{Threonine dehydratase}}$ α-ketobutyrate + $NH_3$

## Decarboxylation

Decarboxylation is the reaction catalyzed by the enzymes, collectively referred to as decarboxylases.

Amino acid $\xrightarrow[\text{PLP}]{\text{Decarboxylase}}$ Amine + $CO_2$

Histidine $\xrightarrow[\text{PLP}]{\text{Histidine decarboxylase}}$ Histamine + $CO_2$

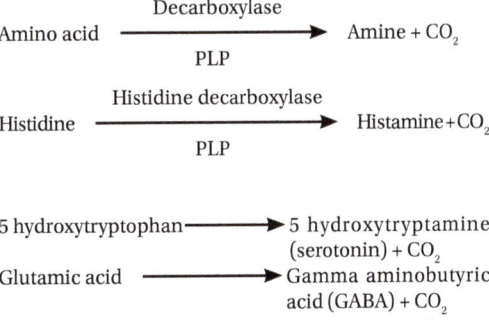

*Importance:* Decarboxylation produces important amines with different biological actions.

## METABOLISM OF INDIVIDUAL AMINO ACIDS

### Glycine

Glycine is a nonessential amino acid. It can be synthesized from:
- Serine
- $CO_2$, $NH_3$, and methylene tetrahydrofolic acid (THFA).

**Synthesis from serine**

Since this reaction is reversible, serine and glycine are interconvertible **(Fig. 11.6)**.

*Synthesis from $CO_2$ and $NH_3$*

The enzyme glycine synthase complex can synthesize glycine from carbon dioxide and ammonia **(Fig. 11.7)**.

*Biological Importance of Glycine*

Glycine is used in the synthesis of the following biologically important compounds.
- Glycine (along with other amino acids) is required for the synthesis of body proteins, enzymes, and polypeptide hormones.
- Synthesis of serine from glycine: The reaction by serine hydroxymethyltransferase **(Fig. 11.6)** is reversible, so as to produce serine.
- Synthesis of **glutathione** (GSH): GSH is a tripeptide containing glutamic acid, cysteine, and glycine.
  - Glutathione is mainly used in reduction reactions
    (Reduced) 2GSH → (Oxidized) GS-SG + $H_2$
    The hydrogen released is used for reducing other substrates. For example,
    Maleylacetoacetate → fumarylacetoacetate
  - Glutathione and RBC membrane integrity
    GSH is present in the RBCs, which is used for inactivation of free radicals formed inside RBC. So GSH will prevent the lysis of RBC membrane.
- Synthesis of **creatine**: Creatine is synthesized from three amino acids, glycine, Arg, and methionine. Creatine phosphate is a form of energy present in the muscles **(Fig. 11.8)**. Creatine phosphate is degraded to **creatinine** which is excreted in urine. Normal serum creatinine level is 0.7–1.4 mg/dL and serum creatine level is 0.2–0.4 mg/dL. Creatinine level in blood is a sensitive indicator of **renal function** (*see* Chapter 20). As the kidney function is decreased, correspondingly blood creatinine level also increases.
  In normal males, negligible amounts of creatine are found in their urine. But in the early phase of **muscular dystrophies**, the blood creatine and urinary creatinine are increased.
  The enzyme creatine kinase (CK), especially the cardiac isoenzyme [CK myocardial band (MB)], is elevated in **myocardial infarction** (*see* Chapter 9).
- Synthesis of **purines** for the production of DNA. The whole molecule of glycine is incorporated in the purine ring.

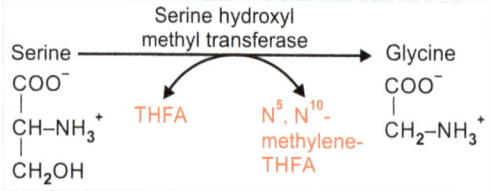

**Fig. 11.6:** Formation of glycine from serine. (THFA: tetrahydrofolic acid)

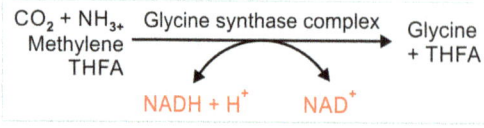

**Fig. 11.7:** Glycine synthesis from $CO_2$. This is a reversible reaction.
(THFA: tetrahydrofolic acid).

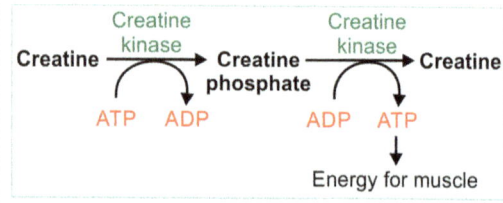

**Fig. 11.8:** Generation of creatine phosphate. (ATP: adenosine triphosphate; ADP: adenosine diphosphate)

## Chapter 11 Amino Acids and Proteins: Metabolism

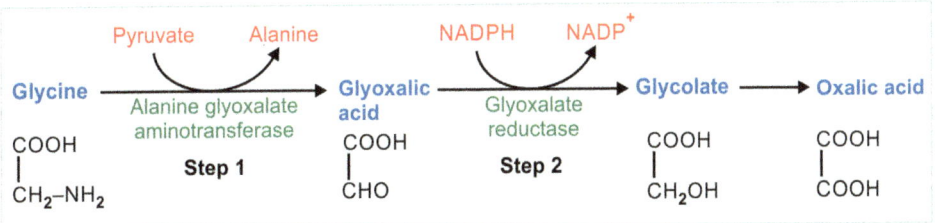

Fig. 11.9: Explanation for oxaluria. Enzyme 1 is inactive in patients.

- Synthesis of **heme**: Heme is the nonprotein part of hemoglobin (*see* Chapter 15).
- Synthesis of glycocholic acid: Cholyl-CoA (activated cholic acid) conjugates with glycine to form glycocholic acid. It is found in bile.
- Detoxification of benzoic acid. Benzoic acid is conjugated with glycine to form benzoyl glycine (**hippuric acid**) and is excreted.

### Catabolism of Glycine

**Formation of glyoxylic acid**

Glycine is transaminated and then undergoes oxidative deamination (**Fig. 11.9**). Increased excretion of oxalates is observed in **primary oxaluria**, which is due to increased production of oxalates. It is an autosomal recessive trait. Normally, the glyoxylic acid formed from glycine by deamination or transamination may be (1) converted back to glycine by transamination or (2) oxidized mostly to formalin and partly to oxalic acid (**Fig. 11.9**). The enzyme alanine glyoxylate aminotransferase (Step 1 in **Fig. 11.9**) is inactive in these patients. So the conversion of glyoxylic acid to glycine does not take place. This leads to increased pool size of glyoxylate and excess production of oxalate. Renal deposition of oxalates would cause nephrolithiasis, renal colic and hematuria. Extrarenal oxalosis may be seen in heart, blood vessels, bone, and other tissues.

**Cleavage of glycine**

The reaction by glycine synthase complex (**Fig. 11.7**) is reversible and this results in the complete breakdown of glycine to $CO_2$, $NH_3$, and methyl THFA.

**Glycine is glucogenic**

Glycine, after being converted to serine, can form glucose (**Fig. 11.10**). A summary of glycine metabolism is shown in **Figure 11.11**.

### Alanine

The metabolism of alanine is as shown in **Figure 11.12**. β-Alanine is a constituent of pantothenic acid and a dipeptide carnosine.

Alanine is a nonessential glycogenic amino acid. Alanine can be formed from pyruvate by transamination (**Fig. 11.3**).

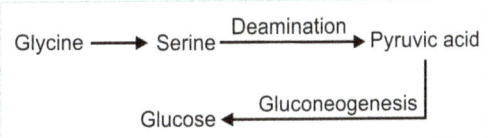

Fig. 11.10: Glucose is formed from glycine.

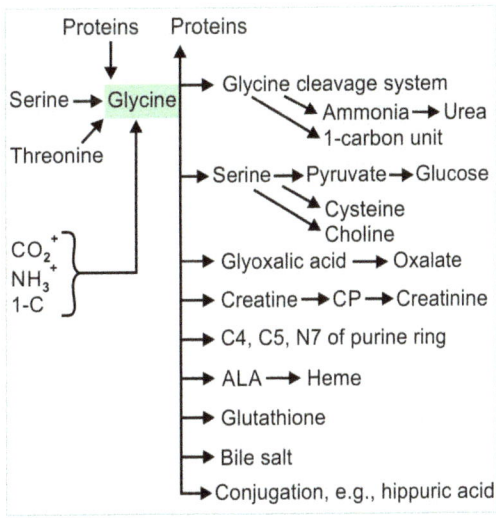

Fig. 11.11: Overview of glycine metabolism.

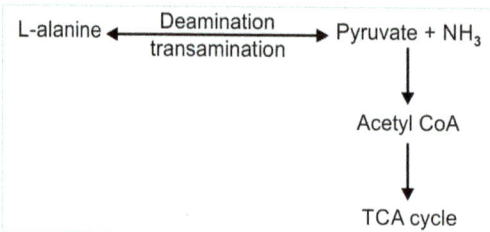

**Fig. 11.12:** Metabolic pathway of alanine.

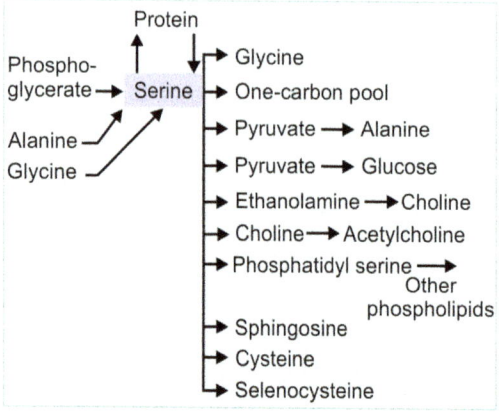

**Fig. 11.13:** An overview of serine metabolism.

$$\text{Pyruvate + Glutamate} \xrightarrow[\text{PLP}]{\text{ALT}} \text{Alanine} + \alpha\text{-KG}$$

It can also be synthesized from other amino acids such as serine, glycine, and cysteine.

The major metabolic role of alanine is to provide pyruvate for gluconeogenesis. Deamination or transamination produces pyruvate that can be readily converted to glucose or oxidized in citric acid cycle. The cycle of glucose transport from liver to muscles and alanine transport from muscles to liver is known as **glucose alanine cycle** (**Cahill cycle**). The glucose alanine cycle is useful under conditions of starvation.

## Serine

Serine is a nonessential glycogenic amino acid (**Fig. 11.13**). Major source of serine in the body is 3-phosphoglycerate. Glycine and serine are interconvertible in the body (**Fig. 11.6**). Serine may also be formed from hydroxypyruvate by transamination with alanine. Important products formed from serine are the following:
- Serine forms the active group of many enzymes (serine proteases).
- Donor of one carbon group.
- Used for the formation of cysteine and alanine.
- Serine is decarboxylated to ethanolamine.
- Used for the synthesis of phospholipid, phosphatidylserine.
- Serine analogues are used as drugs that inhibit nucleotide synthesis, e.g., **azaserine** is an anticancer drug, **cycloserine** is an antituberculosis drug.
- It is a component of proteins. In phosphoproteins, serine serves to attach phosphate groups (e.g., casein).

## Threonine

Threonine is an essential amino acid. It is glucogenic (**Fig. 11.14**). The –OH group of the threonine residues in protein serves to provide a site for phosphorylation.

## Methionine

Methionine is an essential glucogenic amino acid. It contains sulfur. Catabolism of methionine results in the synthesis of cysteine (**Fig. 11.15**).

*Importance of Methionine*

- Methionine is required for the synthesis of body proteins.
- Synthesis of S-adenosylmethionine (SAM), otherwise called active methionine.

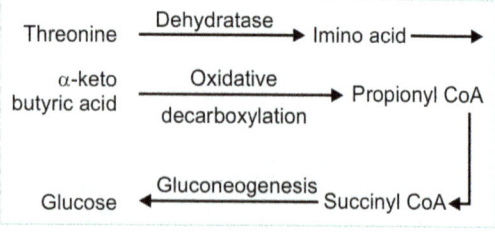

**Fig. 11.14:** Catabolism of threonine.

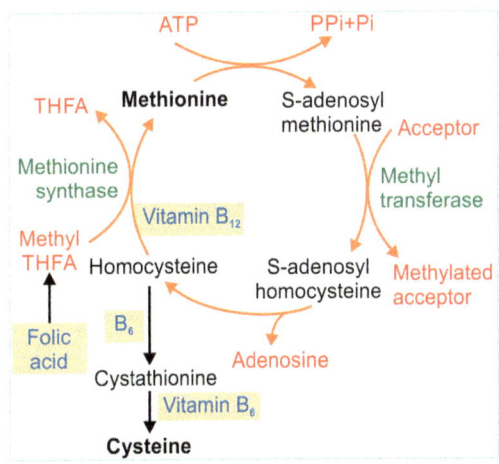

**Fig. 11.15:** Summary of the conversion of methionine to cysteine. Note the role played by the vitamins.

$$\text{Methionine + ATP} \xrightarrow{\text{Methionine adenosyl transferase}} \text{SAM + PPi + Pi}$$

SAM is the methyl group donor for all transmethylation reactions.

For example, noradrenaline + SAM → adrenaline + S-adenosylhomocysteine

## Arginine

Arginine contains guanidinium group. It is a highly basic, **semiessential** amino acid. Arginine is **glucogenic.** In the urea cycle, Arg splits into urea and **ornithine (Fig. 11.2).** Ornithine is then transaminated to glutamate semialdehyde and then to glutamate.
- Arginine, glycine, and methionine are required to form **creatine.**
- Arginine is the precursor of **nitric oxide** (NO) which is an important signal molecule in the body.

## NITRIC OXIDE

It is a toxic pollutant of air and automobile exhausts. However, now it is shown to possess more potential biological functions compared to any other known molecule. Nitric oxide is an uncharged molecule having an unpaired electron, so it is a highly reactive "free radical". Nitric oxide (NO$^\bullet$) is formed from Arg by the enzyme **NO synthase (NOS)**. It contains heme, flavin adenine dinucleotide (FAD), flavin mononucleotide (FMN), and nicotinamide adenine dinucleotide phosphate (NADPH). In urine, NO$^\bullet$ is excreted as **nitrites** and **nitrates**.

## Physiological Actions of Nitric Oxide

### Blood Vessels

Nitric oxide is a potent vasodilator. The normal blood pressure is maintained by NO$^\bullet$ liberated by endothelial NOS. A deficiency of NO$^\bullet$ is associated with hypertension. Excessive production of NO$^\bullet$ results in refractory hypotension, which may be seen in patients with septicemic shock.

### Macrophages

Macrophages contain NOS, which produces NO$^\bullet$ and peroxy nitrite; which are lethal to **microorganisms**.

### Intestinal System

Nitric oxide is a neurotransmitter, especially in gastrointestinal and urogenital tracts. It relaxes smooth muscles and leads to reduced gastrointestinal motility and relaxation of sphincters.

## Nitric Oxide in Diseases and Treatment

### Angina Pectoris

Nitroprusside can directly release NO$^\bullet$. Nitroglycerine (glyceryl trinitrate) when absorbed will dilate coronary arteries and is beneficial in treating angina pectoris.

### Pulmonary Hypertension

Inhalation of NO$^\bullet$ is useful in the treatment of pulmonary hypertension and high-altitude pulmonary edema. NO$^\bullet$ produces pulmonary vasodilatation, without lowering the systemic blood pressure.

## BRANCHED-CHAIN AMINO ACIDS

Valine, leucine, and isoleucine are the essential amino acids. They are required for the synthesis of body proteins. They serve an important role as an alternative source of fuel for the brain especially under conditions of starvation.

**Catabolism (Table 11.2):** The first three steps are common. Beyond the third step, the pathway diverges and each follows a unique pathway. Catabolism of **valine** forms succinyl-CoA which after entering the Krebs tricarboxylic acid (TCA) cycle can also go through gluconeogenesis pathway to give rise to glucose. Thus, valine is glucogenic. Catabolism of **leucine** forms two compounds, acetyl-CoA and acetoacetic acid. Leucine is the only one amino acid that is completely ketogenic. Catabolism of **isoleucine** forms two compounds. One is acetyl CoA which forms ketone body. The other is propionyl-CoA which forms succinyl-CoA and then glucose. Thus, isoleucine is both glucogenic and ketogenic.

## Metabolic Disorder of Branched-Chain Amino Acid Metabolism

**Maple syrup urine disease** (MSUD) is the inborn error of metabolism of valine, leucine, or isoleucine. In this condition, there is absence or deficiency of **α-keto acid dehydrogenase**, which results in the accumulation of α-keto acids of these amino acids and their excretion in urine. These acids give a characteristic smell to the urine. The smell is similar to that of maple syrup (burnt sugar). Hence, this condition is called maple syrup urine disease. The affected children manifest severe **mental retardation**.

## Phenylalanine and Tyrosine

Phenylalanine is an essential, aromatic amino acid. The need for phenylalanine

**Table 11.2:** Catabolism of branched-chain amino acids (valine, leucine, and isoleucine).

| Sl. No. | Reaction and coenzymes | Valine | Leucine | Isoleucine |
|---|---|---|---|---|
| 1. | Transamination to produce branched-chain α-keto acid | α-Keto isovaleric acid | α-Keto isocaproic acid | α-Keto β-methylvaleric acid |
| 2. | Oxidative decarboxylation with the help of CoA, nicotinamide adenine dinucleotide (NAD$^+$) and branched-chain α-keto acid dehydrogenase (lacking in maple syrup urine disease) | Isobutyryl-CoA | Isovaleryl-CoA | α-Methylbutyryl-CoA |
| 3. | Flavin adenine dinucleotide (FAD)-dependent dehydrogenation | Methylacrylyl-CoA | β-methylcrotonyl-CoA | Tiglyl-CoA |
| 4. | Individual reactions | Forms malonyl-CoA | Forms β-hydroxy β-methylglutaryl-CoA (HMG-CoA) | Forms a substituent of acetoacetyl-CoA |
| 5. | End products | $B_{12}$-coenzyme to form succinyl-CoA | HMG-CoA lyase to form acetoacetate and acetyl-CoA | Cleavage to form acetyl-CoA and propionyl-CoA |
| 6. | Final metabolic pathway | Glucogenic only | Ketogenic only | Ketogenic and glucogenic |

becomes minimal, if adequate tyrosine is supplied in the food. This is called the sparing action of tyrosine on phenylalanine. Phenylalanine is converted to tyrosine. This reaction is catalyzed by the enzyme, phenylalanine hydroxylase. Tyrosine is finally broken down to a glucogenic product (fumarate) and a ketone body (acetoacetate). Hence, phenylalanine and tyrosine are partly glucogenic and partly ketogenic (**Fig. 11.16**).

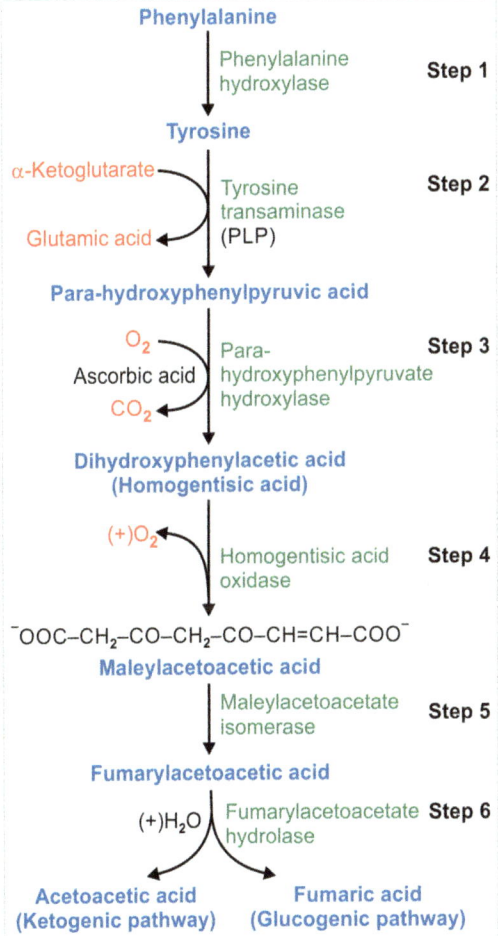

**Fig. 11.16:** Catabolism of phenylalanine and tyrosine.

*Important Specialized Products from Tyrosine*

- **Melanin** pigment gives the black color to the skin and hair. The enzyme necessary for melanin formation is tyrosinase.
- **Norepinephrine and epinephrine** are important hormones produced by adrenal medulla and adrenergic nerve endings. They are synthesized from tyrosine. Norepinephrine is methylated to form epinephrine. S-adenosylmethionine is the methyl donor. Epinephrine and **adrenaline** are the two names of the same hormone. Epinephrine has the following actions:
  – It increases the blood pressure.
  – It increases the rate and force of myocardial contraction.
  – It causes relaxation of smooth muscles of bronchi.
  – It increases glycogenolysis and stimulates lipolysis.
  – Adrenaline is released from adrenal medulla in response to flight, fight, emotions, and exercise.
- **Thyroxin** is a thyroid hormone produced from tyrosine by the thyroid gland. Thyroxin is necessary for the metabolism of all cells.

## Abnormalities in Phenylalanine Metabolism

*Phenylketonuria*

Deficiency of **phenylalanine hydroxylase** is the cause for this disease. So phenylalanine level in blood increases. Unless identified and treated within weeks of child birth, the child will have severe mental retardation. The treatment is to provide a diet containing low phenylalanine. Details of phenylketonuria (PKU) as well as other abnormalities associated with tyrosine metabolism are given in Chapter 12.

## Tryptophan

**Important substances produced from tryptophan are as follows:**

- Alanine (glucogenic).
- Acetoacetyl-CoA (ketogenic).

- **Niacin** and nicotinamide adenine dinucleotide (NAD$^+$) are produced from tryptophan. So tryptophan deficiency leads to niacin deficiency.
- **Serotonin** is an important neurotransmitter in brain. Serotonin level is found to be low in patients with depressive psychosis. Serotonin is involved in mood, sleep, appetite, and temperature regulation. It increases gastrointestinal motility. Summary of tryptophan metabolism is shown in **Figure 11.17**.

*Carcinoid Tumors*

Large quantities of serotonin are produced by **carcinoid tumors**. These tumors develop in small intestine or in the appendix. The patient complains of flushing, sweating, intermittent diarrhea, and often has fluctuating hypertension. Normally, about 1% tryptophan molecules are channeled to serotonin synthesis. But in carcinoid syndrome, up to 60% is diverted to serotonin. Therefore, **niacin deficiency** (pellagra) may also be seen in carcinoid syndrome.

- **Melatonin** is a regulatory neurotransmitter in brain, especially in pineal gland, which regulates the diurnal variations, sleep–wake cycles, and the **biological rhythms**. Melatonin blocks melanocyte-stimulating hormone (MSH) and adrenocorticotrophic hormone (ACTH) secretions. Melatonin is secreted in close association with the light–dark cycle. The light signal on the retina inhibits melatonin secretion, thereby producing wakefulness in the morning and remaining alert during day. Small doses of melatonin can induce sleep, causing a slowing of respiratory rate and decrease in body temperature.

## Fate of Carbon Skeletons of Amino Acids

Those amino acids, which give rise to citric acid cycle intermediates, can be made into glucose. Hence, those amino acids entering the TCA cycle, or at pyruvic acid level, are called **glucogenic** amino acids. This is shown in **Figure 11.18**.

On the other hand, those amino acids which produce acetyl-CoA are called **ketogenic** amino acids. Acetyl-CoA entering into the TCA cycle is completely oxidized. Therefore, there is no net synthesis of glucose from acetyl-CoA. So acetyl-CoA is not entering into the gluconeogenesis pathway. Acetyl-CoA, however, can give rise to ketone bodies. Ketogenic amino acids are shown in **Figure 11.18**.

But phenylalanine, tyrosine, tryptophan, and isoleucine are both **glucogenic and ketogenic.** This is because, during their metabolism, part of the carbon skeleton will enter into some of the TCA cycle intermediates, while the other part will generate acetyl-CoA (**Fig. 11.18**). **Table 11.3** shows the important products formed from various amino acids.

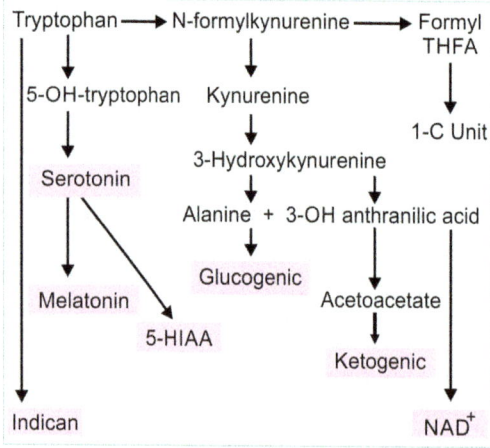

**Fig. 11.17:** Summary of tryptophan metabolism.

## Chapter 11 Amino Acids and Proteins: Metabolism

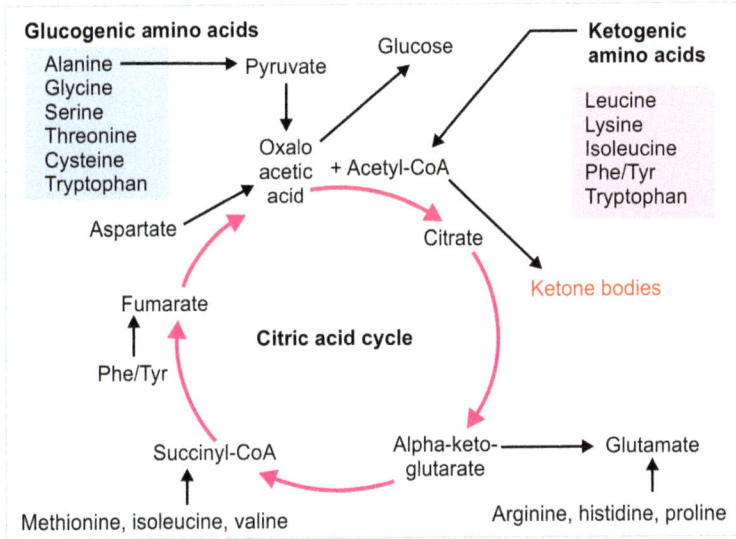

**Fig. 11.18:** Metabolic fates of amino acids. Glucogenic amino acids are shown as blue squares and ketogenic amino acids as red square.

**Table 11.3:** Important products formed from various amino acids.

| Amino acid | Product formed | Importance |
|---|---|---|
| Glycine | Heme | Glycine condenses with succinyl-CoA to form heme, which is used for oxygen carriage |
| | Creatine | Creatine is formed from glycine, arginine, and methionine. Creatine phosphate is the high-energy compound in muscle |
| | Glutathione | It is formed from glutamic acid, cysteine, and glycine. It is an antioxidant in red blood cells (RBCs) |
| | Purine bases | C4, C5, and N7 of purine ring are contributed by glycine |
| | One-carbon group | Glycine cleavage system provides one-carbon groups that are used in DNA synthesis |
| Alanine | Pyruvate | Transamination of alanine to pyruvate is catalyzed by alanine aminotransferase. Alanine is major glucogenic amino acid |
| Serine | One-carbon group | Conversion of serine to glycine generates one-carbon group. It is the major contributor of one carbon groups |
| Methionine | Active methionine | Methionine and ATP forms S-adenosylmethionine, which is the methylating agent for transmethylation |
| | Cysteine | Degradation of methionine produces cysteine |
| Cysteine | SH groups in proteins | Keeps active sites of enzymes |
| | Glutathione | It is formed from glutamic acid, cysteine, and glycine. It is an antioxidant in RBCs |

*Contd...*

Contd...

| Amino acid | Product formed | Importance |
|---|---|---|
| Arginine | Nitric oxide | Nitric oxide synthase acts on arginine releasing NO, which is a potent vasodilator |
| | Creatine | Glycine + arginine + methionine |
| | Ammonia | Glutaminase releases ammonia from glutamine in the renal tubular cells. This is important for excreting H ions as ammonium ions. |
| Aspartic acid | Urea | Urea cycle is important for detoxification of ammonia |
| | Purines, pyrimidines | C and N atoms of aspartate are used for nucleic acids |
| Tyrosine | Epinephrine | Adrenergic activity |
| | Dopamine | Neurotransmitter in brain |
| | T4 and T3 | Stimulators of metabolism and growth |
| | Melanin | Sunscreen pigment |
| Tryptophan | Serotonin | Vasopressor amine; mood regulation |
| | Melatonin | Acetylated serotonin. Important in sleep–wake cycle |
| | Nicotinamide adenine dinucleotide (NAD) | Coenzyme for dehydrogenases |

## SUMMARY OF THE CHAPTER

- Pepsin, trypsin, and chymotrypsin are the important protein hydrolyzing enzymes in the gastrointestinal tract.
- Amino acids are transaminated with α-keto acid to produce another amino acid.
- Glutamic acid is deaminated to produce α-ketoglutaric acid and ammonia.
- Ammonia is finally excreted as urea. Urea is synthesized in the urea cycle.
- Normal urea level in blood is 20–40 mg/dL. It shows an increased level in renal diseases.
- Glycine is used to synthesize serine, creatine, creatinine, purine ring, heme, GSH, bile salts.
- Methionine is activated to SAM, which is used for transmethylation reactions.
- Methionine and cysteine metabolisms are interconnected.
- Glutathione is synthesized by using cysteine.
- Phenylalanine is converted to tyrosine by phenylalanine hydroxylase.
- When this enzyme is absent, it leads to PKU, an inborn error of metabolism, causing severe mental retardation.
- Important specialized products from tyrosine are melanin, epinephrine, and thyroxin.
- Substances produced from tryptophan are alanine (glucogenic), acetoacetyl-CoA (ketogenic), niacin, $NAD^+$, serotonin, and melatonin.

## QUESTIONS FOR SELF-ASSESSMENT

1. Describe the digestion and absorption of proteins.
2. Outline the urea cycle and mention its importance in our body.
3. Name the specialized compounds formed from glycine, phenylalanine, and tryptophan.
4. Name the compounds synthesized from tyrosine.
5. Mention the importance of glycine in our body.
6. Mention the fate of ammonia.
7. Describe the metabolism of phenylketonuria.
8. Mention the fate of carbon skeletons of amino acids.

9. Mention some inborn errors of metabolism related to amino acid metabolism.
10. Mention the metabolic fate of amino acids with the help of a schematic diagram.

## MULTIPLE CHOICE QUESTIONS (MCQs)

11-1. Which of the following compound is not synthesized from glycine?
  a. Heme
  b. Tyrosine
  c. Serine
  d. Creatine

11-2. Pepsin is secreted in:
  a. Pancreas
  b. Stomach
  c. Liver
  d. Intestine

11-3. Which of the following amino acid is both glucogenic and ketogenic:
  a. Phenylalanine
  b. Alanine
  c. Leucine
  d. Lysine

11-4. Which of the following is not an essential amino acid:
  a. Tyrosine
  b. Isoleucine
  c. Histidine
  d. Serine

11-5. Name the hormone that promotes glucose and amino acid uptake by muscle:
  a. Adrenaline
  b. Cortisol
  c. Glucagon
  d. Insulin

11-6. The amino acid methionine without methyl group is:
  a. Cysteine
  b. Homocysteine
  c. Cystine
  d. Formylmethionine

11-7. All of the following are protein hydrolyzing enzymes in gastrointestinal tract, *except:*
  a. Pepsin,
  b. trypsin
  c. chymotrypsin
  d. Lipase

11-8. Which amino acid is oxidatively deaminated in liver?
  a. Aspartic acid
  b. Alanine
  c. Glutamic acid
  d. Valine

11-9. Blood urea level is markedly increased in:
  a. Liver diseases
  b. Renal diseases
  c. Cardiac diseases
  d. Protein intake

11-10. The methyl donor in methyltransfer reactions is:
  a. Methylcobalamin
  b. Methylmalonyl-CoA
  c. *S*-adenosylmethionine
  d. *S*-adenosylhomocysteine

## ANSWER KEYS TO MCQs

11-1. (b)   11-2. (b)   11-3. (d)   11-4. (d)   11-5. (d)   11-6. (b)   11-7. (d)   11-8. (c)
11-9. (b)   11-10. (c)

# CHAPTER 12

# Inborn Errors of Metabolism

At the completion of this chapter, the reader will be able to answer questions on the following topics:

- Phenylketonuria
- Alkaptonuria, albinism, and tyrosinemia
- Glycinuria and primary oxaluria
- Cystinosis
- Hartnup disease
- Maple syrup urine disease
- Histidinemia
- Refsum's disease
- Niemann–Pick disease
- Gaucher's disease
- Galactosemia
- Fructosuria
- Fructose intolerance
- Glycogen storage disease
- Lesch–Nyhan syndrome

Inborn errors of metabolism (IEM) are single-gene disorders with specific enzyme defects in the biochemical pathways of the body. Although these disorders are individually rare, they collectively account for a significant proportion of neonatal and childhood mortality. The explanation for the cause of IEM is based on Garrod's hypothesis, which is schematically shown in **Figure 12.1**.

## SALIENT FEATURES OF INBORN ERRORS OF METABOLISM

- Substrate is converted to a product in the presence of the enzyme. If that enzyme is deficient, metabolism is blocked at that stage, leading to IEM.
- The substrate is converted to the intermediary product, which could not be metabolized further, so the intermediary product is accumulated (B in **Fig. 12.1**). As the final product (C in **Fig. 12.1**) is not produced, the feedback inhibition of A to B is not effective.
- As the substrate is not converted to the final product, there is an accumulation of intermediary substrate (B in **Fig. 12.1**), leading to the opening of alternative pathway, and toxic metabolites (D in **Fig. 12.1**) are produced. This leads to systemic effects, which in turn causes clinical symptoms of IEM.

## INBORN ERRORS ASSOCIATED WITH AMINO ACID METABOLISM

### Phenylketonuria

It is a genetically transmitted disease. Error in the phenylalanine metabolism results in the inability to convert phenylalanine to tyrosine. So the phenylalanine level in blood increases, which leads to phenylalaninemia. This metabolic defect is due to the deficiency or absence of the enzyme **phenylalanine hydroxylase**. This enzyme needs nicotinamide adenine dinucleotide phosphate (NADPH),

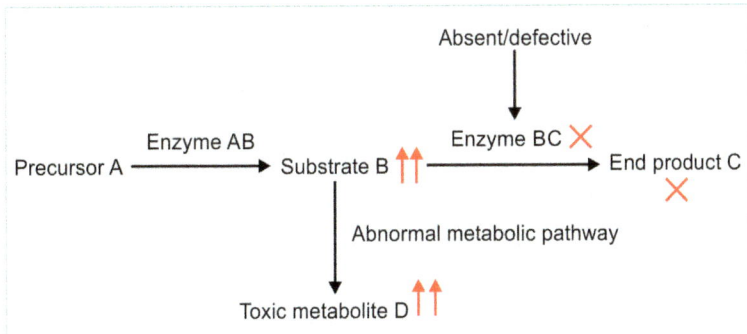

**Fig. 12.1:** Diagrammatic representation of causes and consequences of metabolic disorders.

nicotinamide adenine dinucleotide (NADH), and dihydrobiopterin as coenzymes.

$$\text{Phenylalanine} \xrightarrow[\text{NADPH, NADH, and dihydrobiopterin}]{\text{Phenylalanine hydroxylase}} \text{Tyrosine}$$

Phenylketonuria (PKU) is of four types. Types I, II, and III are due to the deficiency of enzyme (protein part). Type IV is due to the deficiency of dihydrobiopterin reductase.

### Biochemical Abnormalities

A block in the main metabolic pathway mentioned above would lead to an alternate minor pathway. Hence, the abnormal products such as phenylpyruvate, phenyllactate, and phenylacetate are excreted in urine. Therefore, the phenylalanine level in blood increases.

### Clinical Manifestations

- Severe **mental retardation**
- Muscular hypertonia, tremor, fits
- Hypopigmentation
- Eczematous dermatitis

It is important to diagnose the condition within a few weeks of the birth of the child because once mental retardation sets in, it cannot be reversed. Treatment is to restrict phenylalanine in food.

### Alkaptonuria

This metabolic defect is due to the deficiency of the enzyme **homogentisic acid oxidase**. This enzyme is necessary for tyrosine metabolism. So, the pathway gets blocked at the level of reduction of homogentisic acid, which in turn gets accumulated in the tissues and blood. Clinical abnormality involves blackening of urine due to the formation of black pigments and the polymerized products of homogenetisic acid. Another manifestation is **ochronosis** caused due to the accumulation of the dark pigments in cartilage, tendon, joints, ligaments, and sclera of the eye. Many alkaptonuria patients suffer from arthritis due to the deposition of alkapton pigments in joints. Mental retardation is not seen in these cases.

### Albinism

This is due to the deficiency of the enzyme **tyrosinase** in melanocytes. Owing to this, black pigment is not synthesized in the body. Therefore, these patients lack the black pigment melanin in the eyes, skin, and hair. Iris of the eye appears pink and the skin appears white. The most important function of melanin is to protect the body from sunlight. Albinos are sensitive to sunlight. In such patients, increased susceptibility to skin cancer and photophobia are observed.

### Tyrosinemia

This is due to the deficiency of enzyme **hydroxyphenylpyruvic acid oxidase** required for the normal degradation of tyrosine.

## Clinical Manifestations

- Tyrosine level in blood increases.
- Hydroxyphenylpyruvic acid, hydroxyphenylacetic acid, and hydroxyphenyllactic acid appear in urine.
- Defects in the renal tubular reabsorption.
- Hepatosplenomegaly, liver failure, and death at a very early age.

Defect in the enzyme of the metabolic pathway is shown in **Figure 12.2**.

## Glycinuria

In glycinuria, large amounts of glycine are excreted in the urine, which is mainly due to the reabsorption of glycine from the renal tubules. Mental retardation is also observed (*see* Chapter 11).

## Primary Hyperoxaluria

In **hyperoxaluria**, large amounts of oxalic acid are excreted. Normal pathway of oxalic acid production is shown in **Figure 11.9**. In primary hyperoxaluria, the enzyme **alanine-glyoxylate transaminase** is defective, so the conversion of glyoxylic acid to glycine does not take place. This leads to increased pool of glyoxylate and excess production of oxalates. Renal deposition of oxalates would cause nephrolithiasis, renal colic, and hematuria.

## Cystinosis (Cystine Storage Disease)

Cystinosis is a hereditary abnormality of cysteine metabolism. This may be due to the impaired reabsorption of cysteine from urinary tubules and to defective carrier-mediated transport of cysteine. So cysteine is excreted through urine. In the urine, cysteine units are oxidized to form di-cysteine (cystine).

## Clinical Manifestations

- Deposition of cystine crystals in urine, stones in kidney, ureter and urinary bladder (nephrolithiasis), leading to renal colic.
- Accumulation of cystine crystals in lymph nodes, liver, spleen, bone marrow, kidney, cornea, and reticuloendothelial system.
- Cystinosis is usually accompanied by other aminoaciduria, glucosuria, and chronic acidosis.
- Renal functions are seriously impaired leading to uremia and renal failure and eventually death at an early age.

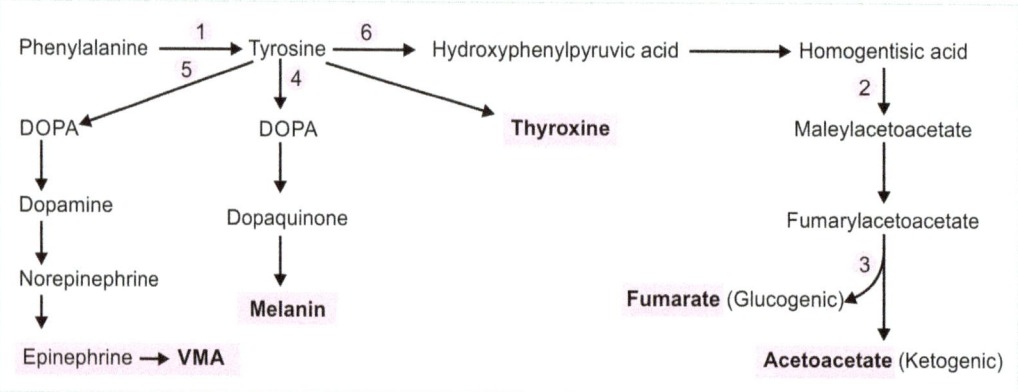

**Fig. 12.2:** Metabolic defects in tyrosine pathway. (1 = phenylketonuria, absence of phenylalanine hydroxylase. 2 = alkaptonuria, absence of homogentisic acid oxidase. 3 = hypertyrosinemia (tyrosinemia type I), absence of fumarylacetoacetate hydroxylase. 4 = albinism, absence of tyrosinase. 5 = tyrosine hydroxylase, key enzyme of epinephrine synthesis. 6 = tyrosinemia type II, absence of tyrosine transaminase.)

## Hartnup Disease (Neutral Aminoaciduria)

Named after the first patient described in literature, Hartnup disease involves defect in the renal and intestinal transport of neutral amino acids, including **tryptophan**. So tryptophan is excreted in large quantities, leading to tryptophan deficiency.

*Clinical Manifestations*

- Urine of the patient contains significantly increased amount of tryptophan and its metabolic product, indole acetic acid.
- Normally, niacin is produced from tryptophan, so tryptophan deficiency leads to niacin deficiency. Niacin deficiency leads to pellagra like skin lesions (erythema, eczematoid skin rash).
- Tryptophan deficiency leads to serotonin deficiency and consequent neurological manifestations, such as cerebellar ataxia and mental confusion.
- Tryptophan is excreted in urine, i.e., 10 times than normal.

## Maple Syrup Urine Disease

Maple syrup urine disease (MSUD) is characterized by the absence of the enzymes required for oxidative decarboxylation of the keto acids derived from the **branched chain amino acids** (valine, leucine, and isoleucine) (*see* Table 11.2).

*Clinical Manifestations*

- These keto acids are accumulated in the blood and excreted in urine. They produce odor in the urine that is similar to the odor of maple syrup or burnt sugar.
- Central nervous system manifestations such as convulsions.
- The patient suffers from significant degree of lethargy.

## Histidinemia

It is associated with deficiency of the enzyme **histidase**, which is required for the normal metabolism of histidine.

*Clinical Manifestations*

- Increased levels of histidine in the blood.
- Histidine and its metabolite (imidazole pyruvic acid) appear in increased amounts in urine.
- Severe **mental retardation**
- Speech defects will appear

## INHERITED DISORDERS OF LIPID METABOLISM

## Refsum's Disease

This disease involves the deficiency of **phytanic acid oxidase** enzyme. So phytanic acid is not converted to pristanic acid. As a result, phytanic acid accumulates in tissue and blood, leading to neurological manifestations.

*Clinical Manifestations*

- Polyneuropathy
- Pigmentary retinitis
- Night blindness
- Severe pain in the knee joints
- Muscular atrophy
- Tachycardia

## Niemann–Pick Disease

In this case, excessive amounts of sphingomyelin are deposited in the spleen, brain, and liver due to the deficiency of the enzyme **sphingomyelinase**.

*Clinical Findings*

- **Mental retardation**
- Enlarged liver and spleen
- Anemia and leukocytosis

## Gaucher's Disease

In this case, glucocerebrosides accumulate in the brain, bone marrow, spleen, and liver. This disease is caused by the deficiency of **glucocerebrosidase**. The concentration of cerrosides is much higher in medullary than in nonmedullary nerve fibers.

*Clinical Manifestations*

- **Severe mental retardation**
- Splenomegaly
- Loss of bone minerals
- Thrombocytopenia

## INBORN ERRORS OF CARBOHYDRATE METABOLISM

## Galactosemia

This inborn error of metabolism is due to the deficiency of enzyme **galactose-1-phosphate uridyltransferase** (*see* **Figs. 5.18 and 5.19**). So galactose-1-phosphate is accumulated in blood and tissues such as liver, spleen, kidney, heart, lens of the eye, and cerebral cortex. This will inhibit galactokinase and leads to the accumulation of galactose in blood.

*Clinical Manifestations*

- Galactose in blood (galactosemia) and in urine (galactosuria).
- Hypoglycemia
- Conjugation of bilirubin is reduced, leading to jaundice.
- Enlargement of liver
- Severe **mental retardation**
- Galactose is reduced to galactitol, which is accumulated in lens of the eye, resulting in cataract.
- Galactose-1-phosphate deposit in renal tubule produces tubular damage.

## Fructosuria

It is a rare congenital disorder caused due to the deficiency of the enzyme hepatic **fructokinase**. In this case, fructose is not converted to fructose-1-phosphate characterized by the inability to utilize fructose completely. These patients excrete fructose in urine. No major clinical manifestations are observed.

## Hereditary Fructose Intolerance

Hereditary fructose intolerance (HFI) is due to the deficiency of **aldolase**-B, an enzyme necessary for fructose metabolism. So fructose-1-phosphate is accumulated, thereby inhibiting glycogen phosphorylase.

*Clinical Manifestations*

- Fructose in urine
- Hypoglycemia
- Accumulation of glycogen in liver
- Hepatomegaly and jaundice
- Death usually by the age of 5 years.

## Glycogen Storage Diseases (Glycogenosis)

These are inborn errors of glycogen metabolism caused due to the deficiency of various enzymes, which are described in **Table 12.1**.

**Table 12.1:** Important glycogen storage diseases.

| Type | Name of the disease | Deficient enzyme | Clinical manifestation |
|---|---|---|---|
| I | Von Gierke disease | Glucose-6-phosphatase | Fasting hypoglycemia, lactic acidosis, ketosis, hyperuricemia, hepatomegaly, cirrhosis, early death |
| II | Pompe disease | Lysosomal maltase | Glycogen accumulates in liver, muscles, and heart; hepatomegaly |
| III | Cori disease (limit dextrinosis) | Debranching enzyme | Highly branched glycogen (limit dextrin) will accumulate in liver and muscle; hepatomegaly |
| IV | Andersen disease (amylopectinosis) | Branching enzyme | Accumulation of glycogen in liver; hepatomegaly and cirrhosis |
| V | McArdle disease | Muscle phosphorylase | Accumulation of glycogen in muscle; patient has exercise intolerance due to its inability to utilize the muscle glycogen |

## INBORN ERRORS OF NUCLEIC ACID METABOLISM

### Lesch–Nyhan Syndrome

It is an inherited disorder of purine metabolism caused due to the deficiency of **hypoxanthine-guanine phosphoribosyltransferase** (HGPRTase).

*Clinical Manifestations*

- **Mental retardation**
- Excessive production of uric acid (hyperuricemia)
- Hypothyroidism
- Hypertension
- Gout
- Increased susceptibility for myocardial infarction

## IMPORTANT GROUPS OF INBORN ERRORS OF METABOLISM

- Disorders of protein metabolism (e.g., aminoacidopathies, organic acidopathies, urea cycle defects).
- Disorder of carbohydrate metabolism (e.g., carbohydrate intolerance disorders, glycogen storage disorders, disorders of gluconeogenesis and glycogenolysis).
- Lysosomal storage disorders (e.g., Gaucher's disease, Niemann–Pick disease).
- Disorder of lipid metabolism (e.g., fatty acid oxidation defects (medium-chain acyl dehydrogenase deficiency), sphingolipidoses).
- Peroxisomal disorders (e.g., Zellweger syndrome, adrenoleukodystrophy).
- Trace metal disorders (Menkes kinky hair syndrome, Wilson's disease).

**Table 12.2** shows the overall clinical findings of IEM.

**Table 12.2:** Analysis of clinical findings of inborn errors of metabolism.

| Clinical findings | Percentage of patients affected |
|---|---|
| Seizure | 10.6 |
| Metabolic acidosis | 7.2 |
| Consanguinity | 6.9 |
| Delayed milestones | 6.0 |
| Hypoglycemia | 4.4 |
| Failure to thrive | 4.0 |
| Abnormal muscle tone | 2.8 |
| Hyperammonemia | 2.3 |
| Behavior disturbances | 1.7 |
| Hepatomegaly | 1.5 |
| Microcephaly | 1.4 |
| Skeletal abnormality | 1.0 |

## SUMMARY OF THE CHAPTER

- Inherited metabolic disorders are a heterogeneous group of genetic conditions mostly occurring in childhood.
- They are individually rare but collectively common, causing substantial morbidity and mortality.
- Advanced techniques were applied to diagnose the disorders of IEM.
- Data analyzed indicates the occurrence of several metabolic disorders in the studied population.
- The need to screen for an inborn error of metabolism arises out of the fact that most cases take to irreversible effects as time progresses.
- Emphasis has to be laid on early detection and prompt management, which could help in alleviating symptoms and preventing complications and consequent incapacitation.
- Deficiency of homogentisic acid oxidase leads to a condition called alkaptonuria, where homogentisic acid is excreted in urine, leading to black urine.
- Absence of tyrosinase will lead to albinism.

## QUESTIONS FOR SELF-ASSESSMENT

1. Define the inborn metabolic disorders.
2. Mention the disorders in tabular form related to carbohydrate, lipid, protein, and Nucleic acid metabolism.
3. Discuss phenylketonuria.
4. Write a short note on alkaptonuria.
5. What are the characteristic features of Lesch–Nyhan syndrome?
6. Mention the features of hereditary fructose intolerance and fructosuria.
7. Describe glycogen storage diseases.
8. What are the features of Refsum's disease?
9. Add a note on Hartnup disease.
10. What are the consequences of MSUD?

## MULTIPLE CHOICE QUESTIONS (MCQs)

12-1. **Inborn errors related to copper metabolism is:**
   a. Paget's disease
   b. Wilson's disease
   c. Keshan's disease
   d. Kashin–Beck disease

12-2. **Clinical symptoms, common to all urea cycle disorders include, all *except*:**
   a. Ammonia intoxication
   b. Protein-induced vomiting
   c. Intermittent ataxia
   d. Dermatitis

12-3. **Ammonia intoxication results in the following symptoms, *except*:**
   a. Impaired brain function
   b. Ataxia
   c. Convulsions
   d. Cataract

12-4. **Maple syrup urine disease:**
   a. Is an inborn error of glycine metabolism
   b. Is the result of deficiency of alpha-keto acid dehydrogenase
   c. Results in increased levels of amino acids in the urine
   d. Does not result in mental retardation

12-5. **Tyrosine would be an essential amino acid in the diet of a child with:**
   a. Lesch–Nyhan syndrome
   b. Defective tyrosine aminotransferase
   c. Classical phenylketonuria
   d. Hyperammonemia type-1

12-6. **Which one of the following deficiencies results in homocystinuria?**
   a. Cystathionine synthase
   b. Phenylalanine hydroxylase
   c. Methyltransferase
   d. Creatine phosphokinase

12-7. **Phenylketonuria results from the deficiency of:**
   a. Keto acid decarboxylase
   b. Tyrosinase
   c. Homogentisate oxidase
   d. Phenylalanine hydroxylase

12-8. **The site of urea synthesis is:**
   a. Liver
   b. Kidney
   c. Brain
   d. Adrenal gland

12-9. **Which of the following amino acids gives rise to alpha-keto acids that accumulate in the urine in maple syrup urine disease?**
   a. Phenylalanine
   b. Valine
   c. Lysine
   d. Tyrosine

12-10. **Malignant carcinoid is a metabolic disorder of:**
   a. Tyrosine
   b. Tryptophan
   c. Histidine
   d. Glycine

## ANSWER KEYS TO MCQs

12-1. (b)   12-2. (d)   12-3. (d)   12-4. (b)   12-5. (c)   12-6. (a)   12-7. (d)   12-8. (a)
12-9. (b)   12-10. (b)

# CHAPTER 13

# Citric Acid Cycle, Biological Oxidation, and Free Radicals

At the completion of this chapter, the reader will be able to answer questions on the following topics:

- Citric acid cycle
- Biological oxidation
- High-energy compounds
- Organization of electron transport chain
- Chemiosmotic theory
- Reactive oxygen species
- Generation of free radicals
- Damage produced by free radicals
- Free radical scavenger enzyme systems
- Clinical significance
- Antioxidants

## CITRIC ACID CYCLE OR TRICARBOXYLIC ACID CYCLE OR KREBS CYCLE

Krebs proposed the original name as tricarboxylic acid (TCA) cycle, because a few intermediaries have three carboxyl (COOH) groups in their structure. Please note that there is no apostrophe to the name Krebs cycle. The first reaction produces citric acid, and hence the cycle is also known as citric acid cycle. This cycle is used to oxidize acetyl coenzyme A (acetyl-CoA) to $CO_2$ and water. Acetyl-CoA is formed from pyruvate as described in the previous paragraph. Other sources of acetyl-CoA are fatty acids and ketogenic amino acids. This cycle also provides precursors for many biosynthetic reactions. All the enzymes of the citric acid cycle are located inside the mitochondria.

In the first reaction of the cycle, oxaloacetate (four carbon atoms) condenses with acetyl-CoA (two carbon atoms) to form a six-carbon compound, the citrate (TCA). The enzyme is citrate synthase (Step 1, **Fig. 13.1**). In the last reaction, oxaloacetate is regenerated.

## Functions of Citric Acid Cycle

- It is the final common oxidative pathway that oxidizes acetyl-CoA to $CO_2$.
- It is the source of reduced coenzymes that provide the substrate for the respiratory chain.
- It acts as a link between catabolic and anabolic pathways (amphibolic role).
- It provides precursors for the synthesis of amino acids and nucleotides.
- Components of the cycle have a direct or indirect controlling effects on key enzymes of other pathways.

## Energetics of Citric Acid Cycle

During the cycle, both carbon atoms of acetyl-CoA are completely oxidized and two molecules of $CO_2$ are removed. Three NADH molecules, each equivalent to three ATPs, one Flavin adenine dinucleotide ($FADH_2$) equivalent to two ATPs, and one molecule of GTP equivalent to one ATP are produced in the cycle. Thus, the oxidation

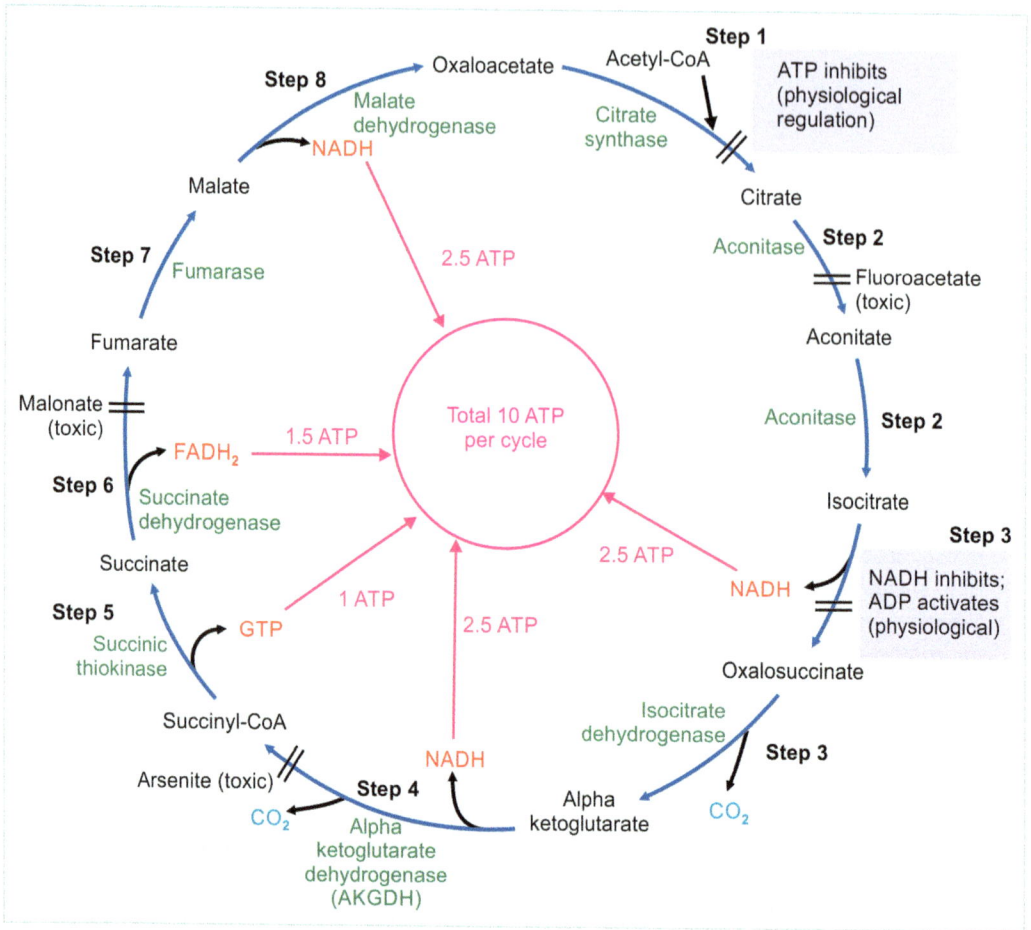

**Fig. 13.1:** Summary of Krebs citric acid cycle with physiological or toxic regulatory substances. Reactions 3 and 4 are carbon dioxide elimination steps. Physiological regulatory steps are as follows: Step 1 (citrate synthase) is physiologically inhibited by ATP. Step 3 (ICDH) is inhibited by nicotinamide adenine dinucleotide (NADH) and activated by ADP. Steps where energy is trapped are marked with the coenzyme and the number of ATP generated during that reaction. A total of 10 ATPs are generated during one cycle. In the electron transport chain (ETC), the NADH produces 2½ ATPs and flavin adenine dinucleotide (FADH) 1½ ATPs.

of a single molecule of acetyl-CoA yields 12 molecules of ATP. (Recent work shows that NADH may produce only **2½** ATPs and FADH only **1½** ATPs; if so, acetyl-CoA yields only 10 ATPs.) Conversion of pyruvate to acetyl-CoA generates one NADH equivalent to three ATPs. Thus, 1 molecule of glucose when completely oxidized through glycolysis and TCA cycle will generate 38 molecules of ATP **(Table 13.1)**. The TCA cycle is governed by the availability of substrates and feedback inhibition. The cycle will be more active when the body needs more energy in terms of ATP and slows down when ATP accumulates.

**Table 13.1:** Energy yield (number of ATP generated) per molecule of glucose in the glycolytic pathway under aerobic conditions (oxygen is available).

| Step No. | Enzyme | Reaction | Source | No. of ATPs gained per glucose molecule |
|---|---|---|---|---|
| Glycolysis 1 | Hexokinase | Glucose to glucose-6-phosphate | | Minus 1 |
| Glycolysis 3 | Phosphofructokinase | Fructose-6-phosphate to Fructose-1,6-bisphosphate | | Minus 1 |
| Glycolysis 5 | Glyceraldehyde-3-phosphate dehydrogenase | Glyceraldehyde-3-phosphate to 1,3-bisphosphoglycerate | 2 NADH | $3 \times 2 = 6$ |
| Glycolysis 6 | 1,3-bisphosphoglycerate kinase | 1,3-bisphosphoglycerate to 3-phosphoglycerate | 2 ATP | $1 \times 2 = 2$ |
| Glycolysis 9 | Pyruvate kinase | Phosphoenolpyruvate to pyruvate | 2 ATP | $1 \times 2 = 2$ |
| | Pyruvate dehydrogenase | Pyruvate to acetyl-CoA | 2 NADH | $3 \times 2 = 6$ |
| Citric acid cycle 3 | Isocitrate dehydrogenase | Isocitrate to oxalosuccinate | 2 NADH | $3 \times 2 = 6$ |
| Citric acid cycle 4 | α-Ketoglutarate DH | α-Ketoglutarate to succinyl-CoA | 2 NADH | $3 \times 2 = 6$ |
| Citric acid cycle 5 | Succinate thiokinase | Succinyl-CoA to succinate | 2 GTP | $1 \times 2 = 2$ |
| Citric acid cycle 6 | Succinate dehydrogenase | Succinate to fumarate | 2 FADH2 | $2 \times 2 = 4$ |
| Citric acid cycle 8 | Malate dehydrogenase | Malate to oxaloacetate | 2 NADH | $3 \times 2 = 6$ |
| | | | | Total = 40 − 2 = 38 |

Note
a. From one molecule of glucose, two molecules of glyceraldehyde phosphate are formed, hence from glycolysis Step No. 5 onward, the molecules are doubled.
b. Calculations were made assuming that nicotinamide adenine dinucleotide (NADH) produces three ATPs and flavin adenine dinucleotide (FADH2) produces two ATPs. This will amount to a net generation of 38 ATPs per glucose molecule when it is completely oxidized in the glycolytic pathway plus citric acid cycle. Recent experiments show that these old values are overestimates, and NADH generates only 2.5 ATPs and FADH2 only 1.5 ATPs. In that case, the net generation will be only 32 ATPs.

## SIGNIFICANCE OF TRICARBOXYLIC ACID CYCLE

*Complete Oxidation of Acetyl Coenzyme A ($CO_2$ Removal Steps)*

During the citric acid cycle, two carbon dioxide molecules are removed in the following reactions:

- Step 3, oxalosuccinate to α-ketoglutarate and Step 4, α-ketoglutarate to succinyl-CoA.
- Acetyl-CoA contains two carbon atoms. These two carbon atoms are now removed as $CO_2$ in steps 3 and 4. Net result is that acetyl-CoA is completely oxidized during one turn of cycle.

*ATP Generating Steps in Tricarboxylic Acid Cycle*

Per turn of the cycle, 10 high-energy phosphates are produced. These steps are marked in **Figure 13.1** and **Table 13.1**.

*Final Common Oxidative Pathway*

The citric acid cycle may be considered as the final common oxidative pathway of all foodstuffs. As shown in **Figure 13.2**, all the major ingredients of food stuffs are finally oxidized through the TCA cycle.

*Integration of Major Metabolic Pathways*

- Carbohydrates are metabolized through glycolytic pathway to pyruvate and then converted to acetyl-CoA which enters the citric acid cycle.
- Fatty acids through β-oxidation are broken down to acetyl-CoA, which then enters this cycle.
- Amino acids after transamination enter into this cycle **(Fig. 13.2)**.
- The integration of metabolisms is achieved at junction points by key metabolites.

*Excess Carbohydrates are Converted as Fat*

Excess calories are deposited as neutral fat in adipose tissue. The pathway is glucose to pyruvate to acetyl-CoA to fatty acid. However, fat cannot be converted to glucose because pyruvate dehydrogenase reaction (pyruvate to acetyl-CoA) is an absolutely irreversible step.

*No Net Synthesis of Carbohydrates from Fat*

Acetyl-CoA entering the cycle is completely oxidized to $CO_2$ by the time the cycle reaches succinyl-CoA. So acetyl-CoA is completely broken down in the cycle. Thus, acetyl-CoA is not present during gluconeogenesis. Therefore, there is no net synthesis of carbohydrates from fat.

*Amphipathic Pathway*

All other pathways such as β-oxidation of fat or glycogen synthesis are either catabolic or anabolic. But the TCA cycle is truly amphibolic (catabolic + anabolic). It is also called amphipathic in nature (Greek word, amphi = both; pathos = feeling). There is a continuous influx (pouring into) and a continuous efflux (removal) of four-carbon units from the TCA cycle. In a traffic circle, many roads converge

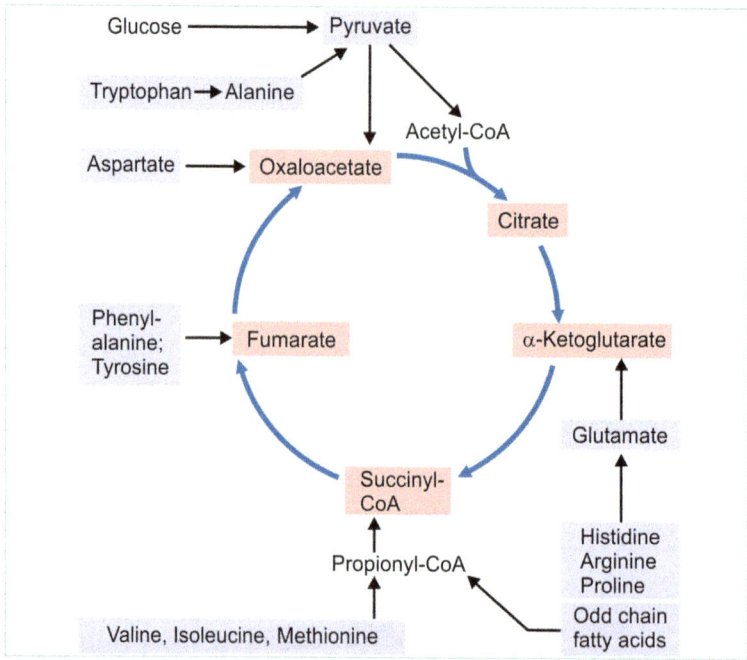

**Fig. 13.2:** Influx of tricarboxylic acid (TCA) cycle intermediates (anaplerotic reactions).

and the traffic is followed toward one way. Since various compounds enter into or leave from TCA cycle, it is sometimes called as the "metabolic traffic circle".

*Anaplerotic Role of Tricarboxylic Acid Cycle*

The citric acid cycle acts as a source of precursors of biosynthetic pathways, e.g., heme is synthesized from succinyl CoA and aspartate from oxaloacetate **(Fig. 13.3)**. To counterbalance such losses, and to keep the concentration of the four-carbon units in the cell, anaplerotic reactions are essential. This is called anaplerotic role of the TCA cycle (Greek word, ana = up; plerotikos = to fill). Anaplerotic reactions are "filling up" reactions or "influx" reactions, which supply four-carbon units to the TCA cycle **(Fig. 13.2)**.

## REGULATION OF THE CITRIC ACID CYCLE

- **Citrate**: The formation of citrate from oxaloacetate and acetyl-CoA is an important part of control (Step 1, **Fig. 13.1)**. The ATP acts as an allosteric inhibitor of citrate synthase. Citrate allosterically inhibits phosphofructokinase (PFK), the key enzyme of glycolysis.
- **Cellular need of ATP**: When the energy charge of the cell is low, as indicated by high level of $NAD^+$ and FAD, the cycle operates at a faster rate. The cycle is tightly coupled to the respiratory chain providing ATP. The Krebs cycle is the largest generator of ATP among the metabolic pathways.

## Inhibitors of Tricarboxylic Acid Cycle

The above-said mechanisms are physiological and regulatory in nature. But the following are toxic or poisonous (nonphysiological) agents that inhibit the reactions, which are shown in **Figure 13.1**.

- Aconitase (citrate to aconitate) is inhibited by fluoroacetate. This is noncompetitive inhibition.
- α-Ketoglutarate dehydrogenase is inhibited by arsenite (noncompetitive inhibition).
- Succinate dehydrogenase (succinate to fumarate) is inhibited by malonate. This is competitive inhibition.

## Stages of Oxidation of Foodstuffs

*First Stage*

Digestion in the gastrointestinal tract converts the macromolecules into small units. For example, proteins are digested to amino acids. This is called **primary metabolism (Fig. 13.4)**.

*Second Stage*

The products of digestion are absorbed, catabolized to smaller components, and ultimately oxidized to $CO_2$. The reducing equivalents are mainly generated in the mitochondria by the final common oxidative pathway, the citric acid cycle. In this process, NADH and $FADH_2$ are generated. This is called **secondary or intermediary metabolism (Fig. 13.4)**.

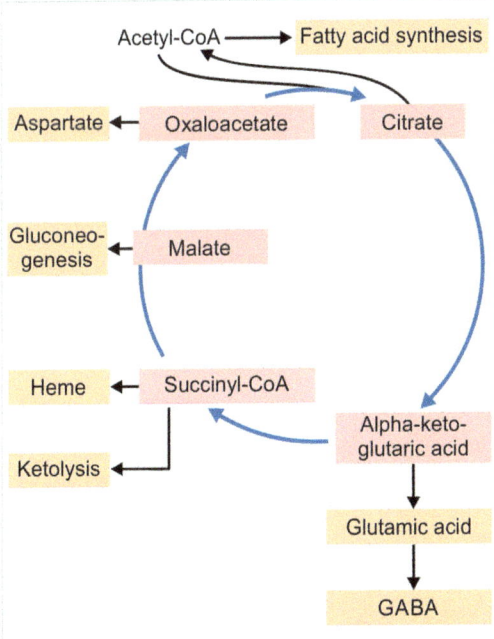

**Fig. 13.3:** Efflux of tricarboxylic acid (TCA) cycle intermediates. (GABA: gamma aminobutyric acid)

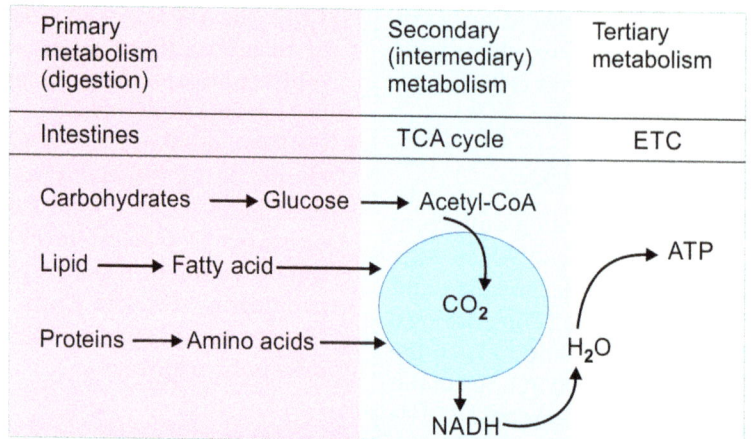

**Fig. 13.4:** Oxidation of foodstuffs in three stages. (NADH: nicotinamide adenine dinucleotide; TCA: citric acid cycle; ATP: adenosine triphosphate; ETC: electron transport chain)

*Third Stage*

These reduced equivalents (NADH and $FADH_2$) enter into the ETC or **respiratory chain**, where energy is released. This is **tertiary metabolism** or **internal respiration** or cellular respiration (**Fig. 13.4**). This energy is then used for synthetic purpose in the body.

## Substrate-Level Phosphorylation

Here energy from a high-energy compound is directly transferred to nucleoside diphosphate to form a triphosphate without the help of ETC, for example:

- Phosphoglycerate kinase (*see* **Fig. 5.7; Step 6**)
- Pyruvate kinase (*see* **Fig. 5.7; Step 9**)
- Succinate thiokinase (**Fig. 13.1;** Step 5).

The ATP thus generated is coupled with a more exergonic metabolic reaction and used for anabolic reactions (**Fig. 13.5**).

## BIOLOGICAL OXIDATION

The transfer of electrons from the reduced coenzymes through the respiratory chain to oxygen is known as biological oxidation. Energy released during this process is trapped as ATP. This coupling of oxidation with phosphorylation is called **oxidative phosphorylation**. In the body, this oxidation is carried out by successive steps of **dehydrogenations**.

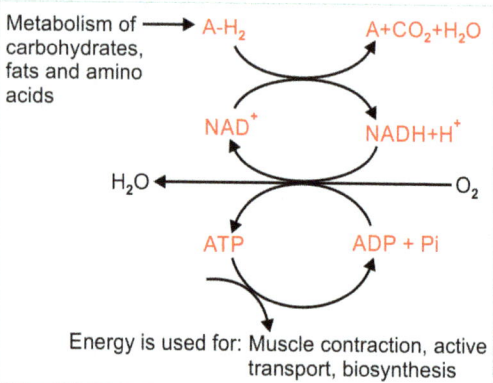

**Fig. 13.5:** ATP generation. Food is catabolized; energy from food is trapped as ATP. It is then used for anabolic reactions. (NADH: nicotinamide adenine dinucleotide; ATP: adenosine triphosphate; ADP: adenosine diphosphate)

## Electron Transport Chain

The electron flow occurs through successive dehydrogenase enzymes together known as ETC. The electrons flow from electronegative potential to electropositive potential. The free energy change between $NAD^+$ and water is very high. If it is released at one stretch, the body cannot utilize it. Hence with the help of ETC assembly, the total energy change is released in small increments, so that energy can be trapped as **chemical bond energy**, i.e., **ATP**.

## Adenosine Triphosphate

The ATP is the **universal currency of energy** within the living cells. The hydrolysis of ATP to ADP releases –7.3 kcal/mol. The energy in the ATP is used to drive all biosynthetic reactions. The structure of ATP is shown in Chapter 21. The ATP captures the chemical energy released by combustion of nutrients and transfers it to synthetic reactions that require energy. The ATPs are formed through ETC.

## HIGH-ENERGY COMPOUNDS

High-energy compounds when hydrolyzed will release a large quantity of energy. The free energy of hydrolysis of an ordinary bond varies from –1 to –6 kcal/mol. For example, glucose-6-phosphate has a free energy of –3.3 kcal/mol. On the other hand, the free energy of high-energy bonds varies from –7 to –15 kcal/mol. Such high-energy compounds are ATP, GTP, creatine phosphate, bisphosphoglycerate; phosphoenolpyruvate, acetyl-CoA, succinyl-CoA, etc.

### Structure of Mitochondrion

The ETC is functioning inside the mitochondria.

The mitochondrion is a subcellular organelle having the outer and inner membranes enclosing the matrix **(Fig. 13.6)**. The **inner membrane** contains the respiratory chain. The knob-like protrusions represent the ATP synthase system **(Fig. 13.6)**.

## Organization of Electron Transport Chain

In the ETC or respiratory chain, the electrons are transferred from NADH to a chain of electron carriers. The four distinct multiprotein complexes are **complexes I, II, III, and IV.** These are connected by two mobile carriers, **co-enzyme Q** and **cytochrome c**. The organization of these components are schematically represented in **Figure 13.7**.

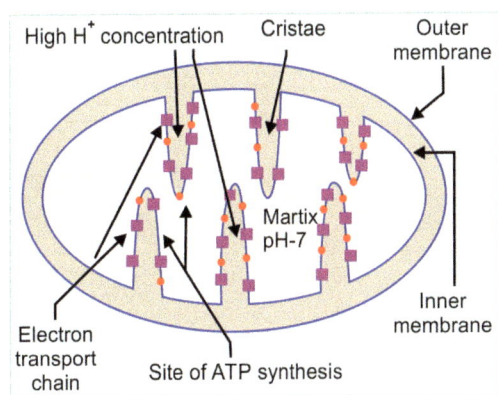

**Fig. 13.6:** Cut section of mitochondria. (ATP: adenosine triphosphate)

### Chemiosmotic Theory

The coupling of oxidation with phosphorylation is termed **oxidative phosphorylation**. The transport of protons from inside to outside of inner mitochondrial membrane is accompanied by the generation of a proton gradient across the membrane. Protons ($H^+$ ions) accumulate outside the membrane, creating an **electrochemical potential** difference. This proton motive force drives the synthesis of ATP by ATP synthase complex. This is explained in **Figure 13.8**.

### Regulation of ATP Synthesis

The availability of ADP regulates the process. When ATP level is low and the ADP level is

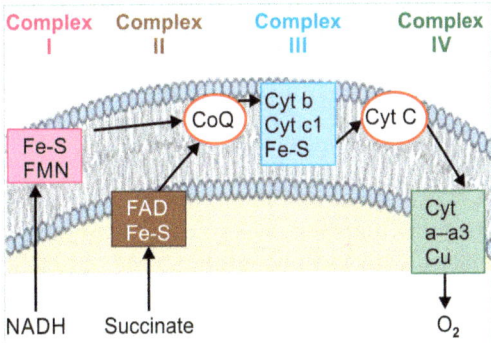

**Fig. 13.7:** Organization of electron transport chain. Summary of electron flow. (NADH: nicotinamide adenine dinucleotide; Cyt: cytochrome; FMN: flavin mononucleotide; FAD: flavin adenine dinucleotide)

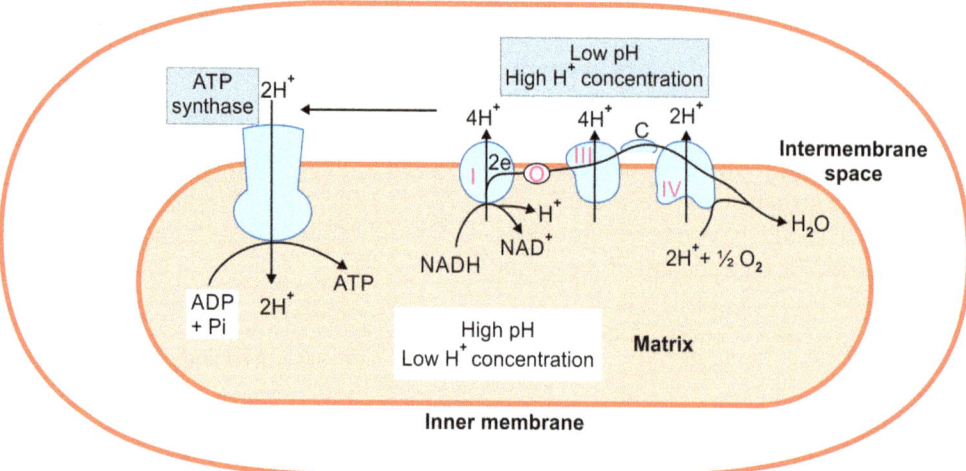

**Fig. 13.8:** Summary of chemiosmotic theory of ATP synthesis. One mitochondrion is depicted, with inner and outer membranes. The electron transport chain (ETC) complexes will push the hydrogen ions from the matrix into the intermembrane space. So intermediate space has more H⁺ (highly acidic) compared to the matrix. Therefore, hydrogen ions tend to leak into the matrix through the ATP synthase complex. Then ATPs are synthesized. The components of ETC are I, II, III, and IV. (NADH: nicotinamide adenine dinucleotide ; ATP: adenosine triphosphate)

high, oxidative phosphorylation proceeds at a rapid rate. This is called respiratory control. The major source of NADH and $FADH_2$ is the citric acid cycle, the rate of which is regulated by the energy charge of the cell.

## FREE RADICALS OR REACTIVE OXYGEN SPECIES

The outermost orbital in an atom or molecule contains two electrons, each spinning in opposite directions. The chemical covalent bond consists of a pair of electrons, each component of the bond donating one electron each.

### Definition of Free Radical

A free radical is a molecule or molecular fragment that contains one or more unpaired electrons in its outer orbital. Free radical is generally represented by a superscript dot (R•).

Oxidation reactions ensure that molecular oxygen is completely reduced to water. The products of partial reduction of oxygen are highly reactive and create havoc in the living systems. Hence, they are also called **reactive oxygen species** or ROS. The following are the members of this group:

- Superoxide anion radical ($O_2^{--•}$)
- Hydroperoxyl radical (HOO•)
- Hydrogen peroxide ($H_2O_2$)
- Hydroxyl radical (OH•)
- Nitric oxide (NO•)

Of these, hydrogen peroxide is not a free radical (i.e., it does not have the superscript dot). However, because of its extreme reactivity, it is included in the group of ROS.

Important characteristics of the ROS are:
- Extreme reactivity
- Short life span
- Generation of new ROS by chain reaction.
- Damage to all tissues in the area.

### Generation of Free Radicals

- They are constantly produced during the normal oxidation of foodstuffs, due to leaks in the ETC in mitochondria. About 1-4% of oxygen taken up in the body is converted to free radicals. Mitochondria are major

sites for production of superoxide ions from the interaction of CoQ and oxygen in the ETC. Hence, a high content of superoxide dismutase (SOD) is needed. Mitochondria also have a high content of glutathione and glutathione peroxidase (POD) for preventing lipid peroxidation.
- **Nicotinamide adenine dinucleotide phosphate (NADPH) oxidase** in the inflammatory cells (neutrophils, eosinophils, monocytes, and macrophages) produces superoxide anion by a process of respiratory burst during phagocytosis. The superoxide is converted to hydrogen peroxide and then to hypochlorous (HClO) acid with the help of **SOD** and **myeloperoxidase** (MPO). The superoxide and HClO ions are the final effectors of bactericidal action. This involves a deliberate production of free radicals by the body **(Fig. 13.9)**. Along with the activation of macrophages, the consumption of oxygen by the cell is increased drastically, and this is called **respiratory burst**. Macrophages also produce NO from arginine by the enzyme **nitric oxide synthase**. This is also an important antibacterial mechanism.
- Ionizing radiation damages tissues by producing hydroxyl radicals, hydrogen peroxide, and superoxide anion.
- Cigarette smoke contains high concentrations of various free radicals.
- Inhalation of air pollutants will increase the production of free radicals.

## Damage Produced by Reactive Oxygen Species

Free radicals are extremely reactive. Their half-life is only a few **milliseconds**. When a free radical reacts with a normal compound, other free radicals are generated. This chain reaction leads to thousands of events

**Peroxidation of polyunsaturated fatty acids (PUFAs)** in plasma membrane leads to loss of membrane functions. Almost all biological macromolecules are damaged by the free radicals. Thus, oxidation of enzyme proteins will lead to loss of function and fragmentation of proteins. Polysaccharides also undergo degradation. DNA is damaged by strand breaks and may directly cause inhibition of protein and enzyme synthesis and indirectly cause cell death or mutation and carcinogenesis.

## Clinical Significance of Free Radicals

- Chronic inflammatory diseases such as **rheumatoid arthritis** are self-perpetuated by the free radicals released by neutrophils. The ROS-induced tissue damage appears to be involved in pathogenesis of chronic ulcerative colitis, **chronic glomerulonephritis**, etc.

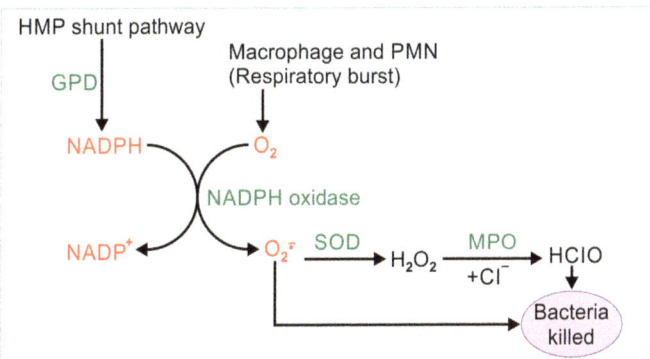

**Fig. 13.9:** Generation of reactive oxygen species (ROS) in macrophages. (HMP: hexose monophosphate; PMN: polymorphonuclear; NADPH: nicotinamide adenine dinucleotide phosphate; MPO: myeloperoxidase; SOD; superoxide dismutase)

- **Respiratory diseases:** Breathing of 100% oxygen for more than 24 hours produces destruction of endothelium and lung edema. This is due to the release of free radicals by activated neutrophils. **Cigarette smoke** contains free radicals. Soot attracts neutrophils to the site which release more free radicals, leading to lung damage.
- **Cataract** formation is related with aging process. Cataract is partly due to photochemical generation of free radicals. Tissues of the eye, including the lens, have high concentration of free radical scavenging enzymes.
- **Reperfusion injury** after myocardial ischemia is caused by free radicals. During ischemia, superoxide is increased. At the same time, the availability of scavenging enzymes is decreased, leading to aggravation of myocardial injury.
- **Atherosclerosis:** Low-density lipoproteins (LDLs) are deposited under the endothelial cells, which undergo oxidation by free radicals. This attracts macrophages. Macrophages are then converted into **foam cells** initiating atherosclerotic plaque formation. Antioxidants offer some protective effect.
- **Carcinogenesis:** Free radicals produce DNA damage and accumulated damages lead to somatic mutations and malignancy.
- **Aging process:** Reactive oxygen metabolites (ROM) play a pivotal role in degenerative brain disorders such as Parkinsonism, Alzheimer's dementia, and multiple sclerosis. Cumulative effects of free radical injury cause gradual deterioration in aging process.

## Free Radical Scavenger Systems

In the first step, superoxide is reacted with the enzyme, **SOD**, to produce hydrogen peroxide. In the next step, this hydrogen peroxide is removed by the enzyme, **POD** to form water (Fig. 13.10). When hydrogen peroxide is generated in excess quantities, the enzyme **catalase** is also used for its removal.

## Antioxidants

Consumption of polyphenol-rich fruits, vegetables, and beverages is beneficial to human health. Dietary polyphenols represent a wide variety of compounds that occur in fruits, vegetables, wine, tea, and chocolate. They contain flavones, flavonols, and catechins. They act as agents having antioxidant, antiaging, anticarcinogenic, anti-inflammatory, and antiatherosclerotic effects. Grape polyphenols can prevent brain damage due to alcohol as well as cerebral ischemia-induced neuronal damage.

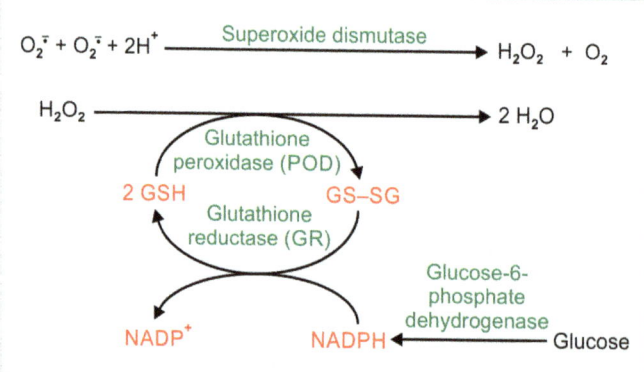

**Fig. 13.10:** Free radical scavenging enzymes. (POD: peroxidase; GSH: glutathione; GR: glutathione reductase)

The following can act as antioxidants, removing the free radicals from the body.
- **Vitamin E** (α-tocopherol), which acts as the most effective naturally occurring antioxidant in tissues.
- **Vitamin C** is the aqueous phase antioxidant.
- **Ceruloplasmin** can act as an antioxidant in extracellular fluid.
- **Caffeine** is another effective antioxidant.
- Food items containing good quantity of antioxidants are (a) spices used in ordinary Indian cooking contain highest quantity of antioxidants; (b) curcumin; (c) fruits and vegetables such as berries, broccoli, spinach, asparagus, and green tea, which contain flavonoids, flavones, isoflavones, and anthocyanins; (d) resveratrol present in grapes; and (e) cysteine, glutathione, carotenoids, and vitamin A are minor antioxidants. The incidence of heart attack is only 50% in vegetarians compared to nonvegetarians, because vegetables contain a lot of such antioxidants.
- The scavenging enzymes described earlier (catalase, POD, and SOD) are powerful antioxidants protecting the tissues.

## Antioxidants Used as Therapeutic Agents

- Vitamin E
- Vitamin C

Caution should be taken in using vitamin E. Excess quantity of vitamin E may act as pro-oxidant and may be deleterious.

## Commercial Use of Antioxidants

Antioxidants are regularly used in food industry to increase the shelf life of products. Commercially used **food preservatives** are vitamin E, propyl gallate, butylated hydroxyanisole (BHA), and butylated hydroxytoluene (BHT). They prevent oxidative damage of oils, particularly those containing PUFA, and prevent rancidity.

## SUMMARY OF THE CHAPTER

- Citric acid cycle is the final common oxidative pathway that oxidizes acetyl-CoA to $CO_2$. It also acts as a link between catabolic and anabolic pathways (amphibolic role).
- Citric acid cycle is the source of reduced coenzymes (that form substrate for respiratory chain) as well as precursors for synthesis of proteins and nucleotides (anaplerotic pathway).
- The sources of acetyl-CoA are pyruvate (from glycolysis) fatty acids (β-oxidation) and ketogenic amino acids.
- All enzymes of the cycle are located inside the mitochondria.
- Three NADH molecules are generated in the cycle at steps 3, 4, and 8. One $FADH_2$ is formed at step 6 and one GTP is formed at step 5.
- Both the carbon atoms of acetyl-CoA are removed as $CO_2$ at steps 3 and 4.
- Ten molecules of ATP are produced per turn of the TCA cycle (1 $FADH_2$ = 1.5 ATP, 3 NADH = (3 × 2.5 = 7.5 ATP, 1 GTP = 1 ATP). It is the main generator of ATP among metabolic pathways.
- α-Ketoglutarate dehydrogenase is the only irreversible step in the TCA cycle.
- Oxaloacetate is the true catalyst, which enters and leaves the cycle unchanged.
- Oxidation of fat (acetyl-CoA) needs the help of oxaloacetate whose major source is pyruvate (carbohydrates). In other words, fats are burnt in the fire of carbohydrates.
- Fat cannot be converted to glucose, as pyruvate to acetyl-CoA is an irreversible step.
- The TCA cycle is regulated by the cellular need of ATP.
- Oxidation of foodstuff occurs in three stages—primary metabolism where macromolecules are converted to smaller units, secondary metabolism where reducing equivalents are formed, and tertiary metabolism where energy is released.
- In substrate-level phosphorylation, energy from high-energy compound is directly

- transferred to nucleoside diphosphate (NDP) to form nucleoside triphosphate (NTP) without the help of ETC.
- Transfer of electrons from reduced coenzymes through respiratory chain to $O_2$ is known as biological oxidation.
- The energy released is trapped as ATP. This coupling of oxidation with phosphorylation is called oxidative phosphorylation.
- Electron flow occurs through successive dehydrogenase enzymes (located in the inner mitochondrial membrane) together known as electron transport chain; the electrons are transferred from higher to lower potential.
- The ETC has four distinct multiprotein complexes, viz. complexes I, II, III, and IV connected by two mobile carriers to CoQ and cytochrome c.
- Cyanide inhibits terminal cytochrome and brings cellular respiration to a standstill.
- Free radicals have important role in our body. They are not always harmful, and macrophages purposely produce free radicals so that bacteria are killed.
- The free radical scavenging enzymes (catalase, POD, and SOD) are powerful antioxidants.
- Antioxidants are regularly used in food industry to increase the shelf life of products.

## QUESTIONS FOR SELF-ASSESSMENT

1. Briefly describes the steps of TCA cycle and mention its importance.
2. Discuss about the energetics of TCA cycle.
3. Why TCA cycle is known as both amphibolic and anaplerotic in nature.
4. What is the importance of biological oxidation?
5. Mention the steps of electron transport chain.
6. Explain the term high-energy compound with examples.
7. Write a note on ATP.
8. Enumerate the role of free radicals in our body.
9. Explain the role of antioxidant enzymes.
10. Mention the role of antioxidants used in therapeutics with examples.

## MULTIPLE CHOICE QUESTIONS (MCQs)

13-1. Glucose has six carbon atoms; and these are removed as carbon dioxide in all the following steps, *except*:
   a. Succinate dehydrogenase reaction
   b. Pyruvate dehydrogenase reaction
   c. Isocitrate dehydrogenase step
   d. α-Ketoglutarate dehydrogenase step

13-2. The location of TCA cycle is:
   a. Cytosol
   b. Golgi apparatus
   c. Endoplasmic reticulum
   d. Mitochondria

13-3. Complete oxidation of one molecule of acetyl-CoA in the TCA cycle generates:
   a. 12 ATP
   b. 38 ATP
   c. 8 ATP
   d. 10 ATP

13-4. Which of the following compound does not contain a high-energy bond?
   a. Fructose-1,6-bisphosphate
   b. 1,3-bisphosphoglycerate
   c. Succinyl-CoA
   d. Creatine phosphate

13-5. The components of ETC are located in the:
   a. Intermediate space of the mitochondrial membrane
   b. Inner mitochondrial membrane
   c. Outer mitochondrial membrane
   d. Cytosolic portion of the cell

13-6. Which of the following processes makes use of free radical effects?
   a. Cell adhesion
   b. Phagocytosis
   c. Contact inhibition
   d. Transcytosis

13-7. The most important antioxidant in the body is:
   a. Vitamin C
   b. Vitamin D

c. Vitamin B$_1$
d. Vitamin E

**13-8. Which is a mineral that is a constituent of glutathione peroxidase:**
a. Copper
b. Iron
c. Selenium
d. Zinc

**13-9. All are functions of antioxidants, *except*:**
a. They may prevent the initiation of chain reactions by removing free radicals.
b. They may scavenge free radicals generated in chain reactions, thereby interrupting the chain sequence.
c. They may remove peroxides, thereby preventing further generation of reactive oxygen species (ROS).
d. They undergo autocatalysis under certain circumstances

**13-10. All the following are types of antioxidant systems, *except*:**
a. Enzyme antioxidant system
b. Vitamin antioxidant system
c. Mineral antioxidant system
d. Nucleic acid antioxidant system

## ANSWER KEYS TO MCQs

13-1. (a)  13-2. (d)  13-3. (a)  13-4. (a)  13-5. (b)  13-6. (b)  13-7. (d)  13-8. (c)
13-9. (d)  13-10. (b)

# CHAPTER 14

# Plasma Proteins, Immunoglobulins, and Tissue Proteins

At the completion of this chapter, the reader will be able to answer questions on the following topics:

- Plasma proteins
- Separation of plasma proteins
- Albumin
- Globulins
- Transport proteins in plasma
- Functions of plasma proteins
- Immunoglobulins
- Immunological mechanisms
- Immunological effector cells
- T cells, B cells, and macrophages
- Antigens and blood group
- HLA and HLA typing
- ELISA technique
- Specialized proteins in tissues
- Collagen
- Elastin
- Keratin
- Myosin
- Actin
- Troponin
- Lens proteins and cataract

## NORMAL VALUES OF PROTEINS IN PLASMA

Total blood volume is about 4.5–5 L in an adult human being. If blood is mixed with an anticoagulant and centrifuged, the cell components [red blood cells (RBC) and white blood cells (WBC)] are precipitated. The supernatant is called plasma. About 55–60% of blood is made up of plasma. Plasma contains hundreds of different proteins. It is a mixture of albumin, glycoproteins, and lipoproteins.

If blood is withdrawn without anticoagulant and allowed to clot, after about 2 hours, liquid portion is separated from the clot. This defibrinated plasma is called serum, which lacks coagulation factors including prothrombin and fibrinogen.

Normal values of plasma proteins are:
- Total proteins: 68 g/dL
- Albumin: 3.5–5 g/dL
- Globulin: 2.5–3.5 g/dL
- Fibrinogen: 200–400 mg/dL
- Albumin/globulin (A/G) ratio: 1.2:1–1.5:1.

Total proteins in serum or plasma are estimated by biuret method and albumin by bromocresol green (BCG) method. Almost all plasma proteins except immunoglobulins are synthesized in liver.

## SEPARATION OF PLASMA PROTEINS

Various methods have been used to separate out the individual proteins in plasma.

a. **Precipitation by ammonium sulfate:** Three proteins have been separated by salting out method using ammonium sulfate. They are albumin, globulin, and fibrinogen.
   - Albumin: Full saturation
   - Globulin: Half saturation
   - Fibrinogen: One-fifth saturation.

b. **Cohn's fractionation (by ethanol):** Various concentrations of ethanol at low temperature are used to separate out fractions of proteins. Each fraction will be a mixture of proteins with one of the protein predominating, for example, fraction I is rich in fibrinogen, while fraction II is rich in gamma-globulins. This method is used in clinical purpose for obtaining purified protein in large scale.

c. **Advanced techniques:** More recent methods include separation of plasma proteins by paper electrophoresis, gel electrophoresis, ultra centrifugation, column chromatography, and so forth. These methods are described in detail in Chapter 22.

## NORMAL VALUES AND INTERPRETATIONS

- The procedure for electrophoresis is described in Chapter 22. In agarose gel electrophoresis, normal serum is separated into the following five bands: albumin, alpha-1-globulin, alpha-2-globulin, beta-globulin, and gamma-globulin. The relative concentrations of these fractions are:
  - Albumin: 55–65%
  - Alpha-1-globulin: 2–4%
  - Alpha-2-globulin: 6–12%
  - Beta-globulin: 8–12%
  - Gamma-globulin: 12–22%.
- Albumin has the maximum and gamma-globulin has the minimum mobility in the electrical field.

## ABNORMAL PATTERNS IN CLINICAL DISEASES

Various abnormalities can be identified in the electrophoretic pattern.

- **Chronic infections:** During chronic infections, the gamma-globulins are increased, but the increase is smooth and wide-based **(Fig. 14.1)**.
- **Multiple myeloma:** In paraproteinemias, a sharp spike is noted and is termed as *M-band*. This is due to the monoclonal origin of immunoglobulins in multiple myeloma **(Fig. 14.1)**. Multiple myeloma is described later.
- **Nephrotic syndrome:** All proteins except very big molecules are lost through urine, and so alpha-2 fraction (containing macroglobulin) will be very prominent.

## ALBUMIN

Albumin is the most abundant protein of plasma, approximately half of the total protein. It has a molecular weight of 69,000. Albumin is a single polypeptide chain having 610 amino acids. Albumin is synthesized mainly in liver. Rate of synthesis is approximately 14 g/day. Plasma albumin is delivered to cell, where it is hydrolyzed to amino acids, and these amino acids are used for the synthesis of cellular proteins. So albumin may be considered as the transport form of essential amino acids to extrahepatic cells.

### Clinical Importance of Albumin Level

- **Hypoalbuminemia** (decrease in albumin concentration)

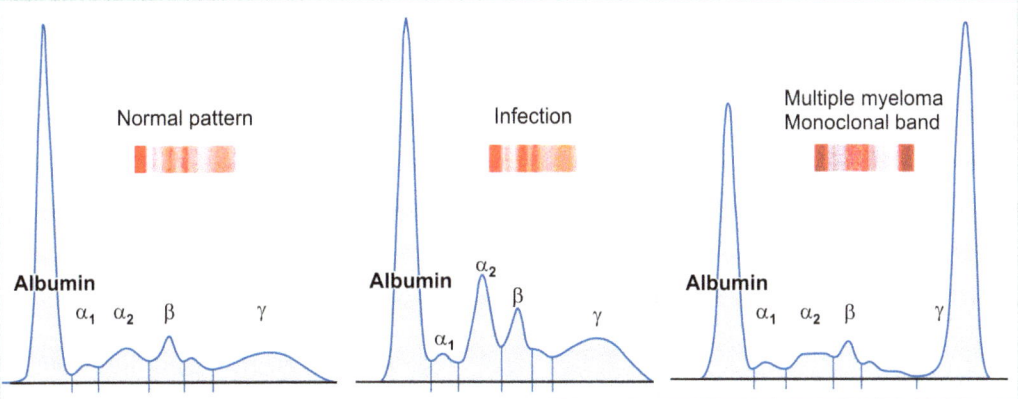

**Fig. 14.1:** Electrophoretogram. Normal and abnormal electrophoretic patterns.

- Concentration of albumin decreases in severe **protein calorie malnutrition** (PCM), where the availability of amino acids is reduced and so albumin synthesis is affected.
- In liver diseases like **cirrhosis**, albumin synthesis is decreased, and so blood level of albumin is lowered. Estimation of albumin is an important liver function test.
- In **nephrotic syndrome**, large amounts of albumin is lost in urine. Detection of albumin in urine is done by **heat and acetic acid test**.
- Decrease in albumin concentration leads to **edema** formation. Edema occurs when total proteins fall below about 5 g% and albumin level below 2.5 g%.
- **A/G ratio:** In all these cases there will be a compensatory increase in globulins, which are synthesized by reticuloendothelial system. A/G ratio is thus altered or even reversed.

## Functions of Albumin

- It contributes about 70–80% of the colloidal **osmotic pressure** of plasma. It is the chief stabilizer of blood volume and regulates the osmotic pressure of vascular compartments by exerting colloidal osmotic pressure.
- It helps in the **transport** of several substances like free fatty acids, bilirubin, calcium ($Ca^{2+}$) and steroid hormones.
- Certain drugs also bind to albumin and are transported to target tissues. For example, sulfonamides, aspirin, penicillin.
- Because of its high concentration in plasma, albumin has maximum buffering capacity.

## GLOBULINS

Globulins are separated by half-saturation with ammonium sulfate. Globulins can be further separated into different fractions, namely, $\alpha_1$-globulins, $\alpha_2$-globulins, $\beta$-globulins, and $\gamma$-globulins. The alpha- and beta-globulins are synthesized in liver, whereas $\gamma$-globulins are synthesized by reticuloendothelial system.

## Alpha-Globulins

They are glycoproteins and are further classified into $\alpha_1$ and $\alpha_2$, depending on their electrophoretic mobility.

*Alpha-1-Acid Glycoprotein (Orosomucoid)*

- It is a reliable indicator of acute inflammation.
- It also serves as a transport protein for hormone progesterone.
- It carries carbohydrate constituents to the sites of tissue repair following injury.

*Alpha-1-Fetoglobulin (Alpha-Fetoprotein)*

- It is present in high concentration in fetal blood.
- It is a diagnostic marker for hepatocellular carcinoma.

*Alpha-1-Antitrypsin*

- It is a protease inhibitor.
- It inhibits proteolytic enzymes such as plasmin, thrombin, trypsin, chymotrypsin, elastase, and so on.

## Alpha-2-Globulins

*Ceruloplasmin*

It is a copper-containing protein and is involved in iron absorption. It is an important antioxidant in plasma.

*Haptoglobin*

It binds free hemoglobin and minimizes the loss of Hb in urine. Haptoglobin is increased in inflammatory conditions.

## Beta-Globulins

- Beta-lipoproteins—transport proteins for lipids.
- Transferrin—transports iron

# Chapter 14  Plasma Proteins, Immunoglobulins, and Tissue Proteins

- Hemopexin—involved in the transport of heme

## TRANSPORT PROTEINS

Body needs the transport of various substances from one tissue to another tissue. Out of these substances, some of them are insoluble in water. Such substances are carried in the blood by certain proteins, which are called transport proteins. They are shown in **Table 14.1**.

## FUNCTIONS OF PLASMA PROTEINS

- **Nutritive:** They are good sources of amino acids; they contribute amino acids for tissue protein synthesis. They may be considered as reserve proteins.
- **Fluid exchange:** The colloid osmotic pressure of plasma protein plays an important role in the distribution of water between blood and tissues.
- **Buffering action:** Proteins are made of amino acids having positively charged amino group and negatively charged carboxyl group. These charged regions can bind with hydrogen ions and hydroxyl ions, thus functioning as buffers.
- **Transport functions (Table 14.1).**

**Table 14.1:** Important transport proteins in plasma.

| Sl. No. | Transport protein in blood | Substance carried by the protein |
|---|---|---|
| 1 | Albumin | Bilirubin, free fatty acid, calcium, and certain drugs |
| 2 | Pre-albumin (Transthyretin) | Thyroid hormones (T3 and T4) |
| 3 | Retinol-binding protein (RBP) | Vitamin A |
| 4 | Transcortin (cortisol-binding globulin [CBG]) | Cortisol and corticosterone |
| 5 | Haptoglobin | Hemoglobin |
| 6 | Hemopexin | Heme |
| 7 | Transferrin | Iron |
| 8 | Lipoproteins | Lipids including cholesterol |

- They provide **viscosity** to blood, which is essential to maintain blood pressure within the normal range.
- **Blood coagulation:** In addition to fibrinogen and prothrombin, plasma contains several enzymes and clotting factors, which participate in the process of blood coagulation.
- **Immunological functions:** Immunoglobulins help in combating against various infections.

## ACUTE PHASE PROTEINS

The level of certain proteins in blood may increase 50–1000 folds in various inflammatory and neoplastic conditions. Such proteins are called acute phase proteins. Important acute phase proteins are described in the following:

### C-Reactive Protein

It is thus named because it reacts with C-polysaccharide of the capsule of pneumococci. It is synthesized in liver. It can stimulate macrophage phagocytosis. When the inflammation has subsided, CRP quickly falls, followed later by erythrocyte sedimentation rate (ESR).

### Ceruloplasmin and Wilson's Disease

Normal blood level of ceruloplasmin is 25–50 mg/dL. This level is reduced to less than 20 mg/dL in Wilson's hepatolenticular degeneration. It is an inherited autosomal recessive condition. Incidence of the disease is 1 in 50,000. Copper is not excreted through bile, and hence copper toxicity. So ceruloplasmin level in blood is decreased. Accumulation in liver leads to hepatocellular degeneration and cirrhosis. Deposits in brain basal ganglia leads to lenticular degeneration and neurological symptoms.

## GAMMA-GLOBULINS (IMMUNOGLOBULINS)

These are proteins having antibody activity. Immunoglobulins are generally abbreviated as "Ig." The antibody reacts with antigen very specifically. This property is widely used in the purification of proteins. This is because of the complementary nature of the three-dimensional structures of antigen and antibody.

The gamma-globulins are proteins synthesized in response to the entry of a foreign substance (antigen) during infection. Thus these proteins provide immunity against infectious diseases. The antibodies will bind with infectious pathogens (antigens) and neutralize them. There are five classes of immunoglobulin (Ig): IgA, IgG, IgM, IgD, and IgE.

## Structure of Immunoglobulins: Heavy and Light Chains

Immunoglobulin (Ig) molecule is made up of two pairs of polypeptide chains, one pair of which is called light chains (L) and the other pair is called heavy chains (H) **(Fig. 14.2)**. The two light chains are identical. Similarly, the two heavy chains are identical. All classes of Ig contain two types of light chains called kappa (κ) and Lambda (λ) chains.

Immunoglobulin molecule is Y shaped. One light chain is held together with one heavy chain by interchain disulfide bridges. The two heavy chains are similarly held together by interchain disulfide bonds. Besides, there are a number of intrachain disulfide bonds **(Fig. 14.2)**.

Amino terminals of light and heavy chains are at one end. These terminals are involved in binding with the antigen. Each chain has a variable amino acid sequence from the amino terminal (variable region, VL, and VH) and a constant amino acid sequence (constant region, CL, and CH) toward carboxy terminal end.

## Immunoglobulin G

Immunoglobulin G (IgG) is the major antibody of secondary immune response. IgG can cross placenta and is the major protective antibody in newborns. IgG has a molecular weight of 1,50,000. The heavy chain is of γ type. IgG is able to activate the complement system and can act as opsonin. It will be present in all extracellular fluids.

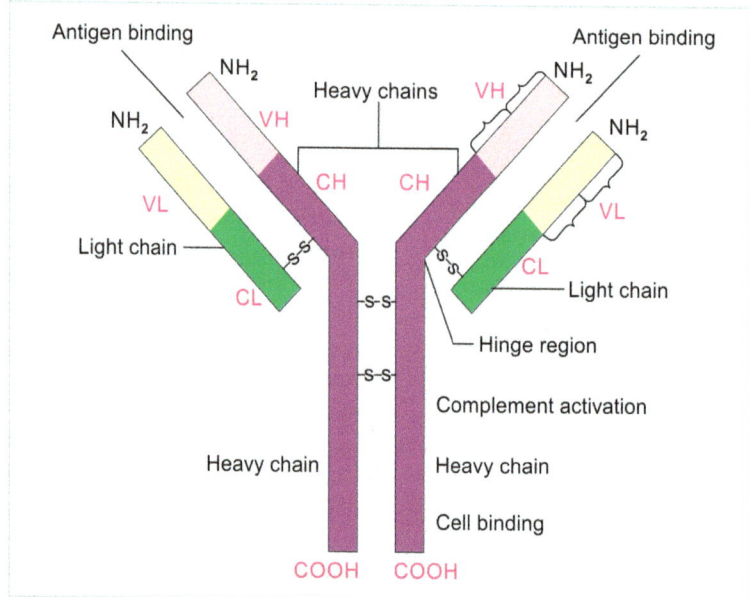

**Fig. 14.2:** Immunoglobulin molecule. Chains are connected by disulfide bridges, shown as —-s-s- linkages. (NH$_2$: amino terminal end; COOH: carboxy terminal end; constant regions are shown as dark; VH: variable heavy region; VL: variable light chain; CH: constant heavy region; CL: constant light region)

## Table 14.2: Characteristics of different immunoglobulin classes.

| | IgG | IgA | IgM | IgD | IgE |
|---|---|---|---|---|---|
| Name of heavy chain | γ | α | μ | δ | ε |
| No. of basic 4-peptide units (2 L + 2 H) | 1 | 2 | 5 | 1 | 1 |
| Additional unit present | — | S and J | J piece | — | — |
| Molecular weight (daltons) | 1,46,000 | 3,85,000 | 9,70,000 | 1,85,000 | 1,90,000 |
| Concentration in normal serum | 800–1200 mg/100 mL | 150–300 mg/100 mL | 50–200 mg/100 mL | 1–10 mg/100 mL | 1.5–4.5 μg /100 mL |

## Immunoglobulin M

They are macroglobulins having five subunits each having four chains. So, there are 10 heavy chains and 10 light chains. It can bind with five antigenic determinant sites simultaneously; so it is very effective in the agglutination of bacteria. It is the predominant class of antibodies in primary immune response. Natural antibodies are also IgM in nature.

## Immunoglobulin A

IgA is the predominant immunoglobulin in body secretions. They are found in external secretions like colostrum, saliva, tears, gastrointestinal fluids, and nasal and bronchial secretions. Secretory IgA provides the primary defense mechanism against local infections, due to its abundance in saliva, tears, and so on. Its main function is to prevent the invasion of microorganisms to susceptible tissues.

## Immunoglobulin D

It is found on the surface of B lymphocytes and also in various body fluids.

## Immunoglobulin E

These antibodies mediate allergic reactions. Upon combination with certain allergens, they trigger the release of mediators such as histamine. This will cause bronchial constriction (asthma), blood vessel dilatation (erythema), and lowering of blood pressure (shock). This is called hypersensitivity reactions to allergens. Characteristics of various immunoglobulin classes are shown in **Table 14.2**. Structural comparison of the various classes are shown diagrammatically in **Figure 14.3**.

## Immunological Alterations in Pathological Conditions

*Multiple Myeloma (Plasmacytoma)*

Immunoglobulin secreting cells are transformed into malignant cells. Thus,

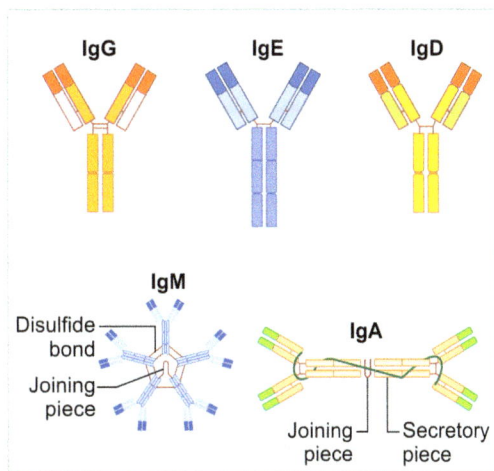

**Fig. 14.3:** Immunoglobulin G (IgG), immunoglobulin E (IgE), and immunoglobulin D (IgD) have one basic unit each; immunoglobulin M (IgM) has five basic units; and immunoglobulin A (IgA) has two basic units. In IgM and IgA, the chains are connected by the joining (J) piece. In IgA, there is an extrasecretory piece.

Ig molecules of the very same type are produced in large quantities. This is seen in electrophoresis as the myeloma band or **monoclonal band** or M-band with a sharp narrow spike. This is seen between beta and gamma regions (**Fig. 14.1**). Multiple myeloma is characterized by paraproteinemia, anemia, lytic bone lesions, and proteinuria. Bone pain and tenderness are the common presenting complaints. Spontaneous pathological fracture of weight-bearing bones, rib, and vertebrae may occur.

*Bence–Jones Proteinuria*

Light chains of immunoglobulins may be seen in the urine of some of the multiple myeloma patients.

*Hypergammaglobulinemia*

An increase in gamma-globulin is seen in leprosy, tuberculosis, malaria, and some other chronic infections.

*Hypogammaglobulinemia*

Decrease or absence of immunoglobulins may be seen in conditions like nephrotic syndrome, myeloma, leukemia, or in certain congenital conditions.

## GENERAL IMMUNITY

Smallpox has been completely eradicated from the world by 1985; this is a triumph of immunology. Three salient features of immunological reactions are **recognition of self** from nonself or foreign substances, **specificity** of the reactions, and **memory** of the response.

When injected with 100 different proteins, an animal will produce 100 different antibodies; this is called specificity. If a person belongs to the A group antigen of RBC, antibodies against B group are seen in circulation. There is an extra N-acetyl group in antigen A; this is the only molecular difference between A and B antigens. Immune system is exquisitely specific to recognize even this small difference at molecular level. If the same antigen is introduced for the second time, body will react immediately; this memory is the basis of vaccination.

## Antigens

Certain components of the cell membranes act as specific antigens. They will be different from person to person in chemical composition and three-dimensional structure. Hence the immunocompetent cells could recognize the self from nonself. Any substance that invokes an immunological response is an **antigen** or immunogen. Antibody response will usually be selective against specific spatial configurations on the antigen, which are called antigenic determinant sites, known as **epitopes.**

## Immune Response

The lymphocytes generated from the bone marrow and passed through and processed by the thymus gland are called **T lymphocytes.** They can directly kill the target cells and are the effector cells for the **cell-mediated immunity** (CMI). The T lymphocytes are found mainly in the paracortical areas of lymph nodes and periarteriolar sheaths in the spleen. In peripheral blood 80% lymphocytes are T cells and 15% are B cells.

Certain other cells originated from bone marrow and processed by the bursa of fabricius in avians are called **B cells.** The bursa-equivalent organs in human beings are gut-associated (including Peyer's patches) and lung-associated lymphoid organs. Immunoglobulins are secreted by **plasma cells** belonging to the B lymphocytes. The B cells govern the **humoral immunity.**

**Clonal selection:** Immunoglobulins of different specificity are available on the B cell surface. When an antigen is introduced, the antigen selects out that particular cell carrying the specific antibody. This results in a series of divisions of that cell, and a clone of cells are produced. These cells are finally differentiated into plasma cells. This is the antigen-dependent **clonal selection.** A particular clone of cells secrete antibodies of the same specificity.

## Effector Mechanisms

The following are the immunological effector mechanisms by which foreign cells are destroyed or particles are removed:

*Cell-mediated Immunity*

The following are the major activities of T lymphocytes:
- **Immunity against infections:** T cells mediate effective immunity against bacteria such as mycobacteria, many viruses, and parasites.
- **Rejection of graft:** When an organ (heart, kidney) is transplanted from one person to another, it is called allograft. Body tries to reject such transplanted organs, mainly by T cell-mediated mechanism.
- **Hypersensitivity:** It is the overreaction of the immune system, often resulting in unwanted tissue destruction. This is responsible for caseation and lung cavity formation in the case of tuberculosis and contact hypersensitivity to chemicals.

*Humoral Immunity*

Antibodies are produced by plasma cells. These are immunoglobulins, described in detail in the following. The antigen–antibody reaction leads to the activation of a complement system, which destroys the foreign cells.

*Macrophages*

Phagocytosis is the nonspecific mechanism by which body tries to eliminate invading organisms. Foreign materials are ingested by the phagocytes and later digested intracellularly. When a foreign particle enters the body, the macrophages phagocytose it, and present the antigens to the lymphocytes.

## Primary and Secondary Immune Responses

When an antigen is injected, antibodies in blood appear within about 10 days, reach a peak level within 20 days, and response declines by about 30 days. The IgM molecules will be predominant in this **primary response.**

When the same antigen is reinjected into the same animal after a few months, the antibody response is **quicker** (within three days), **stronger** (100–1000 times more quantity of antibody), more **avid** (IgG type), and more **prolonged** (response lasts for months). This is the **secondary immune response,** which is due to the memory cells produced in the primary response. This is the basis of immunization. Differences between primary and secondary responses are shown in **Figure 14.4**.

**Active immunity** is induced by immunization with toxoid, or killed or attenuated organisms. Examples are diphtheria, pertussis, tetanus (DPT) vaccine, oral polio vaccine, and hepatitis B vaccine.

In **passive immunity,** protection is given by preformed antibodies. This is used in immunotherapy against diphtheria, tetanus, snake bites, and so on.

## Molecular Structure of Antigens

Blood groups of RBCs express more than many antigens. The most important of them is

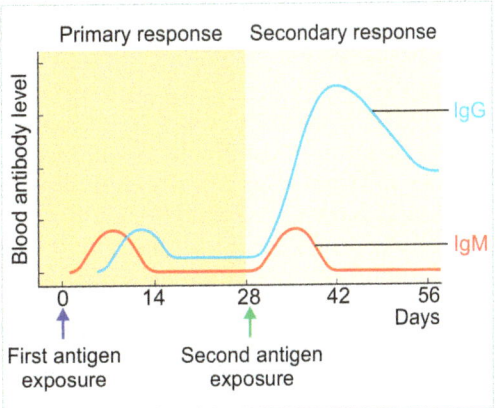

**Fig. 14.4:** Primary and secondary immune responses.

called the ABO system. The RBC of the person may carry antigen A, B, or AB; if none of these antigens are present, it belongs to blood group O. Group A person's serum contains anti-B antibody. Therefore, when RBCs of B group are introduced into the person having blood group A, there will be agglutination of the introduced cells. The ABO system antigens are glycoproteins present on the surface of all cell membranes.

## Human Leukocyte Antigens

The T cell recognition is also closely related with the human leukocyte antigen (HLA) system. When organs are transplanted, the donor and receiver are matched for the HLA system. The genes of major histocompatibility complex (MHC) are involved in the recognition between self and nonself antigens.

The MHC is composed of a set of molecules that play a pivotal role in the immune response against different pathogens and tumors cells. These molecules were described in mice for the first time by Gorer while performing transplantation studies with tumor cell lines injected into inbred strains of mice. In human beings, the MHC genes are present on chromosome 6. In humans, the class I MHC region contains approximately 20 genes. There are HLA-A, HLA-C, HLA-B, HLA-D and HLD-DR loci. All these loci together contain more than 150 alleles. Permutation and combination of them could produce an astronomical number of variations. Hence the antigenic constitution of one person will be entirely different from another one. There are two major types of MHC molecules: class I MHC molecules, which are expressed by nearly all nucleated cells of vertebrate species and consist of a heavy chain linked to alpha-2-microglobulin; and class II MHC molecules consisting of an alpha- and a beta-glycoprotein chain and are expressed only by antigen-presenting cells. These are main transplantation antigens, responsible for the rejection of allograft. Body will try to reject the transplanted organs. To reduce such rejection reactions, various immunosuppressive drugs are used. Basiliximab and daclizumab are monoclonal antibodies, which will inhibit activated T cells.

The B lymphocytes mature within the bone marrow. On the other hand, the T lymphocytes originate from the bone marrow but migrate to the thymus gland to mature. During its maturation within the thymus, the T cell comes to express a unique antigen-binding molecule called the T-cell receptor on its membrane. Unlike membrane-bound antibodies on B cells, which can recognize antigen alone, T-cell receptors can recognize only antigen that is bound to cell-membrane proteins called MHC molecules. MHC molecules that function in this recognition event, which is termed *antigen presentation"* are glycoproteins found in cell.

## HLA Typing

The HLAs are cell surface molecules found on all nucleated cells. They are specialized elements in the family of immune system in recognition of self from nonself in an individual. The HLA is responsible for presenting antigens to T cells and therefore serves as a door to the specific immune system according to the requirement.

Each individual has a unique set of these antigens, half inherited from each parent, and their typing becomes important before organ transplantation. Typing is also used to identify markers for specific diseases, such as HLA $B_{27}$, which is known to be closely associated with ankylosing spondylitis. The HLAs are involved in cancer immunity and therefore in susceptibility and prognosis mainly by presenting certain antigens known as tumor-associated antigens (TAAs). The TAAs are the first contact of malignant cells with adaptive immunity.

Three main processes are used to perform HLA typing:
1. The first is the conventional serological cytotoxicity method where tiny samples of lymphocytes (taken from blood) are added to Terasaki microtiter plates. These plates hold individual wells that contain

different specific antibodies (from either maternal sera or manufactured monoclonal antibodies). The best cells for class II typing are B lymphocytes, and class I typing can be performed with the remaining leukocytes. Magnetic beads are used to purify the required cells from blood.
2. The second potential method used for HLA typing is flow cytometry, particularly for identifying specific alleles. Here fresh nucleated leukocytes are added to monoclonal antibodies that are fluorescence labeled. Cells with surface antigens that bind to the antibody become fluorescent. The flow cytometer detects the fluorescent cells by detecting the light emitted from them as they pass through a laser beam. Flow cytometry takes only about 30 minutes to complete.
3. The third method is DNA polymerase chain reaction (PCR) method of amplification, and then DNA–DNA hybridization is performed.

## ENZYME-LINKED IMMUNOSORBENT ASSAY TEST

Enzyme-linked immunosorbent assay (ELISA) is the abbreviation for **enzyme-linked immunosorbent assay**. The ELISA techniques are widely used not only for hormone measurements but also for detecting other growth factors, tumor markers, bacterial or viral antigens, antibodies against microbes, and any other antigens or antibodies in biological fluids. This test is commonly employed to detect antigens or antibodies present in very small quantities in tissues or blood.

### Antibody Detection by ELISA

This is useful to detect small quantities of antibodies in the blood. A good example is the test for the **detection of HIV antibody.** In patients with AIDS, the human immunodeficiency virus (HIV) produces specific antibody. To detect the HIV antibody, the following method is used.

Antigen from HIV is coated in the wells of a multiwell (microtiter) plate. Patient's serum is added and incubated. If it contains the antibody, it is fixed. The wells are washed. This is to remove excess antibodies in the serum.

Next a second antibody (antibody against human immunoglobulin) conjugated with horseradish peroxidase (HRP) is added. Then a color reagent containing hydrogen peroxide and diaminobenzidine is poured over. If a brown color develops, it means that the antibody was originally present in the patient's serum **(Fig. 14.5)**. Here the **color developed is proportional to the antibody.** So, the quantity of the antibody can be calculated. HIV antibody is an example, any antibody could be detected by using the specific antigen.

### Antigen Detection by ELISA Method

At first, the specific antibody is fixed to the well of a microtiter plate (Step 1, **Fig. 14.6**). A good example is the assay of thyroid hormone T4. The patient's serum is added in the well, and incubated for 30 minutes at 37°C. By this time, antigen (T4 in this example) present in the serum is fixed on the antibody (Step 2,

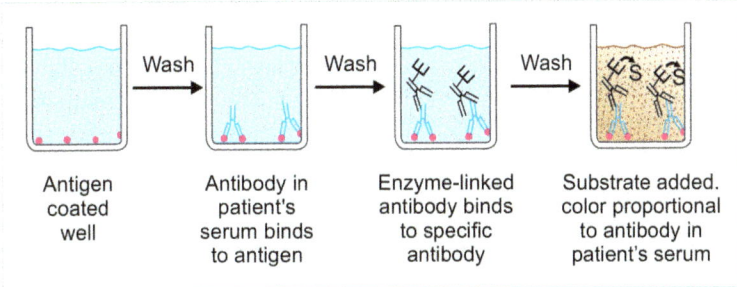

Antigen coated well → Antibody in patient's serum binds to antigen → Enzyme-linked antibody binds to specific antibody → Substrate added. color proportional to antibody in patient's serum

**Fig. 14.5:** Indirect enzyme-linked immunosorbent assay to detect antibody.

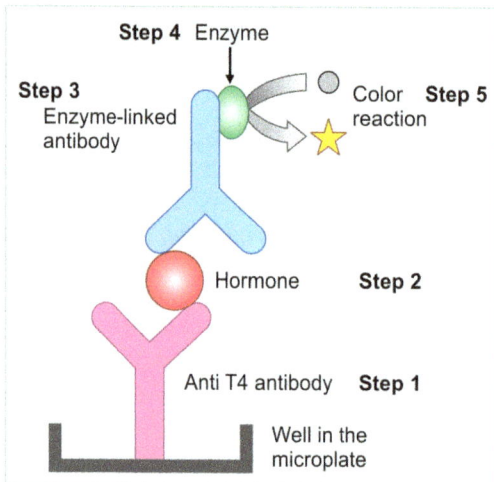

**Fig. 14.6:** Antigen detection by ELISA. The five sequential steps are explained in the text.

**Fig. 14.6).** Excess antigen and other unwanted proteins are washed out.

Then, specific antibody (antibody against T4, tagged with enzyme HRP) is added (**Step 3, Fig. 14.6**). If the antigen is already fixed, the antibody–enzyme conjugate will be fixed in the well (**Step 4, Fig. 14.6**). Then a color reagent containing hydrogen peroxide ($H_2O_2$) and diaminobenzidine (DAB) are added (**Step 5, Fig. 14.6**). The hydrogen peroxide will oxidize the DAB to produce a color. The intensity of the color thus produced will be proportional to the antigen (T4) present in the patient's blood. So, quantitation is easy.

## SPECIAL PROTEINS

### Collagen

The major structural protein found in connective tissue is the collagen. It is the most abundant protein in the body. About 25–30% of the total weight of protein in the body is collagen. It serves to hold together the cells in the tissues. It is the major fibrous element of tissues like bone, teeth, tendons, cartilage, and blood vessels.

### Structure of Collagen

The **tropocollagen** is made up of three polypeptide chains, each having about 1,000 amino acid residues. About 33% of the amino acids is **glycine,** that is, every third residue is glycine. Other abundant amino acids are proline and hydroxyproline. The hydroxylated amino acid residues are of special functional significance.

The collagen fiber has **triple-stranded, quarter staggered** arrangement. This arrangement helps in mineralization. The collagen fibers are strengthened by covalent cross-links between lysine and hydroxylysine residues. The older the collagen, the more the extent of cross-linkages. The process continues, especially in **old age,** so that the skin, blood vessels and other tissues become less elastic and more stiff, contributing to a great extent to the medical problems of the old people. The synthesis and maturation of collagen are depicted in **Figure 14.7**.

### Functions of Collagen

- To give support to organs.
- To provide alignment of cells, so that cell anchoring is possible.
- In blood vessels, if collagen is exposed, platelets adhere and thrombus formation is initiated.

### Abnormalities in Collagen

**Osteogenesis imperfecta** is an inherited disease. It is the result of a mutation which results in the replacement of a single glycine residue by cysteine in collagen. This change disrupts the triple helix. So, unfolding of the helix takes place and fibrillar array cannot be formed. The result is **brittle bones** leading to multiple fractures and skeletal deformities.

### Elastin

Elastin is a protein found in the connective tissue and is the major component of elastic fibers. The elastic fibers can stretch and

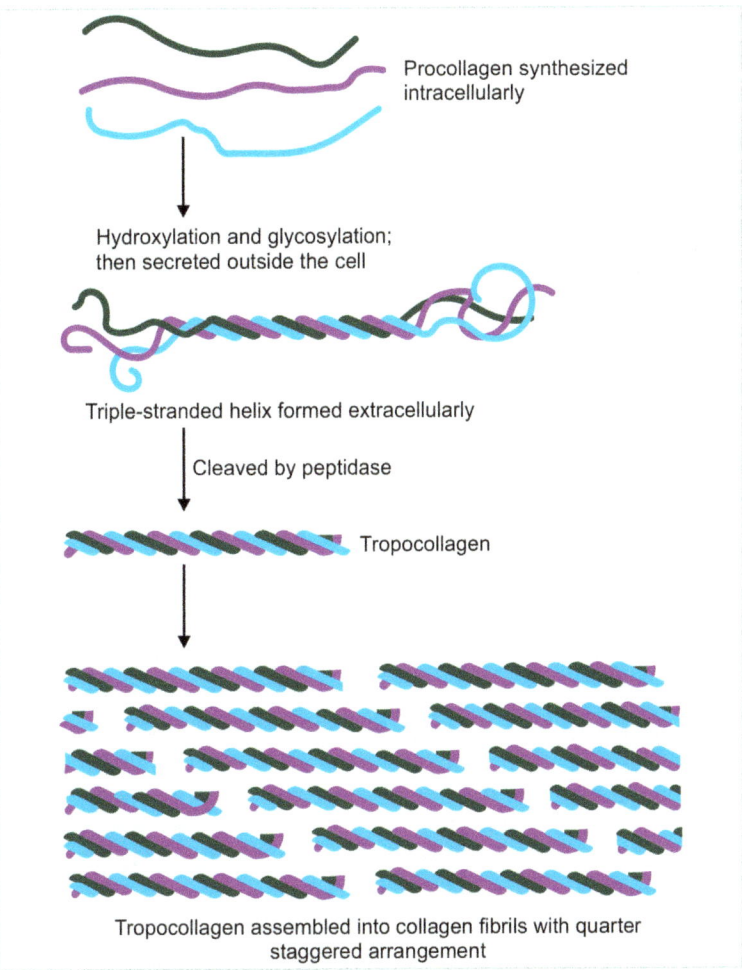

**Fig. 14.7:** Maturation of procollagen into tropocollagen and assembly of collagen fibrils. Note the quarter staggered arrangement of collagen fibril. Each row moves one-fourth length over the last row; the fifth row repeats the position of the first row.

then resume their original length. They have high tensile strength. They are found in the ligaments as well as in the walls of the blood vessels, especially large vessels like aorta. One-third of the residues are glycine. Proline is present in large amounts. When elastin matures, **desmosine** cross-links are formed from four lysine residues. Once mature, elastin is very stable; the turnover rate is very low.

**Pseudoxanthoma elasticum:** It is an inherited defect in the formation of elastin. Clinical manifestations are similar to Ehlers–Danlos syndrome.

### Keratins

Keratins are proteins present in hair, skin, nails, horn, hoof, and so forth. They mainly have the alpha-helical structure. Each fibril has three polypeptide chains, and each bundle has about 10–12 fibrils. It has cysteine-rich polypeptide chains, which are held together by disulfide bonds. The keratin present in hair has significantly more number of disulfide bonds, which give the mechanical strength.

## MUSCLE PROTEINS

Striated muscle is made up of multinucleated cells bound by plasma membrane called **sarcolemma**. Sarcomere is the functional unit of muscle. The myofibrils are immersed in a cytosol that is rich in glycogen, ATP, creatine phosphate, and glycolytic enzymes.

There are variable combination of thick and thin filaments. The thick filament is primarily **myosin**, and the thin filament contains **actin**, **tropomyosin**, and **troponin**. Thick and thin filaments slide past each other during the muscle contraction, so that the muscle shortens by as much as one-third of its original length. However, the lengths of the thick and thin filaments do not change during muscle contraction.

### Myosin

Myosin molecules assemble into filaments. Myosin acts as the enzyme ATPase. Myosin binds to actin polymer, which is the major component of the thin filaments. Myosin molecules are large (about 540 kD), each with six polypeptide chains, two identical heavy chains, and four light chains.

### Actin

It is the major protein of the thin filaments. The muscle contraction results from the interaction of actin and myosin to form actomyosin, with energy provided by ATP. When the two thin filaments that bind the crossbridges of a thick filament are drawn toward each other, this could result in the process of contraction of muscle fibers.

### Troponins

The muscle contraction is modulated by troponin and tropomyosin. In the resting muscle, the $Ca^{++}$ is within the sarcoplasmic reticulum. The nerve impulse releases $Ca^{++}$ from the sarcoplasmic stores and increases its cytosolic concentration about 10 times. The action of calcium is brought about by two proteins, troponin and tropomyosin, located in the thin filament.

Troponin is a marker for myocardial infarction. Its level in serum is increased within 4 hours of myocardial infarction and remains high for about 7 days. Troponins are described in Chapter 9, under the heading "Cardiac Biomarkers."

## LENS PROTEINS

India has the maximum number of blind persons in the world. Cataracts and opacities of cornea are the cause for 70% of blindness. The eyes of older people and of diabetics are prone to cataract formation. Being avascular, lens relies on the aqueous humor for the provision of oxygen and essential metabolites. These normal lens cells possess the usual protein-synthesizing machinery. Lens tissue has a very active HMP shunt pathway, and has the maximum concentration of NADPH. Lens also contains a high quantity of ascorbic acid. They scavenge the free radicals and maintain the transparency of lens.

### Crystallins

Major lens proteins are alpha-, beta-, and gamma-crystallins. They undergo no replacement throughout the life of the individual. There is no turnover of these proteins. The proteins at the center of the lens are as old as the individual. The orderly arrangements of the molecules make the lens proteins transparent.

### Cataracts

When there is change in the three-dimensional structure of lens proteins, the lens becomes opaque. In diabetes mellitus, when the blood glucose level is increased, lysine residues of these proteins are glycated. This leads to increased susceptibility for the aggregation of the proteins, resulting in opalescence and cataract.

In the lens, the enzyme aldose reductase reduces monosaccharides to corresponding sugar alcohols, glucose to sorbitol, and galactose to galactitol. These polyols do not readily cross cell membranes and hence accumulate, causing osmotic swelling and consequent disruption of cell architecture. Thus **diabetes mellitus** (increased glucose in blood) and galactosemia (high galactose level) cause cataract.

## SUMMARY OF THE CHAPTER

- Total plasma protein content is 6–8 g/dL of which albumin is 3.5–5 g/dL and the rest is globulin. Almost all plasma proteins are synthesized in the liver except immunoglobulins.
- In agar gel electrophoresis, albumin has maximum mobility while gamma-globulin has minimum mobility.
- In chronic infection, gamma-globulins are increased smoothly, while in paraproteinemias, M-band is seen. The alpha-2 fraction is increased in nephrotic syndrome, while albumin is decreased in liver cirrhosis, malnutrition, and nephrotic syndrome.
- Albumin contributes to colloid osmotic pressure of plasma, has buffering capacity, and is a transport medium for various hydrophobic substances.
- Hypergammaglobulinemia is seen in conditions of hypoalbuminemia, chronic infection, and paraproteinemias.
- The transport proteins in blood are albumin, prealbumin (transthyretin), retinol-binding protein (RBP), thyroxine-binding globulin (TBG), transcortin, and haptoglobin.
- The levels of certain proteins in blood may increase 50–100-fold in various inflammatory and neoplastic conditions. Such proteins are called acute phase proteins. For example, C-reactive protein (CRP), ceruloplasmin, haptoglobin, alpha-1-acid glycoprotein, alpha-1-antitrypsin, and fibrinogen.
- Proteins that are decreased in blood during inflammatory response are called negative acute phase proteins. For example, albumin, transthyretin, and transferrin.
- The collagen fiber has triple-stranded, quarter staggered arrangement. This arrangement helps in mineralization.
- The muscle contraction is modulated by troponin and tropomyosin.
- Major lens proteins are alpha-, beta-, and gamma-crystallins. They undergo no replacement throughout the life of the individual.

## QUESTIONS FOR SELF-ASSESSMENT

1. Define plasma proteins with examples.
2. Mention the functions of albumin.
3. What is the significance of albumin and globulin ratio?
4. Classify immunoglobulins and mention the various types.
5. What are acute phase proteins? Mention their importance.
6. Write a note on muscle proteins.
7. What are the major functions of collagen?
8. Write a note on crystallins.
9. What are the main functions of troponin?
10. Name the negative acute phase proteins.

## MULTIPLE CHOICE QUESTIONS (MCQs)

14-1. **All are major plasma proteins, *except*:**
 a. Albumin
 b. Troponin
 c. Alpha-2-globulin
 d. Beta-globulin

14-2. **Plasma albumin performs the following functions:**
 a. Osmotic function
 b. Transport function
 c. Detoxification
 d. Buffering functions

14-3. **All the following are acute phase reactant proteins, *except*:**
 a. C-reactive protein (CRP)
 b. High-density lipoprotein (HDL)
 c. Ceruloplasmin
 d. Haptoglobin

14-4. Regarding the structure of immunoglobulin all of the following statements are true, *except:*
   a. Heavy chains are linked to carbhydrates
   b. Amino acid sequence of the variable regions are responsible for binding with antigen
   c. Light chains contain unusual amino acids
   d. Light chains are of two types

14-5. Which protein is not present in plasma?
   a. Albumin
   b. Fibrinogen
   c. Hemoglobin
   d. Globulins

14-6. The collagen defect present in scurvy is:
   a. Decreased protein stability due to decreased hydroxylation of proline and lysine residues
   b. Substitution of valine for proline and lysine residues in the collagen sequence
   c. Increased formation of Schiff base cross-links
   d. Decreased protein stability due to increased glycosylation

14-7. The protein present in highest concentration in plasma is:
   a. Fibrinogen
   b. Gamma-globulin
   c. Albumin
   d. Alpha-globulin

14-8. Regarding multiple myeloma identify the correct statement:
   a. Multiple myeloma results from a polyclonal proliferation of lymph node plasma cells
   b. Multiple myeloma often presents with eye pain
   c. Hypercalcemia develops in 50% of patients
   d. Most patients have a serum called alpha proteinemia

14-9. All are examples of structural proteins, *except:*
   a. Collagen
   b. Fibrillin
   c. Elastin
   d. Troponin

14-10. The two main contractile proteins found in skeletal muscle are:
   a. Actin and myosin
   b. Troponin and tropomyosin
   c. Myosin and tropomyosin
   d. Actin and tropomyosin

### ANSWER KEYS TO MCQs

14-1. (b)  14-2. (c)  14-3. (b)  14-4. (c)  14-5. (c)  14-6. (a)  14-7. (c)  14-8. (c)
14-9. (d)  14-10. (a)

# CHAPTER 15

# Hemoglobin

At the completion of this chapter, the reader will be able to answer questions on the following topics:

- Structure of heme
- Biosynthesis of heme
- Catabolism of heme
- Jaundice
- Hyperbilirubinemias
- Porphyrias
- Abnormal hemoglobins
- Sickle cell anemia
- Thalassemias
- Transport of oxygen
- Transport of carbon dioxide

## HEME METABOLISM

Blood is made of cell components and plasma. Cell components include red blood cell (RBC), white blood cell (WBC), and platelets. Red blood cells or erythrocytes are biconcave discs having a life span of about 120 days. RBCs contain hemoglobin, which is a conjugated protein having heme as the prosthetic group and the protein globin. Heme is present not only in hemoglobin but also in myoglobin, cytochromes, and certain enzymes like catalases and peroxidases.

## Structure of Heme

Heme is a **tetrapyrrole**; it has four pyrrole rings surrounding a central **iron** atom. **Porphyrins** are cyclic compounds formed by the fusion of four pyrrole rings linked by methenyl bridges. The rings are denoted as I, II, III, and IV. The bridges are α, β, γ, and δ (alpha, beta, gamma, and delta; **Fig. 15.1**).

## Biosynthesis of Heme

Heme can be synthesized in almost all tissues.

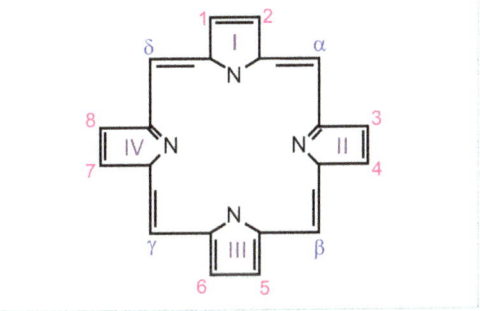

**Fig. 15.1:** Porphyrin ring. The pyrrole rings are numbered from I to IV and the bridges are named from alpha to delta. The possible sites of substitutions are denoted from 1 to 8.

**Step 1:** Condensation of amino acid glycine with succinyl CoA to form amino levulinic acid (ALA)

$$\text{Glycine + succinyl CoA} \xrightarrow[\text{PLP}]{\text{ALA synthase}} \text{α-amino levulinic acid}$$

This is a rate-limiting step. This needs PLP as the coenzyme.

**Step 2:** Condensation of two molecules of ALA to form porphobilinogen.

$$\text{2ALAs} \xrightarrow{\text{ALA dehydratase}} \text{porphobilinogen}$$

**Step 3:** Four porphobilinogen molecules condense to form linear tetrapyrrole. This cyclizes to form uroporphyrinogen III.

$$\text{Porphobilinogen} \xrightarrow[\text{Uroporphyrinogen III cosynthase}]{\text{Uroporphyrinogen I synthase}} \text{Uroporphyrinogen}$$

The pathway continues with a number of modifications to the substituent groups (i.e., groups attached to the outside of the ring structure), finally forming protoporphyrin. At this point, ferrous iron is inserted into the central cavity of the ring system and converts it into heme.

$$\text{Protoporphyin} \xrightarrow[\text{Fe}^{2+}]{\text{Heme synthase}} \text{Heme}$$

This heme gets attached to proteins (apoproteins) to form biologically functional proteins.

$$\text{Heme + protein (globin)} \longrightarrow \text{Hemoglobin}$$

Summary of heme synthesis pathway is shown in **Figure 15.2**.

## Regulation of the Heme Synthesis

ALA Synthase is the rate-limiting enzyme. It is regulated by:
- **Repression mechanism:** Heme, the end product, acts as a repressor of gene for the synthesis of ALA synthase. So the enzyme is not synthesized, and the reaction is decreased. But this mechanism will take a few days for effective operation.
- **Feedback inhibition:** The end product heme directly inhibits the enzyme molecules. So the activity is reduced with immediate effect.

## Disorders of the Heme Synthesis

**Porphyrias** are a group of inborn errors of metabolism associated with the biosynthesis of heme (in Greek *porphyria* means "purple"). This is mainly because of the deficiency of enzymes of heme synthesis. There is increased production and excretion of porphyrins and/or their precursors [aminolevulinic acid (ALA) + porphobilinogen (PBG)]. Porphyrias are not associated with anemia. Porphyrias may be classified into hepatic and erythropoetic porphyrias. Most of the porphyrias are inherited, whereas some are acquired. A classic example of the hepatic variety is acute intermittent porphyria.

### Acute Intermittent Porphyria

It is inherited as an autosomal dominant trait. In this trait, PBG-deaminase is deficient. This leads to a secondary increase in activity of ALA synthase, since the end-product inhibition is not effective. The levels of ALA and PBG are elevated in blood and urine. Urine is colorless when voided, but the color is increased on standing due to photo-oxidation of PBG to porphobilin. Symptoms appear intermittently. Hence it is at times called the "little imitator." Most commonly, patients present with acute abdominal pain. The patients often land up with the surgeon as a case of acute abdomen, and on several instances exploratory laparotomies are done.

### Acquired Porphyrias

Porphyria can result from lead poisoning. Most of the paints contain lead more than the permitted levels. Children suck painted toys, and they get the poison. The toxic effect of lead is due to the inhibition of ferrochelatase. So, there is decreased levels of heme with consequent increased activity of ALA synthase.

## Catabolism of Heme

End products of heme catabolism are bile pigments. The RBCs are broken down by the reticuloendothelial system to liberate hemoglobin. The protein globin and its constituent amino acids will be reutilized. Iron of heme is also reutilized. However, the porphyrin part of heme is degraded to bile pigments mainly **bilirubin (Fig. 15.3)**. In an adult, about 250–300 mg of bilirubin is formed

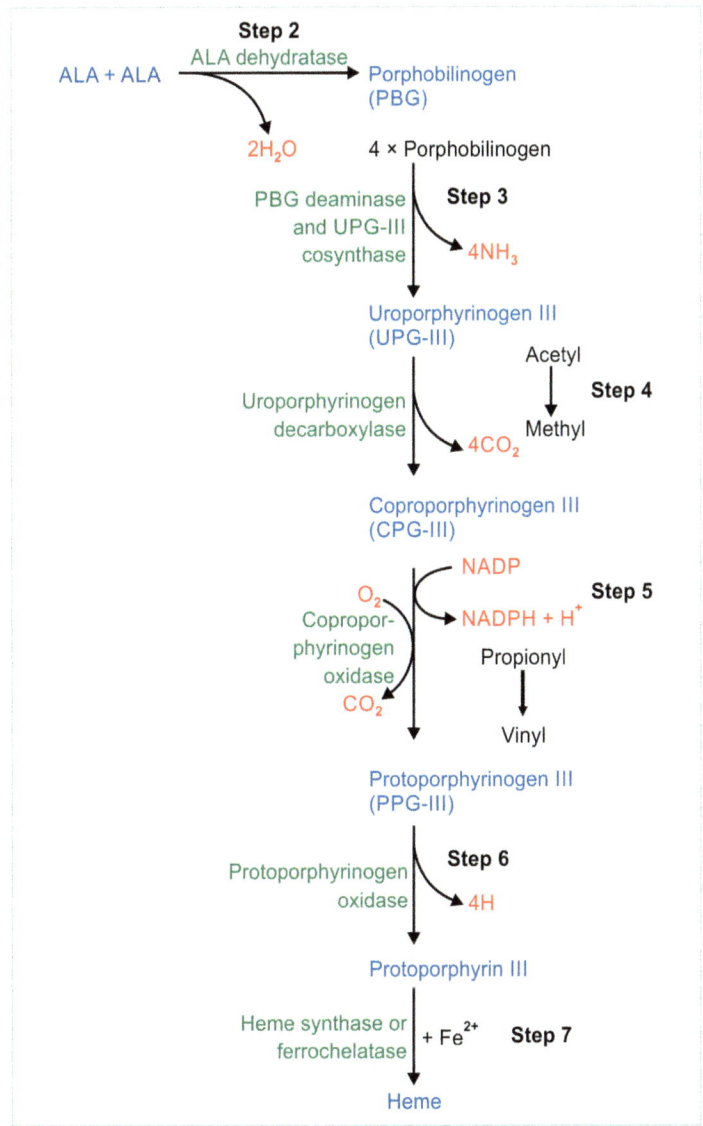

**Fig. 15.2:** Steps of heme synthesis.

every day from hemoglobin and other heme-containing compounds. Bile pigments are excreted through bile.

After bilirubin is formed by the breakdown of heme in the reticuloendothelial system (RES), its disposal occurs in the liver. Bilirubin formed in RES is insoluble in water, so for transport in the blood, bilirubin is loosely bound to albumin. In the liver cells, bilirubin is converted into water-soluble form **bilirubin diglucuronide** by conjugating with glucuronic acid **(Fig. 15.4)**.

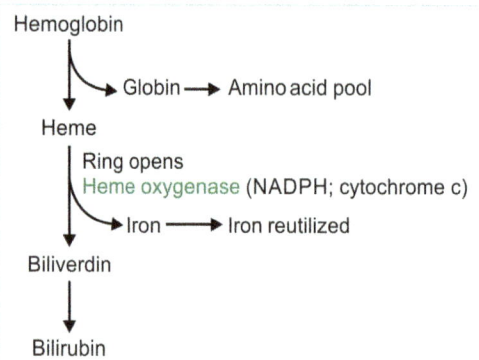

**Fig. 15.3:** Heme is broken down to bilirubin. Heme oxygenase system requires nicotinamide adenine dinucleotide phosphate hydrogen (NADPH) and molecular oxygen.

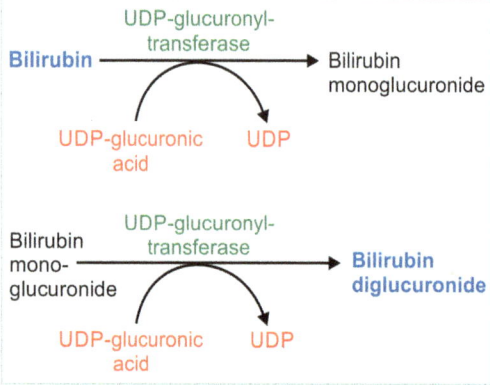

**Fig. 15.4:** Formation of conjugated bilirubin. (UDP: uridine diphosphate)

The conjugated bilirubin (bilirubin diglucuronide) is secreted into bile by active transport. Bilirubin is acted upon by the intestinal bacteria, deconjugated, and reduced to **urobilinogen**. A major portion of urobilinogen is reduced to stercobilinogen and is excreted through feces.

A part of urobilinogen is absorbed from large intestine into blood; this reaches again into liver and is re-excreted to bile. This is called **enterohepatic circulation** of bile pigments. A small part of this urobilinogen is taken up from blood by the kidney and is excreted through urine **(Fig. 15.5)**.

## Clinical Significance of Bilirubin

Normal plasma bilirubin level is less than 1 mg/dL (range 0.2–0.8 mg/dL). The unconjugated bilirubin is about 0.2–0.6 mg/dL; the rest will be conjugated bilirubin. Properties of these two types of bilirubins are shown in **Table 15.1**.

When the total bilirubin level exceeds 1 mg/dL, then the condition is called **hyperbilirubinemia**. This may be due to (i) production of more bilirubin than the liver can excrete, (ii) failure of damaged liver to excrete bilirubin produced in normal amounts, or (iii) block in excretion of bile.

In all these cases, when bilirubin concentration exceeds 2 mg/dL, it diffuses back into the tissues to give yellow coloration to tissues especially sclera, conjunctiva of eye, skin, and mucous membrane. This condition is called **jaundice**. Jaundice may be of three types depending on the cause:

1. **Hemolytic jaundice (pre-hepatic jaundice):** This is due to excessive destruction of RBC. Common causes are (a) congenital spherocytosis, (b) autoimmune hemolytic anemias, and (c) toxins. Since the cause is increased production of bilirubin and not an abnormality in conjugation or excretion, it is characterized by the increase in **unconjugated bilirubin.**
2. **Hepatic jaundice (hepatocellular jaundice):** This is caused by damage to liver cells. Damage may be due to viral hepatitis or due to poisons like carbon tetrachloride, chloroform, and so on. Increased unconjugated bilirubin in blood.
3. **Obstructive jaundice (post-hepatic jaundice):** This is caused by an obstruction to bile duct either by bile stones or by cancer. **Conjugated bilirubin** in blood is increased, and it is excreted through urine. Urobilinogen in urine is decreased or absent. Since no pigments are entering into the gut, the feces become clay colored **(Table 15.2)**. These three types of jaundices are compared in **Figure 15.6**.

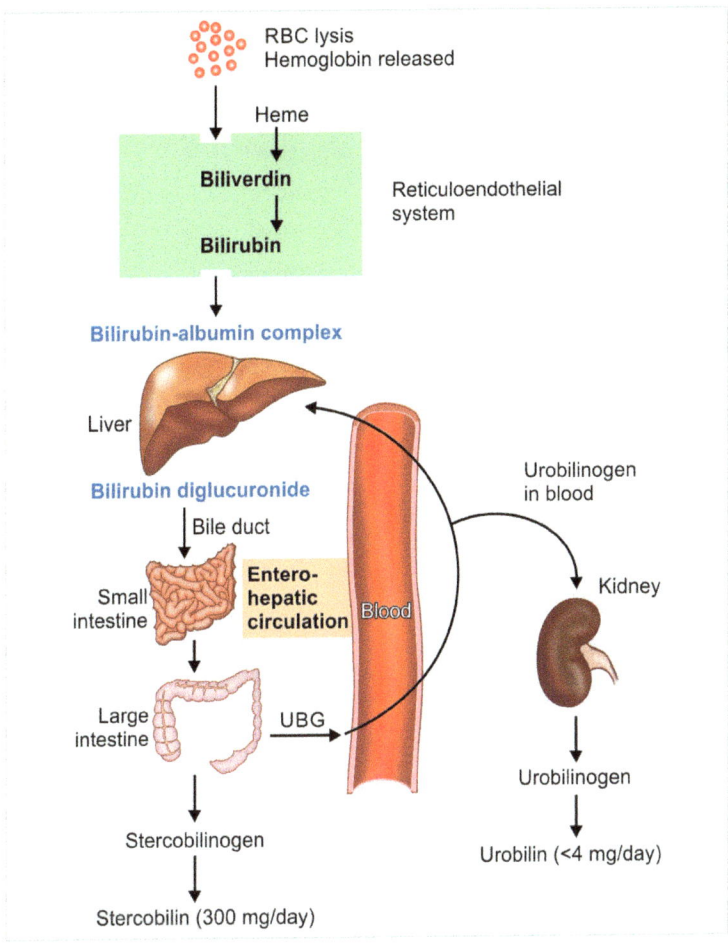

**Fig. 15.5:** Production and excretion of bilirubin.
(RBC: red blood cell; UBG: urobilinogen)

**Table 15.1:** Properties of conjugated and free bilirubin.

| | Free bilirubin | Conjugated bilirubin |
|---|---|---|
| In water | Insoluble | Soluble |
| In alcohol | Soluble | Soluble |
| Normal plasma level | 0.2–0.6 mg/dL | 0–0.2 mg/dL |
| In bile | Absent | Present |
| In urine | Always absent | Normally absent |
| Van den Bergh test | Indirect positive | Direct positive |

In clinical studies of jaundice, bilirubin is quantitatively estimated by **van den Bergh test**. This test is based on coupling of diazotized sulfanilic acid and bilirubin to produce reddish purple compound. **Ehrlich's test** is done to detect urobilinogen in urine and **Fouchet's test** to detect bilirubin in urine.

Other causes of hyperbilirubinemia:
- **Physiological jaundice:** Seen in newborns. Results from excessive hemolysis. This disappears quickly, but if this condition prolongs, then the bilirubin may cross the blood–brain barrier and get deposited in the brain, leading to a condition called Kernicterus.

**Table 15.2:** Differences between three types of jaundices.

|  | Hemolytic jaundice | Hepatocelluar jaundice | Obstructive jaundice |
|---|---|---|---|
| Other name | Pre-hepatic | Hepatic | Post-hepatic |
| Cause | Hemolysis | Infection | Stone in bile duct |
| Blood, free bilirubin | Increased | Increased | Normal |
| Blood, conjugated bilirubin | Normal | Increased | Increased |
| Blood, alkaline phosphatase | Normal | Increased | Very high |
| Urine, conjugated bilirubin | Nil | Nil | Present |
| Urine, urobilinogen | Increased | Nil | Nil |
| Feces | Normal | Normal | Clay colored |

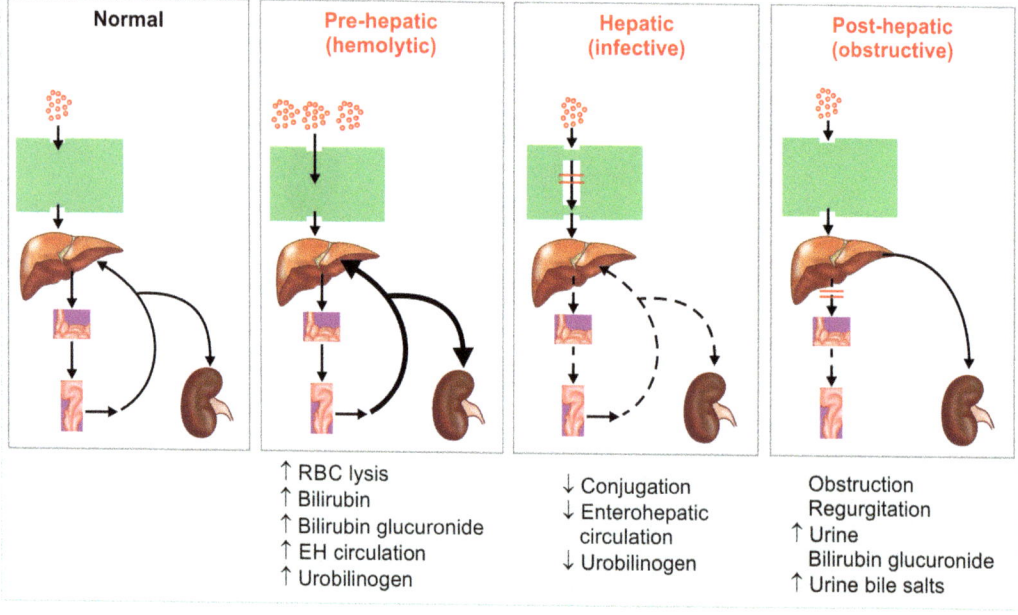

**Fig. 15.6:** Comparison of different types of jaundice.

- **Crigler–Najjar syndrome:** Due to metabolic defect in conjugation, that is, absence of enzyme UDP glucuronyl transferase.
- **Gilbert syndrome:** Defect in the uptake of bilirubin by liver cells.
- **Dubin–Johnson syndrome:** Defect in the secretion of conjugated bilirubin into the bile.

# HEMOGLOBIN

Hemoglobin has four polypeptide chains associated with noncovalent bonds. $HbA_1$, which is the major hemoglobin found in RBCs in adults, has two identical alpha chains and two identical beta chains, that is, $\alpha_2, \beta_2$. Each peptide chain has one heme and one $Fe^{++}$. Each heme can combine with one molecule of

$O_2$. Thus, one hemoglobin molecule can carry four molecules of $O_2$ or eight atoms of oxygen. The other forms of hemoglobin are $HbA_2$ ($\alpha_2$, $\delta_2$) (two alpha, two delta) and HbF ($\alpha_2$, $\gamma_2$) (two alpha, two gamma). In the circulation of a normal adult, $HbA_1$ constitutes 95%, $HbA_2$ has 4.5%, and HbF concentration is 0.5%.

## Modified Hemoglobin

When glucose concentration increases in blood, $HbA_1$ and glucose undergo nonenzymatic reaction to form **glycated hemoglobin** ($HbA_{1c}$). Normally this fraction is less than 5.5% of total hemoglobin, but in diabetic patients, the amount of $HbA_{1c}$ is increased proportionate to the average blood glucose concentration. In untreated diabetes mellitus, the $HbA_{1c}$ is more than 6.5% of total hemoglobin. Please refer to Chapter 6 for further details on $HbA_{1c}$.

## HEMOGLOBIN DERIVATIVES

### Carboxyhemoglobin (Carbon Monoxy-Hb)

Hemoglobin binds with carbon monoxide (CO) to form carboxy-Hb. The affinity of CO to Hb is 200 times more than that of oxygen. It is then unsuitable for oxygen transport. CO poisoning is a major occupational hazard for workers in mines. Breathing the automobile exhaust in closed space is the commonest cause for CO poisoning. One cigarette liberates 10–20 mL carbon monoxide into the lungs. Clinical symptoms manifest when carboxy-Hb levels exceed 20%. Symptoms are breathlessness, headache, nausea, vomiting, and pain in the chest. At 40–60% saturation, death can result. Treatment is to administer oxygen.

### Methemoglobin

When the ferrous ($Fe^{++}$) iron is oxidized to ferric ($Fe^{+++}$) state, methemoglobin (met-Hb) is formed. Met-Hb cannot release oxygen and therefore it can no longer function. Small quantities of met-Hb formed in the RBCs are readily reduced back to the ferrous state by met-Hb reductase enzyme systems using nicotinamide adenine dinucleotide hydrogen (NADH) and nicotinamide adenine dinucleotide phosphate hydrogen (NADPH).

## ABNORMAL HEMOGLOBINS (HEMOGLOBIN VARIANTS OR HEMOGLOBINOPATHIES)

Alterations in the amino acid sequence of polypeptide chain can produce altered conformation and hence abnormal function and diseases. These clinical conditions are together called **hemoglobinopathies**.

## Sickle Cell Hemoglobin (or HbS)

Sickle cell hemoglobin is made up of $\alpha_2$ and $\beta_2$ chains. There is no change in the amino acid sequence in the $\alpha$ chain, but in the $\beta$ chain, the amino acid at the sixth position (normal glutamic acid) is replaced by valine in HbS. Due to this substitution, a sticky patch is generated on the surface of $\beta$ chain, leading to the polymerization of HbS to form fibrous precipitates. This leads to lysis of RBC and hence anemia.

## HbM

In this variant, there is substitution of histidine residues of HbA by tyrosine to form HbM. This substitution forms a very tight complex with ferric iron and leads to an increase in amount of met-Hb (oxidized form of Hb).

## HbC (Cooley's Anemia)

In this variant, glutamic acid in the sixth position of normal $\beta$ chain is replaced by lysine. This variant also produces sickling.

## HbD

It results from the replacement of $\beta_{121}$ glutamic acid by glutamine. It does not produce sickling. However, HbSD is a severe condition, where HbD copolymerizes with HbS and produces sickling.

### HbE

This variant results from the replacement of $\beta_{26}$ glutamic acid by lysine.

### Thalassemia

Thalassemia is a group of inherited diseases characterized by severe anemia in early life, due to abnormal hemoglobin and premature destruction of RBCs.

Normally the rate of synthesis of α and β chains of Hb must be identical. In thalassemia, there is deficiency or absence of either α or β chains.

*Alpha Thalassemia*

In α-thalassemia, there is absence of α chains; so the Hb will have four β chains. In β-thalassemia, there is deficiency of β chains. The common form is β-thalassemia.

*Beta Thalassemia*

Beta thalassemia is more common than the alpha variety. Beta type is characterized by a decrease or absence of synthesis of beta chains. As a compensation, gamma or delta chain synthesis is increased.

*Thalassemia Syndromes*

All cases of thalassemias are characterized by the deficit of HbA synthesis. Hypochromic microcytic anemia is seen. In homozygous state, clinical manifestations are severe and hence are called **thalassemia major**, for example, Cooley's anemia. In heterozygous conditions, the clinical signs and symptoms are minimal; they are called **thalassemia minor**.

## TRANSPORT OF OXYGEN BY HEMOGLOBIN

The main function of hemoglobin is carriage of oxygen from lungs to tissues. There are four heme residues per Hb molecule, one for each subunit in Hb. The iron atom of heme occupies the central position of the porphyrin ring. The reduced state is called ferrous ($Fe^{++}$) and the oxidized state is ferric ($Fe^{+++}$). In hemoglobin, iron always remains in the ferrous state. The oxygen atom directly binds to the iron atom.

In the lungs, when Hb takes up oxygen, then it is called **oxygenated hemoglobin**. Even in the presence of oxygen, the iron is in the ferrous (reduced) state. Only in reduced state, Hb can take up oxygen in lungs and release oxygen into tissues. In the rare condition, when iron is oxidized to ferric state, then it is called met-hemoglobin (met-Hb or oxidized Hb); it cannot release oxygen and is therefore useless physiologically. Please note that oxygenation and oxidation are different. In the peripheral tissues, when oxygen is released from Hb, then it is called **deoxy-hemoglobin (Fig. 15.7)**.

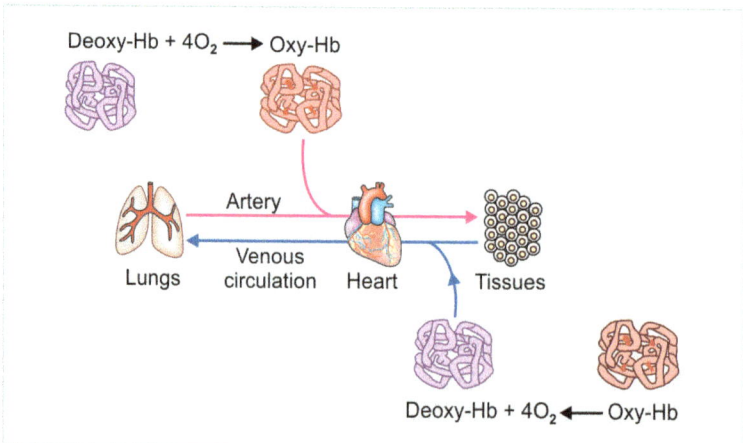

**Fig. 15.7:** In tissues, oxy-Hb releases oxygen to become deoxy-Hb. In lungs, the reverse takes place.

The special feature of hemoglobin is its ability to load and unload oxygen at physiological $pO_2$ (partial pressure of oxygen).
- At the oxygen tension in the pulmonary alveoli, the Hb is **97% saturated** with oxygen. Normal blood with 15 g/dL of Hb can carry 20 mL of $O_2$ per deciliter of blood.
- In the tissue capillaries, where $pO_2$ is only 40 mm of Hg, the Hb is about **60% saturated**. So physiologically, 40% of oxygen is released (**Fig. 15.8**).

## TRANSPORT OF CARBON DIOXIDE

At rest, about 200 mL of $CO_2$ is produced per minute in tissues. The $CO_2$ is carried by the following ways:

### Dissolved Form

About 10% of $CO_2$ is transported as dissolved form.

$$CO_2 + H_2O \rightarrow H_2CO_3 \rightarrow HCO_3^- + H^+$$

The hydrogen ions thus generated are buffered by the buffer systems of plasma.

### Isohydric Transport of Carbon Dioxide

Isohydric transport constitutes about 75% of the transported form of $CO_2$. It means that there is minimum change in pH during the transport. The $H^+$ ions liberated in the previous reaction are buffered by the deoxy-Hb.

### Carriage as Carbaminohemoglobin

The rest 15% of $CO_2$ is carried as carbaminohemoglobin, without much change in pH. A fraction of $CO_2$ that enters into the red cell is bound to Hb as a carbamino complex.

$$R\text{-}NH_2 + CO_2 \dashrightarrow R\text{-}NH\text{-}COOH$$

## SUMMARY OF THE CHAPTER

- Hemoglobin—the oxygen transporter—is a conjugated protein with heme as prosthetic group and globin polypeptide. It is a tetramer with two alpha and two beta chains.
- Heme is a derivative of porphyrin, which is a cyclic compound formed by the fusion of four pyrrole rings linked by methenyl bridges and has an iron atom at its center.
- Regulation of heme synthesis is by repression of ALA synthase by heme. Glucose prevents ALA synthase induction. Barbiturates induce the enzyme. ALA synthase is also allosterically inhibited by hematin.
- Porphyrias are the class of metabolic disorders associated with heme synthesis. Acute intermittent porphyria (AIP) occurs due to deficiency of porphobilinogen (PBG) deaminase. Congenital erythropoietic porphyria is due to imbalance in activities of uroporphyrinogen I synthase and cosynthase. Acquired porphyrias result from lead poisoning.
- Normal plasma bilirubin levels range from 0.2 to 1.0 mg/dL and can be detected by van den Bergh test.
- Congenital hyperbilirubinemias include Gilbert's disease, where bilirubin uptake is defective; Criggler–Najjar syndrome, where conjugation is defective; and Dubin–Johnson syndrome, where defect is in the excretion of conjugated bilirubin.
- Acquired hyperbilirubinemias or jaundice may be physiological jaundice, hemolytic

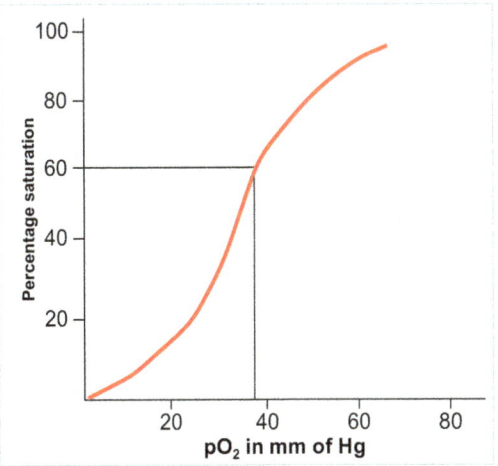

**Fig. 15.8:** Oxygen dissociation curve (ODC).

jaundice, hepatocellular jaundice, and obstructive jaundice.
- Hemoglobin (Hb) is a globular protein containing two alpha two beta (HbA), two alpha two gamma (HbF), and two alpha two delta (HbA2).
- The ability of Hb to load and unload $O_2$ at physiological $pO_2$ is shown by the $O_2$ dissociation curve and has a sigmoid shape.
- HbS is sickle cell Hb, resulting from a Glu–Val substitution at the sixth position on the beta chain.
- Thalassemia is caused due to the functional abnormality of alpha or beta chains of Hb. Homozygous states exhibit thalassemia major, while heterozygous states exhibit thalassemia minor.
- Arterial blood is 97% saturated at $pO_2$ of 100 mm of Hg and can carry about 40 mm of Hg. In venous blood, where $pO_2$ is about 40 mm of Hg, Hb is only 60% saturated. Net effect is a release of 40% $O_2$ to tissues.
- Transport of $CO_2$ by hemoglobin is called isohydric transport, which helps in buffering and the elimination of the volatile acid carbonic acid. In the tissues where $pCO_2$ is high, $CO_2$ diffuses into RBCs and combines with $H_2O$ to form carbonic acid catalyzed by carbonic anhydrase. The $H^+$ formed by the dissociation of $H_2CO_3$ is buffered by HHb, which can take up $H^+$ when oxygen dissociates.
- A fraction of $CO_2$ is transported as carbamino-Hb.
- In anemic children and children with beta chain defects, HbF level remains high as a compensatory mechanism.

## QUESTIONS FOR SELF-ASSESSMENT

1. Mention the steps of heme synthesis. Mention the rate-limiting step.
2. Describe heme catabolism.
3. Define porphyrias.
4. What are hemoglobinopathies.
5. Explain the role of bilirubin in our body.
6. Explain the characteristic features of the oxygen dissociation curve.
7. What are different types of thalessemias?
8. What are the clinical consequences of hyperbilirubinemias?
9. Mention the functions of hemoglobin.

## MULTIPLE CHOICE QUESTIONS (MCQs)

15-1. Rate-limiting enzyme in heme synthesis is:
  a. Heme synthase
  b. ALA dehydrase
  c. Uroporphyrinogen synthase
  d. ALA synthase

15-2. A patient with acute intermittent porphyria should be given:
  a. High-carbohydrate diet
  b. Barbiturates
  c. Aspirin
  d. Saline infusion

15-3. All the following manifestations are seen in sickle cell anemia (HbS disease), *except:*
  a. Pain and swellings in joints
  b. Hemolytic anemia
  c. Sickled cells in peripheral circulation
  d. Inclusion bodies in RBCs

15-4. Methemoglobinemia is found in all the following conditions, *except:*
  a. Ingestion of nitrites
  b. Carbon monoxide poisoning
  c. Presence of HbM
  d. Poisoning by aniline dyes

15-5. All of the following are features of isohydric transport, *except:*
  a. $CO_2$ is transported in plasma as bicarbonate
  b. Carbonic anhydrase is the major enzyme involved
  c. Chloride shift is a result of $CO_2$ transport
  d. $CO_2$ is bound to amino terminal group

15-6. The polypeptide chains of $HbA_1$ are:
  a. $\alpha_2\beta_2$
  b. $\alpha_2\gamma_2$
  c. $\alpha_2\delta_2$
  d. $\alpha_2\varepsilon_2$

15-7. Unconjugated hyperbilirubinemia results in:
   a. Dubin–Johnson syndrome
   b. Crigler–Najjar syndrome
   c. Rotor's syndrome
   d. Metabolic syndrome

15-8. The porphyria that lacks photosensitivity is:
   a. Acute intermittent porphyria
   b. Variegate porphyria
   c. Porphyria cutanea tarda
   d. Erythropoietic protoporphyria

15-9. In obstructive jaundice, prothrombin time:
   a. Remains normal
   b. Decreases
   c. Becomes normal when vitamin K is administered
   d. Increases when vitamin K is administered

15-10. Carbon dioxide is carried in blood with all the forms, *except:*
   a. Dissolved form
   b. Isohydric transport
   c. Carboxyhemoglobin
   d. Carbaminohemoglobin

## ANSWER KEYS TO MCQs

15-1. (d)   15-2. (a)   15-3. (d)   15-4. (b)   15-5. (d)   15-6. (a)   15-7. (a)   15-8. (c)
15-9. (c)   15-10. (c)

# CHAPTER 16

# Vitamins

At the completion of this chapter, the reader will be able to answer questions on the following topics:

- Vitamin A
- Vitamin D and calcitriol
- Vitamin E
- Vitamin K
- B complex group of vitamins
- Thiamine (vitamin $B_1$) and thiamine pyrophosphate
- Riboflavin (vitamin $B_2$) and flavin adenine dinucleotide (FAD)
- Niacin (nicotinic acid), nicotinamide adenine dinucleotide (NAD), and NADP
- Pyridoxine (vitamin $B_6$) and pyridoxal phosphate
- Pantothenic acid and coenzyme A
- Biotin
- Folic acid
- Vitamin $B_{12}$
- Vitamin C

## VITAMINS IN GENERAL

Vitamins may be defined as organic compounds occurring in small quantities in different natural foods and is necessary for the growth and maintenance of good health in human beings. In their absence in food different deficiency diseases may be developed. Vitamins are required for the proper utilization of carbohydrates, lipids, and proteins. These compounds are not related chemically; but they are considered as a group because of the similarity in their biological functions. Vitamins are generally divided into two major groups:

1. **Fat-soluble vitamins:** They are usually found associated with the lipids of natural foods. They are vitamins A, D, E, and K.
2. **Water-soluble vitamins:** This includes B complex group of vitamins and vitamin C. These two types of vitamins are compared in **Table 16.1**.

**Table 16.1:** Comparison of two types of vitamins.

| Criteria for comparison | Fat-soluble vitamins | Water-soluble vitamins |
|---|---|---|
| Solubility in fat | Soluble | Not soluble |
| Water solubility | Not soluble | Soluble |
| Absorption | Along with lipids require bile salts | *Absorption is simple |
| Carrier proteins | Present | *No carrier proteins |
| Storage | Stored in liver | *No storage |
| Deficiency | Manifest only when stores are depleted | *Manifest rapidly as there is no storage |
| Toxicity | Hypervitaminosis may result | Unlikely, since excess is excreted |
| Major vitamins | A, D, E, and K | B and C |

*Vitamin $B_{12}$ is an exception.

# VITAMIN A

## Chemistry

Vitamin A group is a common general structure made up of a β-ionone ring with polyisoprene side chain. Three different compounds with vitamin A activity are **retinol** (vitamin A alcohol), **retinal** (vitamin A aldehyde), and **retinoic acid** (vitamin A acid) **(Fig. 16.1)**. The retinal may be reduced to retinol by retinal reductase. This reaction is readily reversible. Retinal is oxidized to retinoic acid, which cannot be converted back to the other forms **(Fig. 16.2)**.

**Beta-carotene** is otherwise called provitamin A. Plant foods contain β-carotene, which can be oxidatively cleaved in the intestine to form two molecules of retinal.

## Absorption

Vitamin A is present in diet as retinyl esters. It is hydrolyzed in the intestinal mucosa to retinol and free fatty acid. Beta-carotene is cleaved by dioxygenase to form retinal, which is then reduced to retinol by retinal reductase. In the intestinal epithelium, this retinol is re-esterified with long-chain fatty acid and incorporated into chylomicrons, which enter into lymphatics as retinyl esters and stored in liver.

## Transport

- Retinol is released from liver and transported after binding to the specific transport protein called retinol-binding protein (RBP).
- Retinol is converted to retinal, which is essential for visual function, and retinoic acid for growth and epithelial function.

## Biological Role

Vitamin A (retinal) is a component of the visual pigments of rods and cones present in retina. So, vitamin A is essential for vision.

## Role of Vitamin A in Vision

*Rhodopsin Cycle/Wald's Visual Cycle*

Retinal is the prosthetic group of visual pigments like rhodopsin (11-cis retinal bound to the protein **opsin**). When light falls on retina, rhodopsin undergoes a series of photochemical isomerization. The 11-cis retinal group of visual pigments is isomerized by light into all-trans-retinal. Consequently, these pigments get hydrolyzed into all-trans-retinal and opsin. This photolysis of rhodopsin triggers a nerve impulse that is transmitted by the optic nerve to the brain. The 11-cis retinal is regenerated and forms rhodopsin after combining with opsin **(Fig. 16.3)**.

**Dark Adaptation Mechanism**

Bright light depletes stores of rhodopsin in the rods of retina. Therefore, when a person shifts suddenly from bright light to a dimly lit area, there is difficulty in seeing, for example, when entering a cinema theater. After a few minutes, rhodopsin is resynthesized and vision is improved. This period is called **dark adaptation time.** The dark adaptation time is increased in persons with vitamin A deficiency. Red light bleaches rhodopsin to a lesser extent; so doctors use red glasses during fluoroscopic X-ray examination of the patients.

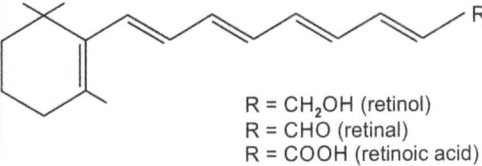

R = CH₂OH (retinol)
R = CHO (retinal)
R = COOH (retinoic acid)

**Fig. 16.1:** Structure of vitamin A.

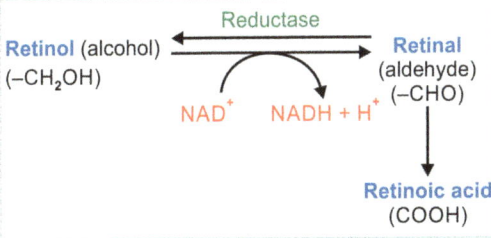

**Fig. 16.2:** Interconversion of vitamin A molecules. (NAD: nicotinamide adenine dinucleotide; NADH: nadh nicotinamide adenine dinucleotide hydride)

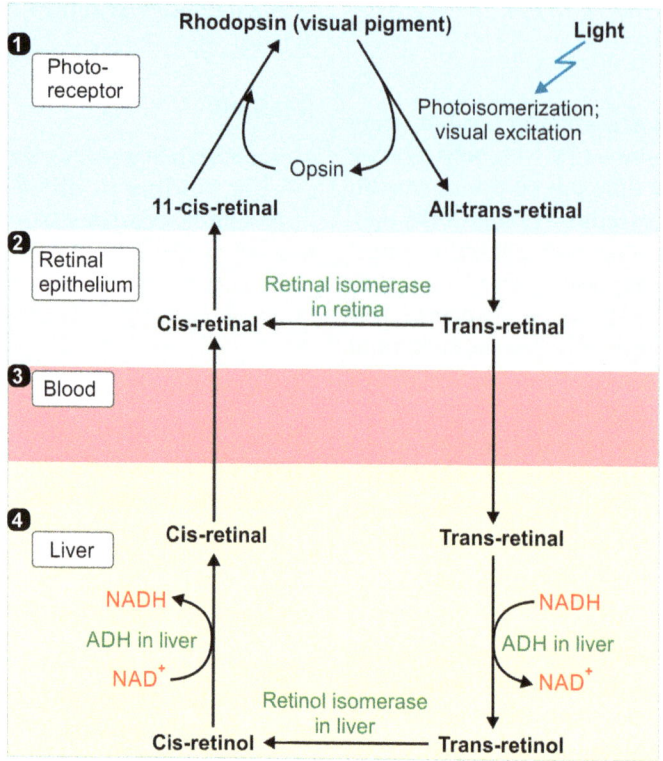

**Fig. 16.3:** Wald's visual cycle. (1) represents reactions in the photoreceptor matrix. (2) represents reactions in the retinal pigment epithelium. (3) depicts blood. (4) shows reactions taking place in the liver. (NAD: nicotinamide adenine dinucleotide; NADH: nicotinamide adenine dinucleotide hydride; ADH: alcohol dehydrogenase)

**Rods and Cones**

In the retina, there are two types of photosensitive cells, the rods and the cones. The rods are responsible for perception in dim light. Rhodopsin present in rods is made up of 11-cis-retinal + opsin. Deficiency of cis-retinal will lead to increase in dark adaptation time and night blindness.

Cones are responsible for vision in bright light as well as color vision. They contain the photosensitive protein called **conopsin**. There are three types of cones; each is characterized by a different conopsin that is maximally sensitive to either blue, green, or red. In the cone also, 11-cis-retinal is the chromophore. Reduction in the number of cones or the cone proteins will lead to **color blindness**.

## Other Biological Functions of Vitamin A

- Retinoic acid is required for normal growth and maintenance of epithelium of various tissues.
- Retinol is required for normal reproduction. It supports spermatogenesis in males. It prevents fetal resorption in females.
- β-carotene is a natural antioxidant. It is useful in preventing heart attacks and cancer.
- **Immunological system:** Retinoic acid influences the differentiation of T lymphocytes. When vitamin A is deficient in the diet, the population of hematopoietic stem cells decreases.
- **Effect on skin:** Vitamin A is necessary for the maintenance of normal epithelium

and skin. The alterations in skin may cause increased occurrence of generalized infection.

## Deficiency Manifestations of Vitamin A

- **Night blindness (nyctalopia):** Visual activity is diminished in dim light; the dark adaptation time is increased.
- **Bitot's spot:** Greyish-white triangular plaques (thickenings) adherent to the conjunctiva.
- **Xerophthalmia:** In severe vitamin A deficiency, dry, thick and wrinkled cornea and conjunctiva are seen. Microorganisms often infect these tissues.
- **Keratomalacia:** Softening and ulceration of cornea, which may lead to total blindness.
- **Keratinization (dry and rough epithelium):** Keratinization of epithelial tissue occurs in skin, respiratory tract, alimentary tract and genitourinary tract. Deterioration of epithelial surface leads to increased incidence of infectious diseases.
- **Acne** affects 85% of teenagers. Androgens and estrogens are associated with the acne. Acne may be a feature in many endocrine disorders, including polycystic ovary disease and Cushing syndrome. **Isotretinoin,** a synthetic variant of vitamin A, is known to reduce the sebaceous secretions, hence it is used to prevent acne formation during adolescence.
- Retarded growth and skeletal abnormalities.

## Sources of Vitamin A

Vitamin A is available in fish liver oils (cod liver oil and shark liver oil), egg yolk, cream, butter, milk. β-carotene is present in yellow and green fruits and vegetables, green leafy vegetables, carrot, papaya, mango, and so on.

## Recommended Dietary Allowance of Vitamin A

For adults, 750–1,000 μg (microgram) of vitamin A or 6,000 μg of β-carotene is recommended. Normal level of vitamin A in blood is 25–50 μg/dL.

## Hypervitaminosis A

Increased intake of vitamin A for prolonged periods will cause hypervitaminosis A. Anorexia, irritability, head ache, peeling of skin, drowsiness, and vomiting are the manifestations.

## VITAMIN D (CHOLECALCIFEROL)

### Chemical Structure

In the case of vitamin D, two forms are seen: (a) **Ergosterol** in plants is **provitamin** of vitamin $D_2$ (Provitamins are precursors of vitamins. They are converted into vitamins). (b) 7-dehydrocholesterol in the skin is the provitamin of vitamin $D_3$. These two provitamins are converted into the corresponding vitamin by sunlight.

### Sources of Vitamin D

Vitamin $D_2$ is found in plants and vitamin $D_3$ in animal tissues like fish, fish oil, egg yolk, and liver. Milk contains moderate quantity of the vitamin. The current recommendation is to fortify dairy products with vitamin D. Vitamin $D_2$ and $D_3$ are also formed when their precursors (provitamins) are exposed to sunlight. (a) Ergosterol, a plant sterol when irradiated is converted to vitamin $D_2$. Since $D_2$ is formed from ergosterol it is called **ergocalciferol**. (b) 7-dehydrocholesterol present in the epidermis of human skin is converted to vitamin $D_3$ by sunlight. Chemically $D_3$ is called **cholecalciferol**.

### Recommended Dietary Allowance of Vitamin D

The recommended daily allowance of vitamin D is (1) for children = 10 mg (400 IU)/day; (2) for adults = 5 mg (200 IU)/day; (c) during pregnancy and lactation = 10 mg/day; and for persons above the age of 60 = 600 IU per day.

### Absorption of Vitamin D

Being a fat-soluble vitamin, bite salts are needed for the absorption of vitamin D from food.

## Activation of Vitamin D

Vitamin $D_3$ (cholecalciferol) is first transported to liver, where hydroxylation at 25th position occurs to form 25-hydroxycholecalciferol (25-HCC). This reaction in liver requires the enzyme hepatic 25-hydroxylase, which contains cytochrome P-450 and nicotinamide adenine dinucleotide phosphate (NADPH) as coenzymes. The 25-HCC is the major transport and storage form. In plasma, it is bound to "vitamin D-- binding protein," an α-2 globulin.

In the kidney, it is further hydroxylated at first position. The 1-alpha-hydroxylase is located in proximal convoluted tubules. This needs NADPH as coenzyme. Thus 1,25-dihydroxycholecalciferol is generated. Since it contains three hydroxyl groups at positions 1, 3, and 25, it is also called **calcitriol,** which is the active form of vitamin D **(Fig. 16.4)**. Calcitriol acts as a hormone. Since calcitriol is generated from vitamin D, this vitamin is considered as a **pro-hormone.**

## Biochemical Effects of Vitamin D

Function of 1,25-dihydroxy vitamin $D_3$ (calcitriol) is to maintain an adequate serum calcium level. The sites of action are: (a) intestinal villi cells, (b) bone osteoblasts, and (c) distal tubular cells of kidney. These effects are explained in the following:

### Vitamin D and Intestinal Absorption of Calcium

**Calcitriol** promotes the absorption of calcium from the intestine. The mechanism of action is typical of steroid hormone. It enters the intestinal cell and binds to a cytoplasmic receptor. The calcitriol–receptor complex then moves to the nucleus, where it interacts with DNA. This results in the synthesis of messenger RNA (mRNA) and this in turn synthesizes a calcium-binding protein called **calbindin.** Due to the increased availability of the calcium-binding protein, the absorption of calcium is increased **(Fig. 16.5)**.

### Deficiency Manifestations of Vitamin D

Vitamin D deficiency causes rickets in children and osteomalacia in adults.

### Effect of Vitamin D on Bone

Mineralization of the bone is increased by increasing the activity of **osteoblasts**. Calcitriol stimulates osteoblasts, which secrete alkaline phosphatase, and also regulates the synthesis of calcium-binding protein called **osteocalcin** So calcium and phosphate ions in the bone increase. These ions enhance the mineral deposition in bones.

### Effect of Vitamin D on Renal Tubules

Calcitriol increases the reabsorption of calcium and phosphorous by renal tubules,

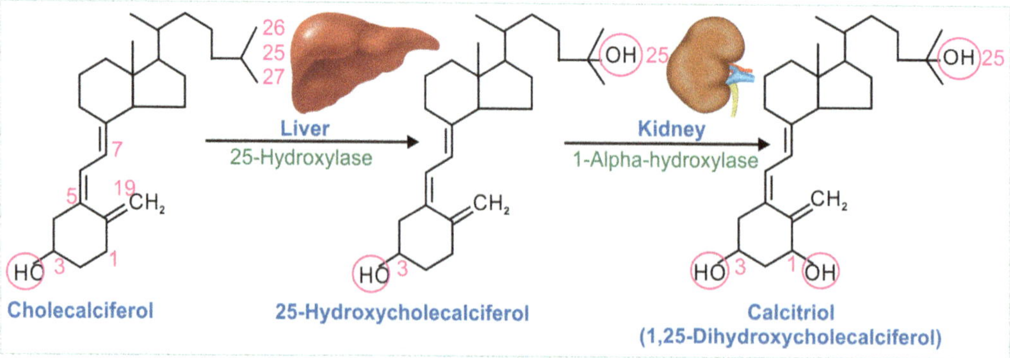

**Fig. 16.4:** Generation of calcitriol.

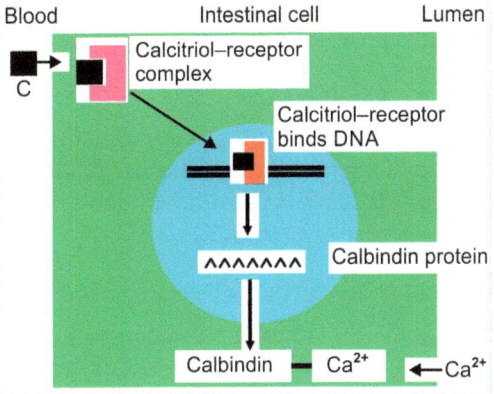

**Fig. 16.5:** Calcitriol increases calcium absorption. (DNA: deoxyribonucleic acid)

therefore, both minerals are conserved (PTH conserves only calcium). So, blood level of calcium is tend to increase under the influence of vitamin D.

## Serum Vitamin D

Reference level of serum 25(OH) D is 30 ng/mL. The level of 20-29 ng/mL is considered as insufficient, while the level 11-20 ng/mL is deficient and the level less than 10 ng/mL is the mark of severe vitamin D deficiency. Toxic level is more than 150 ng/mL.

## Vitamin D Deficiency

About 50-80% of elderly and 20-50% of children have vitamin D deficiency. The deficiency is more common in obese people, because vitamin D is stored in adipose tissue and not released for utilization.

### Causes for Vitamin D Deficiency

- Deficiency of vitamin D can occur in people who are not exposed to sunlight properly, for example, inhabitants of northern latitudes, in winter months, people who are bedridden for longer periods, or those who cover the whole body (e.g., with burka).
- Nutritional deficiency of calcium or phosphate may also produce similar clinical picture.
- Malabsorption of vitamin D (obstructive jaundice and steatorrhea). High phytate content in diet may also reduce the absorption of vitamin.
- Abnormality of vitamin D activation. Liver and renal diseases may retard the hydroxylation reactions.

**Rickets** is a disease of growing bone. In people with vitamin D deficiency, during the stage of bone growth, the deposition of minerals (calcification) in the newly formed matrix fails to occur, but matrix formation continues. This results in soft, easily bent bones. Deformities occur because soft bones cannot withstand the weight of the body. This results in bow legs, knock knees, pigeon breast, and rachitic rosary (a beaded appearance of the ribs). Rickets is characterized by **hypocalcemia** and hypophosphatemia; serum alkaline phosphatase level is also increased.

In people with **osteomalacia**, bones are softened due to insufficient mineralization and increased osteoporosis.

## Other Diseases Related with Vitamin D Deficiency

Low levels of vitamin D in pregnancy are associated with gestational diabetes, pre-eclampsia, and small size of infants. Therapeutic doses given to children with rickets have been found to correct the anemia, hypocellularity of the bone marrow, and increased susceptibility to infection. Vitamin D-deficient patients present a higher risk of cardiovascular disease than the general population. Various studies have shown the association between decreased dihydroxycholecalciferol (DHCC) concentrations in blood and the risks of cardiovascular disease, stroke, cancer, fractures, and mortality.

## Hypervitaminosis D

Because of the body's ability to store excess vitamin D, toxicity occurs from vitamin D overdosage. Symptoms include weakness, headache, nausea, and deposition of calcium in soft tissues and blood vessels of kidney.

## VITAMIN E

- Chemical name: α-tocopherol
- Sources: vegetable oils and nuts
- RDA: 10 mg/day

### Biochemical Role (Functions) of Vitamin E

It is an important natural **antioxidant**. As an antioxidant it:
- Prevents peroxidation of polyunsaturated fatty acids (PUFAs) in membrane lipids.
- Protects RBCs from hemolysis.
- Decreases oxidation of low-density lipoproteins (LDLs), thereby reduces the risk of myocardial infarction.
- Slows down aging process and delays the onset of cataract.
- Acts in cooperation with selenium to minimize oxidative degeneration of tissues.

### Deficiency Manifestations of Vitamin E

In adults deficiency is rare. In premature infants RBCs become susceptible to lysis by peroxides, and hemolysis may occur in the absence of vitamin E.

### Recommended Dietary Allowance of Vitamin E

RDA for males is 10 mg/day, for females 8 mg/day. The requirement increases with higher intake of PUFAs.

### Sources of Vitamin E

**Vegetable oils** are rich sources of vitamin E, for example, wheat germ oil, sunflower oil, safflower oil, and cottonseed oil. Fish liver oils are devoid of vitamin E.

## VITAMIN K

### Chemical Structure

They are naphthoquinone derivatives. Vitamin $K_1$ (phylloquinone) and vitamin $K_2$ (menaquinone) are naturally occurring; while menadione (vitamin $K_3$) is the synthetic vitamin.

### Sources of Vitamin K

Green leafy vegetables are good sources. Intestinal bacteria can synthesize vitamin K, and so deficiency is very uncommon. (Milk is a poor source of vitamin K).

### Recommended Dietary Allowance of Vitamin K

For adults, 50–100 µg/day is recommended.

### Functions of Vitamin K

It is necessary for **blood clotting**. Vitamin K is required for the synthesis of **prothrombin**, and other blood-clotting factors II, VII, IX, and X. These proteins are synthesized as inactive precursors and require vitamin K-dependent carboxylation to form mature clotting factors.

### Deficiency Manifestations of Vitamin K

As intestinal bacteria synthesize vitamin K, deficiency manifestations are rare in adults. However, hemorrhagic disease in premature infants may be seen. This is manifested as profuse bleeding after minor injuries. Delayed clotting time and prolonged prothrombin time are the laboratory findings.

### Antagonists of Vitamin K

Warfarin and dicumarol are vitamin K antagonists. They are used as anticoagulants for treating clot formation.

A summary of fat-soluble vitamins are shown in **Table 16.2**.

## WATER-SOLUBLE VITAMINS

Water-soluble vitamins include vitamin C and members of the vitamin B complex group, namely, thiamine, riboflavin, niacin, pyridoxine, biotin, folic acid, pantothenic acid, and cobalamins. The major function of these vitamins is to serve as coenzymes for the enzyme systems of intermediary metabolism **(Table 16.3)**.

**Table 16.2:** Summary of fat-soluble vitamins.

| Name of vitamin | Major function | Deficiency disease/symptoms | Requirement per day or recommended daily allowance (RDA) |
|---|---|---|---|
| Vitamin A | Visual cycle | Night blindness, xerophthalmia | 1 mg |
| Vitamin D | Calcium absorption, mineralization | Rickets, osteomalacia | 10 µg |
| Vitamin E | Antioxidant | Hemolytic anemia in newborns | 10 mg |
| Vitamin K | Clotting | Hemorrhage, delay in clotting | 60 µg |

**Table 16.3:** B complex vitamins and their coenzymes.

| | B complex vitamins | Coenzyme form |
|---|---|---|
| 1 | Thiamine (vitamin $B_1$) | Thiamine pyrophosphate (TPP) |
| 2 | Riboflavin (vitamin $B_2$) | Flavin mononucleotide (FMN) and flavin adenine dinucleotide (FAD) |
| 3 | Nicotinic acid (niacin) | Nicotinamide adenine dinucleotide (NAD) |
| 4 | Pyridoxine (vitamin $B_6$) | Pyridoxal phosphate (PLP) |
| 5 | Pantothenic acid | Coenzyme A (CoA-SH or CoA) |
| 6 | Biotin | Biotin |
| 7 | Folic acid | Tetrahydrofolic acid (FH4) |
| 8 | Cyanocobalamin (vitamin $B_{12}$) | Methylcobalamin, 5-deoxyadenosyl cobalamin |

# Thiamine (Vitamin $B_1$) (Anti-beriberi Factor; Aneurin)

*Structure of Vitamin $B_1$*

It contains pyrimidine ring and a thiazole ring held by a methylene bridge. It is a sulfur-containing vitamin. In older literature, it was called as anti-beriberi factor.

## Source of Vitamin $B_1$

Whole grains are good sources (polishing of grains removes about 80% of thiamine). Other sources are oilseeds, nuts, and meat.

*Biological Functions of Vitamin $B_1$*

The coenzyme thiamine pyrophosphate (TPP) is formed from vitamin $B_1$. This formation needs energy in the form of adenosine triphosphate (ATP).

**Reactions that Involve Thiamine Pyrophosphate**

**Oxidative decarboxylation:** TPP is required for the conversion of pyruvate to acetyl-CoA by the pyruvate dehydrogenase enzyme. It is also required for the reaction of α-ketoglutarate to succinyl-CoA by α-ketoglutarate dehydrogenase (α-KGDH) complex.

$$\text{Pyruvate} \xrightarrow[\text{TPP}]{\text{PDH complex}} \text{Acetyl-CoA}$$

$$\text{α-ketoglutarate} \xrightarrow[\text{TPP}]{\text{α-KGDH}} \text{Succinyl-CoA}$$

TPP is also the coenzyme for transketolase, an enzyme in the hexose monophosphate (HMP) shunt pathway (*see* Chapter 5).

Ribose-5-(P) + Xylulose-5-(P) $\xrightarrow[\text{TPP}]{\text{Transketolase}}$ Sedoheptulose-7-(P) + Glyceraldehyde-3-(P)

*Recommended Dietary Allowance of Thiamine*

- Children: 1.2 mg/day
- Adults: 1.5 mg/day and increases during pregnancy and lactation. The requirement also increases with increased carbohydrate intake.

*Deficiency of Thiamine*

Deficiency of thiamine leads to beriberi. Symptoms include anorexia (loss of appetite), weakness, nausea, and so on. Beriberi may be manifested as

**Dry beriberi:** Here central nervous system manifestations are predominant. There will be peripheral neuritis and muscle wasting

**Wet beriberi:** Cardiovascular manifestations are predominant, leading to edema of legs and face, palpitation, and tachycardia.

**Wernicke-Korsakoff syndrome** (cerebral beriberi), which includes symptoms such as hallucinations.

**Polyneuritis:** It is common in chronic alcoholics. Alcohol utilization needs large doses of thiamine. Alcohol inhibits intestinal absorption of thiamine, leading to thiamine deficiency.

## Riboflavin (Vitamin B$_2$)

*Chemistry/Structure*

Riboflavin has the isoalloxazine ring to which a ribitol (a carbohydrate) is attached.

*Sources of Vitamin B$_2$*

Milk, milk products, egg, liver, and meat are good sources. Cereals, fruits, and vegetables are moderate sources.

*Biological Functions*

Riboflavin is important in energy metabolism. Flavin mononucleotide (**FMN**) and flavin adenine dinucleotide (**FAD**) are the two coenzyme forms of riboflavin (**Fig. 16.6**). Riboflavin is converted to FMN and FAD with the help of ATP. FAD and FMN participate in many oxidation–reduction reactions of intermediary metabolism.

Flavin adenine dinucleotide is involved in pyruvate dehydrogenase, α-ketoglutarate dehydrogenase, succinate dehydrogenase, and fatty acyl CoA dehydrogenase, and so on.

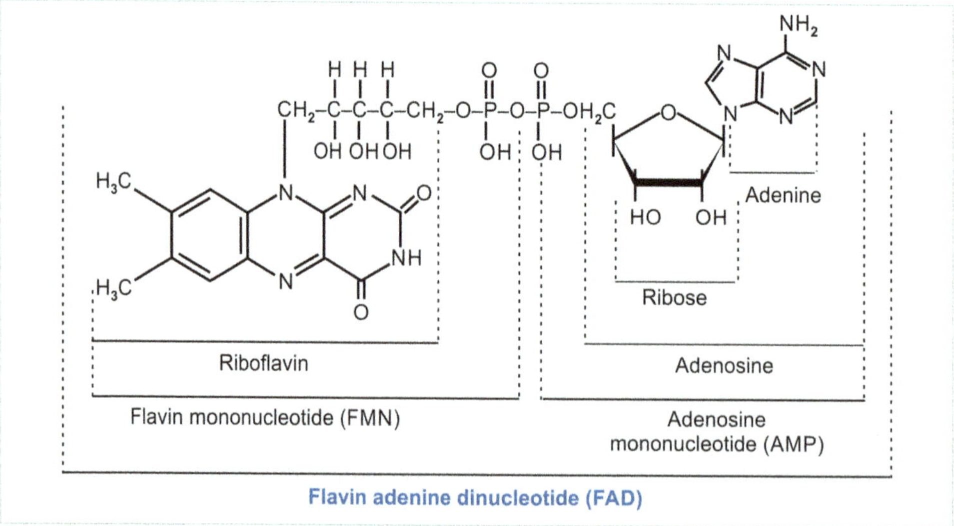

**Fig. 16.6:** Coenzymes FMN and FAD.

Flavin mononucleotide is involved in the electron transport chain, which is required to trap energy in the form of ATP. FMN-dependent enzyme is L-amino acid oxidase.

*Recommended Dietary Allowance of Vitamin $B_2$*

About 2 mg/day.

*Symptoms of Vitamin $B_2$ Deficiency*

Deficiency is uncommon because riboflavin is synthesized by intestinal bacterial flora. Deficiency affects epithelial tissues, and symptoms include **cheilosis** (fissuring at the corners of mouth), **circumcorneal vascularization,** and **glossitis** (magenta-colored tongue).

## Niacin (Nicotinic Acid)

It is also known as pellagra preventing factor.

*Chemistry/Structure*

It contains a pyridine ring (pyridine-3-carboxylic acid). The amide form is niacinamide (nicotinamide; **Fig. 16.7**).

*Sources of Niacin*

Liver, dried yeast, whole grains, milk, fish, legumes, and peanuts. It can be synthesized in the body from tryptophan (60 mg of tryptophan is equivalent to 1 mg of niacin)

*Biological Functions of Niacin*

- Nicotinamide adenine dinucleotide (NAD$^+$) and nicotinamide adenine dinucleotide phosphate (NADP$^+$) are the active forms of this vitamin **(Fig. 16.8)**. NAD$^+$ and NADP$^+$

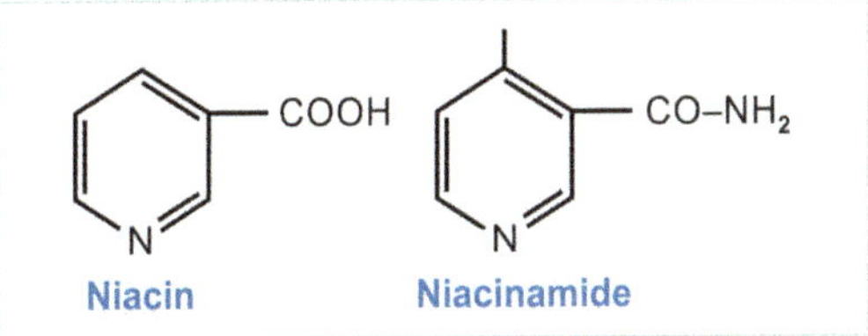

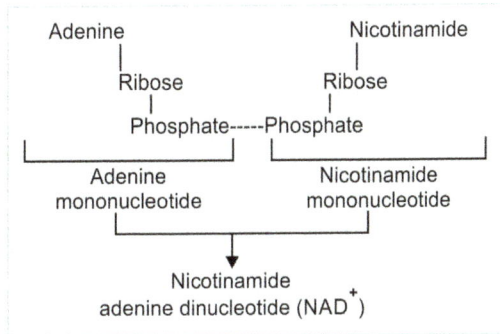

**Fig. 16.8:** Structure of NAD$^+$.

are vital for oxidation–reduction reactions catalyzed by dehydrogenases.
- NAD$^+$-dependent enzymes are lactate dehydrogenase, pyruvate dehydrogenase, alcohol dehydrogenase, and branched-chain α-keto acid dehydrogenase.
- NADP$^+$-dependent enzyme is glucose-6-phosphate dehydrogenase.
- NADPH is involved in reductive biosynthesis. NADPH-dependent enzymes are HMG-CoA reductase, dihydrofolate reductase, and phenylalanine hydroxylase.

*Recommended Dietary Allowance of Niacin*

About 20 mg/day; the requirement is linked to the total energy expenditure of the individual.

*Dietary Sources of Niacin*

The richest natural sources of niacin are dried yeast, unpolished rice, liver, peanut, whole cereals, legumes, meat, and fish. About half of the requirement is met by the conversion of tryptophan to niacin. About 60 mg of tryptophan will yield 1 mg of niacin.

*Causes for Niacin Deficiency*

**Dietary deficiency of tryptophan:** Deficiency is seen among people whose staple diet is maize and **sorghum** (jowar) as in Central and Western India.

**Deficient synthesis:** Kynureninase, an important enzyme in the pathway of tryptophan, is pyridoxal phosphate dependent.

**Fig.16.7:** Structure of niacin and niacinamide.

So conversion of tryptophan to niacin is not possible in pyridoxal deficiency.

**Isoniazid (isonicotinylhydrazide or INH):** It is an anti-tuberculosis drug, which inhibits pyridoxal phosphate formation. Hence there is block in the conversion of tryptophan to $NAD^+$.

**Carcinoid syndrome:** The tumor utilizes major portion of available tryptophan for the synthesis of serotonin; so tryptophan is unavailable.

*Deficiency Manifestations of Niacin*

Deficiency of niacin leads to **pellagra**, which involves skin, gastrointestinal tract, and central nervous system. Symptoms include dermatitis, diarrhea, and dementia.

## Pyridoxine (Vitamin $B_6$)

*Chemistry/Structure*

Pyridoxine exists in three states: pyridoxine (alcohol), pyridoxal (aldehyde), and pyridoxamine (amine). The active form of vitamin $B_6$ is pyridoxal phosphate (PLP) **(Fig. 16.9)**. It is synthesized by pyridoxal kinase utilizing ATP.

*Sources of Pyridoxine*

Whole grains, dried yeast, egg, milk, meat, fish, and green leafy vegetables.

*Biological Functions of Pyridoxine*

Pyridoxal phosphate (PLP) participates in essential enzymatic reactions of amino acid catabolism:

- **Transamination:** These reactions are catalyzed by amino transferases (transaminases), which employ PLP as the coenzyme. For example,
  alanine + alpha ketoglutarate →
  pyruvate + glutamic acid
  (enzyme alanine transaminase).
- **Decarboxylation:** All decarboxylation reactions of amino acids require PLP as coenzyme, for example, glutamate → GABA (gamma aminobutyric acid), which is an inhibitory neurotransmitter. Hence, in $B_6$ deficiency, especially in children, **convulsions** may occur.
- PLP is required for the synthesis of aminolevulinic acid (ALA), the precursor for heme synthesis (*see* Chapter 15).
- It is required for the synthesis of niacin from tryptophan (*see* Chapter 11).

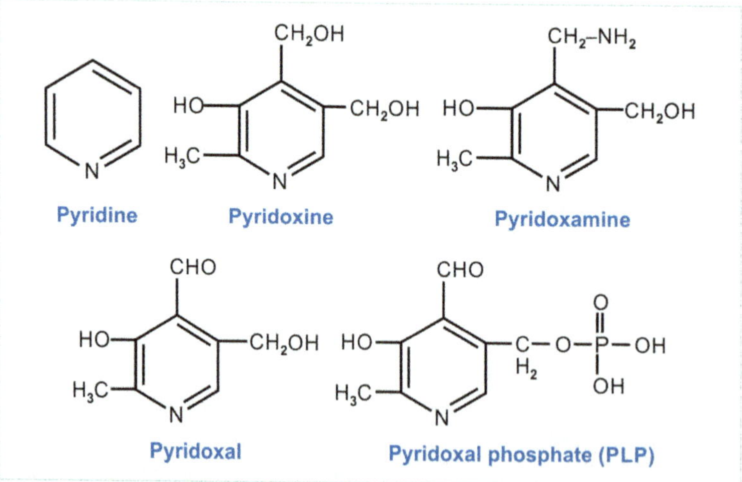

**Fig. 16.9:** Structure of $B_6$-related compounds.

- Glycogen phosphorylase enzyme requires PLP (*see* Chapter 5).

*Recommended Dietary Allowance of Pyridoxine*

Recommended dietary allowance (RDA) is 1-2 mg/day. This requirement is increased with increased protein intake and also during pregnancy, lactation, or old age.

*Deficiency Manifestations of Pyridoxine*

Pyridoxine is required for the synthesis of neurotransmitters like serotonin, GABA, and epinephrine. Deficiency of pyridoxine leads to decreased synthesis of these neurotransmitters. This leads to symptoms like convulsion, depression, irritability, and peripheral neuritis. Moreover, symptoms of pellagra are also seen in $B_6$ deficiency due to the defect in the synthesis of niacin. Hypochromic microcytic anemia may occur, due to inhibition of heme synthesis.

## PANTOTHENIC ACID

## Chemistry/Structure

It consists of two components—pantoic acid and β-alanine held together by a peptide linkage.

*Sources of Pantothenic Acid*

It is widely distributed in plants and animals. Egg, liver, meat, milk, vegetables, and grains are good sources.

*Biological Functions of Pantothenic Acid*

The coenzyme form of pantothenic acid is coenzyme A (CoA), which is involved in:
- Pyruvate + CoA + $NAD^+$ →
  Acetyl-CoA + $CO_2$ + NADH
  (Enzyme is pyruvate dehydrogenase)
- Alpha-ketoglutarate to succinyl-CoA, enzyme is ketoglutarate dehydrogenase complex.
- Fatty acid oxidation and fatty acid synthesis
- Acetyl-CoA + Choline → Acetyl choline + CoA (enzyme is acetyl choline synthase)
- Pantothenate is also a part of acyl carrier protein (ACP) involved in fatty acid synthesis.

*Recommended Dietary Allowance of Pantothenic Acid*

Recommended dietary allowance is 5-10 mg/day

*Sources of Pantothenic Acid*

It is widely distributed in plants and animals. Moreover, it is synthesized by the normal bacterial flora in intestines. Therefore, deficiency is very rare. Yeast, liver, and eggs are good sources.

*Deficiency Manifestations of Pantothenic Acid*

Grierson-Gopalan syndrome or **Burning Foot Syndrome** is manifested as paresthesia (burning, lightning pain) in lower extremities and staggering gait due to impaired co-ordination and sleep disturbances.

## BIOTIN

## Chemistry/Structure

Biotin is a heterocyclic sulfur-containing vitamin. It has an imidazole ring.

*Sources of Biotin*

It is widely distributed in both animal and plant kingdoms. Egg yolk, liver, kidney, and milk are rich sources of biotin. It is also synthesized by intestinal bacterial flora.

*Biological Function of Biotin*

Biotin serves as a carrier of $CO_2$ in carboxylation reactions (carbon dioxide fixation reactions). Biotin is the prosthetic group for carboxylases so it is involved in the following reactions.

**Acetyl-CoA carboxylase:** This is the rate-limiting reaction in the biosynthesis of fatty acids.

Acetyl-CoA + $CO_2$ + ATP → Malonyl-CoA + ADP + Pi

**Pyruvate carboxylase:** This reaction is important in two aspects. First, it provides the oxaloacetate, which is the catalyst for TCA cycle. Second, it is an important enzyme in the gluconeogenic pathway.

Pyruvate + $CO_2$ + ATP →
$\quad\quad$ Oxaloacetate + ADP + Pi

*Recommended Dietary Allowance of Biotin*

Recommended dietary allowance is 30–50 µg/day.

*Deficiency Manifestations of Biotin*

Deficiency is uncommon. Symptoms include anemia, loss of appetite, nausea, dermatitis, and so forth.

*Sources of Biotin*

Normal bacterial flora of the gut will provide adequate quantities of biotin. Moreover, it is distributed ubiquitously in plant and animal tissues. Liver, yeast, peanut, soybean, milk, and egg yolk are rich sources.

## Folic Acid

*Chemistry/Structure*

It consists of a pteridine ring attached to p-amino benzoic acid and conjugated with one or more glutamic acid residues.

*Biological Function of Folic Acid*

The active form of folic acid is tetrahydrofolic acid (THFA or $FH_4$). It is formed from folic acid by reduction involving NADPH.

Tetrahydrofolic acid is actively involved in **one-carbon metabolism**. Donors of one-carbon fragments are serine, glycine, histidine, and tryptophan. These one carbon groups are taken up by THFA, which transfers them to intermediates during the synthesis of amino acids, purines, and pyrimidines. Methyl group of methylene-THFA is used for the **methyl transfer reactions** during the synthesis of choline, epinephrine, creatine, and so on.

*Recommended Daily Allowance of Folic Acid*

Recommended dietary allowance is 300–500 µg/day. The requirement is increased during pregnancy and lactation.

*Sources of Folic Acid*

Green leafy vegetables, liver, and whole grain cereals.

*Causes for Folate Deficiency*

**Pregnancy:** Folate deficiency is commonly seen in pregnancy, where requirement is increased.

**Defective absorption:** Absorption of folate is deficient in sprue, celiac disease, gluten-induced enteropathy, resection of jejunum, and short-circuiting of jejunum in gastroileostomy.

**Hemolytic anemias:** As the requirement of folic acid becomes more, deficiency is manifested in hemolysis.

**Dietary deficiency:** Absence of vegetables in food for prolonged periods may lead to deficiency of folic acid.

*Deficiency Manifestations of Folic Acid*

Deficiency is more common during pregnancy. Folic acid is supplemented to pregnant women to avoid fetal malformations like spina bifida and other neural tube defects. Deficiency results in **megaloblastic anemia (Figs. 16.10A and B)**, caused by decreased synthesis of purines and pyrimidines, leading to an inability of cells to make DNA and divide, which is manifested as **immature-looking nucleus** and mature eosinophilic cytoplasm in the bone marrow cells. **Reticulocytosis** is often seen with folate deficiency. These abnormal RBCs are rapidly destroyed in the spleen. This **hemolysis** leads to the reduction of life span of RBCs. Reduced generation and increased destruction of RBCs result in anemia. **Leukopenia** and thrombocytopenia are also manifested.

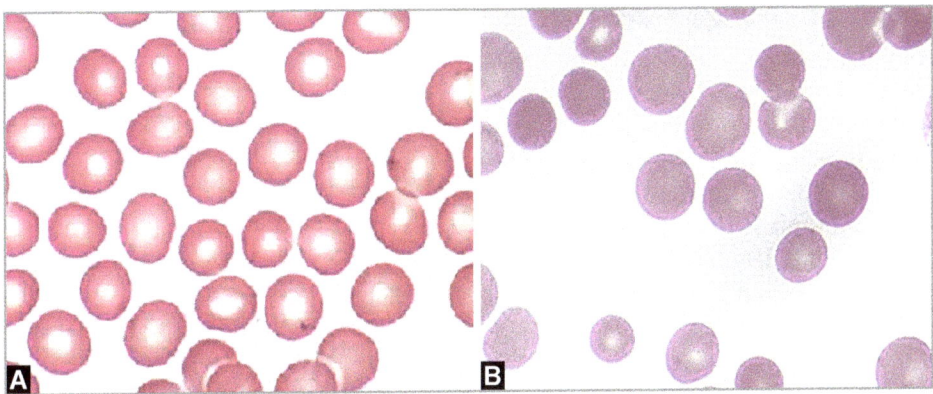

**Figs. 16.10A and B:** Peripheral blood picture. (A) Normal RBCs; (B) Macrocytic anemia due to folic acid deficiency.

## Clinical Applications

**Birth defects:** Folic acid deficiency during pregnancy may lead to neural tube defects such as spina bifida in the fetus. Administration of folic acid to mother prevents birth defects and fetal malformations. So, supplementation of folic acid from early pregnancy is a common practice.

**Antagonists** of folic acid are used for various therapeutic purposes. **Methotrexate** (folic acid analogue) competitively inhibits dihydrofolate reductase and is used for the remission of acute leukemia in children.

**Sulfonamides** are anti-folic acid drugs used as antibacterial agents.

## COBALAMIN

## Vitamin $B_{12}$

### Chemistry/Structure

It contains four pyrrole rings (corrin ring) linked directly with a cobalt atom and held in the center by coordination bonds with the nitrogens of pyrrole rings. Cobalamin is attached with cyanide to form cyanocobalamin (commercial preparation of vitamin $B_{12}$).

### Sources of Vitamin $B_{12}$

Liver, meat, egg, and curd. It is synthesized only by microorganisms. It is not present in plants.

### Absorption and Storage of Vitamin $B_{12}$

Vitamin $B_{12}$ from diet binds to the intrinsic factor (IF) produced by the gastric mucosa. This complex binds to specific receptors on the surface of the mucosal cells of ileum. It is then released into circulation and is stored in the liver. (It is the only water-soluble vitamin that can be stored in the body).

### Biological Function of Vitamin $B_{12}$

The coenzyme forms of cobalamin are 5'-deoxy-adenosyl cobalamin and methyl cobalamin. It is required in humans for two essential enzymatic reactions:
1. Synthesis of methionine **(Fig. 16.11)**
2. Isomerization of methyl malonyl-CoA to succinyl-CoA

### Recommended Daily Allowance of Vitamin $B_{12}$

RDA is 1–3 µg/day.

### Causes of $B_{12}$ Deficiency

**Nutritional:** The only source for $B_{12}$ in vegetarian diet is curd/milk, and lower-income

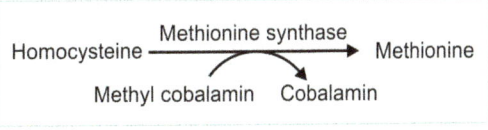

**Fig. 16.11:** Synthesis of methionine needs cobalamin.

group may not be able to afford it.

**Decrease in absorption:** Absorptive surface is reduced by gastrectomy, resection of ileum, and malabsorption syndromes.

**Addisonian pernicious anemia:** It is very rare in India but common in European countries. It is an autoimmune disease with a strong familial background. Antibodies are generated against IF. So, IF becomes deficient, leading to defective absorption of $B_{12}$.

**Gastric atrophy:** Although it is true that Addisonian pernicious anemia is rare in India, similar atrophy of gastric epithelium leading to deficiency of IF and decreased $B_{12}$ absorption is common in India. In chronic **iron deficiency** anemia, there is generalized mucosal atrophy.

**Pregnancy:** Increased requirement of vitamin $B_{12}$ during pregnancy is another common cause for vitamin $B_{12}$ deficiency in India.

### Manifestations of Vitamin $B_{12}$ Deficiency

Deficiency is characterized by **megaloblasts**, low hemoglobin levels and decreased number of erythrocytes. Hematological picture of $B_{12}$ deficiency will be similar to folic acid deficiency. In addition, patients with cobalamin deficiency also show neurological symptoms like progressive demyelination, degeneration of nerve fibers, and paresthesia of limbs. These neurological features together are called **subacute combined degeneration**. Deficiency of vitamin $B_{12}$ leads to folic acid deficiency. So, while treating this type of anemia, vitamin $B_{12}$ and folic acid are administered together.

A summary of vitamin B complex group is given in **Table 16.4**.

## Vitamin C

### Chemical Name

The chemical name of vitamin C is L-ascorbic acid. Its structure resembles that of glucose. The active forms are L-ascorbic acid and dehydro ascorbic acid.

### Sources of Vitamin C

Gooseberry is the richest source. Citrus fruits, guava, and green leafy vegetables also contain vitamin C. Milk is a very poor source of vitamin C.

**Table 16.4:** Summary of vitamin B complex group.

| Vitamin | Coenzyme form | Major biochemical function | Deficiency disease | Daily requirement |
| --- | --- | --- | --- | --- |
| Thiamine ($B_1$) | TPP | Oxidative decarboxylation | Beriberi | 1.2–1.8 mg |
| Riboflavin ($B_2$) | FMN, FAD | Dehydrogenation | Cheilosis, glossitis | 1–2 mg |
| Niacin | $NAD^+$, $NADP^+$ | Dehydrogenation | Pellagra | 12–20 mg |
| Pyridoxine ($B_6$) | PLP | Transamination | Peripheral neuritis | 1–2 mg |
| Pantothenic acid | CoA | Formation of CoA derivatives | Muscle cramps | 5–10 mg |
| Biotin | Biotin | $CO_2$ fixation | Anemia | 30–50 µg |
| Folic acid | THFA | One-carbon carriage | Macrocytic anemia | 300–500 µg |
| Vitamin $B_{12}$ | Methylcobalamin | Trans-methylation | Megalocytic anemia | 1–3 µg |

(TPP: thiamine pyrophosphate; FMN: flavin mononucleotide; FAD: flavin adenine dinucleotide; NAD: nicotinamide adenine dinucleotide; NADP: nicotinamide adenine dinucleotide phosphate; PLP: pyridoxal phosphate; THFA: tetrahydrofolic acid)

*Recommended Daily Allowance of Vitamin C*

75 mg/day. Requirement is increased during pregnancy, lactation, old age (100 mg/day) and also in smokers.

*Absorption and Excretion of Vitamin C*

It is readily absorbed from the GI tract. Excess is excreted in urine as vitamin C or as its metabolite, oxalate.

*Functions of Vitamin C*

- It is a strong reducing agent. It is necessary for the normal production of supporting tissues such as osteoid, dentine, and intercellular cement substances of capillaries. It is required in the synthesis of collagen, as a coenzyme in the hydroxylation of proline and lysine residues.
- It is required in tyrosine metabolism: para-hydroxyphenylpyruvate → homogentisate.
- It is necessary for the synthesis of serotonin from tryptophan.
- It is required in bile acid synthesis: cholesterol → 7-α-OH cholesterol, 7-α-hydroxylase
- It helps in the synthesis of steroid hormones.
- Vitamin C enhances iron absorption from the intestine by reducing $Fe^{3+}$ → $Fe^{2+}$ and helps in the reduction of met-hemoglobin → Hb.
- It is required for the reduction of folate → THF, thereby it helps in the maturation of RBC.
- It is an important antioxidant as it reduces the occurrence of cataract formation, atherosclerosis, and certain types of cancer.
- It is essential for wound healing and helps in the synthesis of antibodies.

*Deficiency Manifestations of Vitamin C*

- Deficiency of vitamin C causes **scurvy**. It is characterized by painful, swollen, and spongy gums that bleed on slight pressures. Capillaries are fragile. Hemorrhage may occur in conjunctiva and retina. These symptoms are due to abnormal or defective synthesis of collagen and intercellular cement substances. Osteoid of bone is abnormal. Hence bones are weak and brittle.
- Delayed wound healing due to synthesis of abnormal collagen.
- Moderate microcytic hypochromic anemia is seen.
- Immunocompetence is decreased.

## SUMMARY OF THE CHAPTER

- Vitamin A is a fat-soluble vitamin whose active form is present only in animal tissues, but provitamin A (beta-carotene) is present in plant tissues.
- Retinols are polyisoprenoid compounds with vitamin A activity, having the beta ionone ring system.
- Active forms of the vitamin A include retinol, retinal, and retinoic acid. The two important isomers are all-trans-retinal and 11-cis retinal.
- Vitamin A is transported with the help of retinal-binding protein (RBP), and this retinal–RBP complex has specific receptors in various tissues.
- Rods are for vision in dim light, and cones are for color vision.
- Decrease in the number of cones/cone proteins leads to color blindness.
- Vitamin D is derived from 7-dehydro-cholesterol by the action of UV rays.
- Vitamin D deficiency results in rickets and osteomalacia.
- Vitamin E is tocopherol. It is absorbed along with fats with the help of bile salts.
- Vitamin E is the most important antioxidant in tissues.
- Vitamin K is absorbed in intestine along with chylomicrons. They are also synthesized by intestinal flora.
- Vitamin K is involved in blood coagulation. Vitamin K is required for post-translational modification of coagulation factors.
- The coenzyme form of thiamine is thiamine pyrophosphate (TPP).
- Deficiency of thiamine leads to beriberi.

- Coenzyme forms of riboflavin are flavin mononucleotide (FMN) and flavin adenine dinucleotide (FAD).
- Examples of FAD-dependent enzymes are succinate dehydrogenase and acyl-CoA dehydrogenase.
- Examples of NAD⁺-dependent enzymes are lactate dehydrogenase, glyceraldehyde-3-phosphate dehydrogenase, and pyruvate dehydrogenase.
- Niacin is synthesized from tryptophan.
- Active form of pyridoxine is pyridoxal phosphate (PLP).
- PLP is essential for the transamination and decarboxylation reactions of amino acids.
- ALA synthase in heme biosynthesis is also a PLP-dependent enzyme. Hence anemia is common in B6 deficiency.
- Isonicotinylhydrazide (INH or isoniazid), used as an anti-tuberculosis drug, can produce pyridoxine deficiency.
- Coenzyme A (CoA) contains pantothenic acid.
- Important CoA derivatives are acetyl-CoA, succinyl-CoA, HMG-CoA, and acyl-CoA.
- Biotin acts as a coenzyme for carboxylation reactions. For example, acetyl-CoA carboxylase, propionyl-CoA carboxylase, and pyruvate carboxylase.
- THFA (tetrahydro folic acid) is the carrier of one-carbon groups.
- Folate antagonists are sulfonamides, aminopterin, and amethopterin.
- Absorption of vitamin $B_{12}$ requires the intrinsic factor (IF). Transcobalamin, a glycoprotein, is the specific carrier of vitamin $B_{12}$.
- $B_{12}$-containing enzymes in the human body are methylmalonyl-CoA isomerase and homocysteine methyltransferase.
- $B_{12}$ deficiency leads to pernicious anemia.
- Man, higher primates, guinea pigs, and bats are the only species that cannot synthesize ascorbic acid (vitamin C).
- Vitamin C deficiency causes scurvy. It is characterized by abnormal collagen, ecchymoses, hemorrhage, and anemia.

## QUESTIONS FOR SELF-ASSESSMENT

1. Mention the active form and functions of vitamin A.
2. Mention the sources and functions of vitamin E.
3. Mention the antioxidant property of vitamin E.
4. What is the role of vitamin K in blood coagulation process?
5. What is the active form of vitamin D? Mention its role in calcium homeostasis.
6. What are the deficiency symptoms of thiamine.
7. Name the coenzymes related to riboflavin.
8. Mention the role of hematopoietic and non-hematopoietic vitamins.
9. Mention the role of folic acid in our body.
10. Mention the sources and major functions of vitamin C.

## MULTIPLE CHOICE QUESTIONS (MCQs)

16-1. All the following are the symptoms of vitamin A deficiency, *except:*
a. Night blindness
b. Beriberi
c. Keratinization of lacrimal glands
d. Keratomalacia

16-2. Beriberi is due to the deficiency of:
a. Niacin
b. Thiamine
c. Riboflavin
d. Vitamin $B_{12}$

16-3. Regarding the major functions of vitamin E, all of the following statements are correct, *except:*
a. Antioxidant
b. Normal reproduction in animals
c. Helps in vision
d. Protects cellular membrane

16-4. The deficiency symptom of vitamin C is:
a. Scurvy
b. Dermatitis
c. Pellagra
d. Beriberi

16-5. All the following vitamins are produced from precursors in the human body, *except:*
   a. Niacin
   b. Ascorbic acid
   c. Vitamin D
   d. Vitamin A

16-6. One of the following enzymes does not require coenzyme A (CoASH):
   a. Pyruvate dehydrogenase complex
   b. Alpha ketoglutarate dehydrogenase complex
   c. Phosphofructokinase
   d. Thiokinase

16-7. Deficiency manifestation of vitamin $B_6$ leads to the following, *except:*
   a. Hypochromic microcytic anemia
   b. Glossitis
   c. Pigmented scaly dermatitis
   d. Megaloblastic anemia

16-8. Pellagra is due to the deficiency of:
   a. Niacin
   b. Thiamine
   c. Riboflavin
   d. Vitamin $B_{12}$

16-9. Which of the following vitamins is required for oxidative decarboxylation?
   a. Pyridoxal phosphate
   b. Thiamine pyrophosphate
   c. Biotin
   d. Riboflavin

16-10 Among the following most of them are vitamin K dependent clotting factors, *except:*
   a. Prothrombin
   b. Factor XII
   c. Factor VII
   d. Factor X

## ANSWER KEYS TO MCQs

16-1. (b)   16-2. (b)   16-3. (c)   16-4. (a)   16-5. (b)   16-6. (c)   16-7. (d)   16-8. (a)
16-9. (a)   16-10. (b)

# CHAPTER 17

# Minerals

At the completion of this chapter, the reader will be able to answer questions on the following topics:

- Calcium
- Phosphorus
- Magnesium
- Sodium
- Potassium
- Chloride
- Sulfur
- Iron
- Copper
- Iodine
- Zinc
- Fluoride
- Selenium

A large number of minerals are found in the human body. They are supplied through the diet. Based on the daily requirement, minerals may be classified into **major elements** or **macronutrients** (if the requirement is more than 100 mg/day) and **microelements** or **trace elements** (if the requirement is less than 100 mg/day). The principal macronutrients include calcium, phosphorus, magnesium, sodium, potassium, chloride, and sulfur. The trace elements are listed in **Table 17.1**. The following minerals are **toxic** and should be avoided: aluminum, lead, cadmium, and mercury.

**Table 17.1:** Important minerals.

| Major elements | Trace elements |
| --- | --- |
| • Calcium | • Iron |
| • Magnesium | • Iodine |
| • Phosphorus | • Copper |
| • Sodium | • Manganese |
| • Potassium | • Zinc |
| • Chloride | • Molybdenum |
| • Sulfur | • Selenium |
|  | • Fluoride |

## CALCIUM

Total body calcium is about 1–1.5 kg. Nearly 99% of this is present in bones and teeth and remaining 1% is extracellular, mainly in blood.

### Sources of Calcium

Milk and milk products like cheese are good sources of calcium. Egg, fish, vegetables, and nuts are moderate sources.

### Daily Dietary Allowance Requirement of Calcium

- Adults: 500 mg/day
- Children: 1,000 mg/day
- Pregnant and lactating Women: 1,500 mg/day
- Old age: 1,500 mg/day (to prevent osteoporosis)

### Absorption of Calcium

Calcium is absorbed from the first part of small intestine. It is an active process requiring energy and a carrier protein. Calcium absorption is increased by **calcitriol** (active form of vitamin D, described in Chapter 16), acidity (gastric juice), and basic amino acids. Maximum absorption is seen when the ratio of Ca:P is between 1:2 and 2:1. Calcium absorption is decreased by **phytic acid** (seen in cereals), **oxalates** (present in vegetables), excess fatty acids and phosphates, all of which form insoluble salts of calcium.

## Serum Level of Calcium

Normal level of calcium in serum is 9–11 mg/dL. About 4–4.5 mg/dL of calcium is in the ionized form, which is metabolically active. The remaining calcium is present as protein–calcium complex, which is non-diffusible.

## Regulation of Calcium Level in Blood

The normal level of calcium in the blood is maintained by (1) vitamin D, (2) parathyroid hormone, and (3) calcitonin. The actions of these three factors are summarized in **Table 17.2**. Interrelationship of these factors to maintain the normal calcium level in blood is depicted in **Figure 17.1**.

*Vitamin D*

- **Calcitriol** (1,25-dihydroxy cholecalciferol) is the metabolically active form of vitamin D.
- Calcitriol increases the absorption of calcium from the intestine by inducing the formation of calcium-binding protein **calbindin**.
- Calcitriol also increases the mineralization of bones by increasing the activity of osteoblast cells of bone.
- Vitamin D also increases the reabsorption of calcium from the kidneys and decreases the excretion of calcium through the kidneys.

*Parathyroid Hormone*

- Parathyroid hormone (PTH) is secreted by the four parathyroid glands that are embedded in the thyroid gland. The main action of the parathyroid hormone is to increase the blood calcium level.
- The PTH causes demineralization of the bones. It increases the number and activity of osteoclast cells of the bone, which in turn helps in solubilizing calcium from the bones.
- In the kidneys, PTH causes decreased renal excretion of calcium and also increases the excretion of phosphate.
- PTH also stimulates the activity of enzyme 1-hydroxylase which is necessary for activation of vitamin D to calcitriol. Calcitriol increases calcium absorption from the intestine.

| Table 17.2: Comparison of the actions of the three major factors affecting serum calcium. | | | |
|---|---|---|---|
| | Vitamin D | PTH | Calcitonin |
| Blood calcium | Increased | Drastically increased | Decreased |
| Main action | Absorption from gut | Demineralization | Opposes demineralization |
| Calcium absorption from gut | Increased | Increased | |
| Bone resorption | Decreased | Increased | Decreased |
| Deficiency manifestation | Rickets | Tetany | -- |
| Effect of excess | Hypercalcemia+ | Hypercalcemia++ | Hypocalcemia |

(PTH: parathyroid hormone)

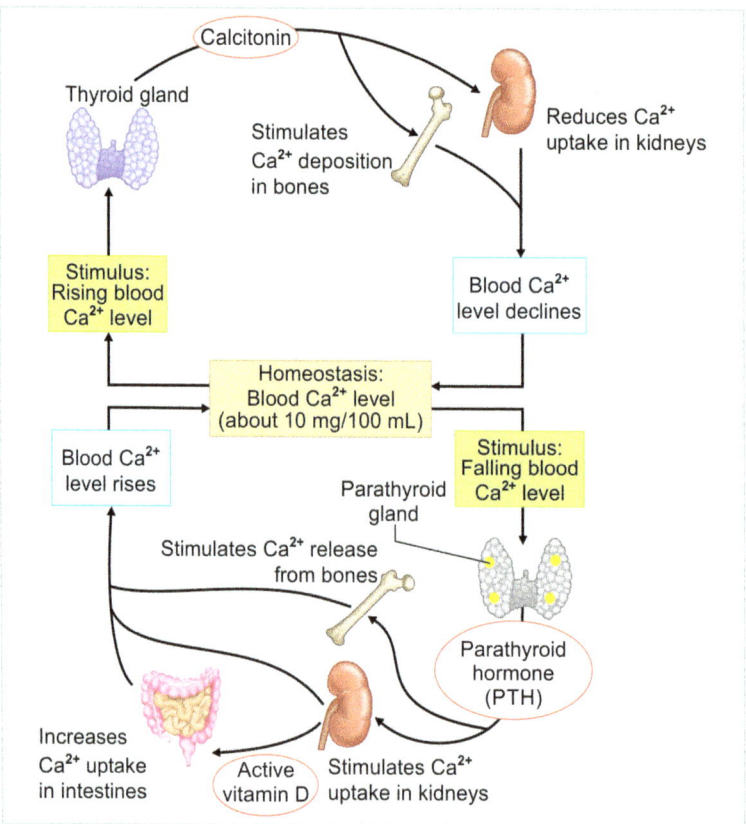

**Fig. 17.1:** Calcium homeostasis. When serum calcium is low, PTH is stimulated, resulting in increased calcium release from bone and decreased renal calcium excretion. PTH also stimulates increased production of calcitriol, which acts to increase the absorption of calcium from intestine.

- When the calcium level is low, PTH secretion is stimulated, which increases the blood calcium level.

*Calcitonin*

- It is a hormone secreted by the parafollicular cells of the thyroid gland. The main action of this hormone is to decrease blood calcium level.
- Calcitonin inhibits by inhibiting the demineralization of bones.
- It inhibits the activity of osteoclast cells and also increases the activity of osteoblast cells, thus causing calcium deposition in the bones.
- The secretion of these hormones is controlled by the serum calcium level. But when blood calcium level is increased, PTH secretion will be inhibited and secretion of calcitonin will be stimulated.
- When the calcium level is low, calcitonin secretion is inhibited.

## Biological Role of Calcium

- The major quantity of calcium is used for bone and teeth formation.
- Calcium helps in activation of enzymes either directly or through a calcium-binding regulatory protein called **calmodulin**. The calcium–calmodulin complex activates enzymes such as glycogen synthase, pyruvate kinase, adenyl cyclase, and so on **(Table 17.3)**. Enzymes activated by calcium include pancreatic lipase, rennin, and so on.

| Table 17.3: Selected list of enzymes activated by Ca$^{++}$ and mediated by calmodulin. |
|---|
| Adenyl cyclase |
| Ca$^{++}$-dependent protein kinases |
| Ca$^{++}$–Mg$^{++}$–ATPase |
| Glycogen synthase |
| Myosin kinase |
| Phospholipase C |
| Pyruvate dehydrogenase |

- Calcium mediates excitation and contraction of skeletal, cardiac, and smooth muscles.
- Calcium is necessary for the transmission of nerve impulses from the pre-synaptic to post-synaptic region.
- Calcium decreases neuromuscular excitability.
- Calcium is necessary for blood coagulation.
- Calcium is necessary for the adhesion of cells in a tissue.

*Hypocalcemia (Calcium Deficiency)*

When serum calcium level is less than 8 mg/dL, it may lead to mild **tremors**. If serum calcium is further lowered, it can lead to **tetany**, where the neuromuscular irritability is increased, thus causing spasm of muscles. It is mainly manifested as carpopedal spasm (**Fig. 17.2**) and laryngeal spasm.

A chronic dietary deficiency of calcium may cause loss of bone mass (bone resorption). It may lead to osteoporosis in old age, especially in postmenopausal women. Calcium deficiency is mainly seen in: (a) inadequate dietary intake and (b) hypoparathyroidism. Calcium supplementation in combination with vitamin D is recommended.

*Hypercalcemia (Calcium Toxicity)*

An increase in blood calcium level is known as hypercalcemia. It is usually due to hyperparathyroidism. Clinical manifestations include osteoporosis and bone resorption. Strength of bone reduces, thus increasing the risk of bone fracture. X-rays of the skull and bones show punched-out areas of bone resorption.

Excretion of calcium increases causing hypercalciuria. This may lead to deposition of calcium in urinary tract, causing urinary calculi. Calcification (deposition of calcium) may be seen in renal tissues, arterial walls, muscle tissues, and so on.

## PHOSPHORUS

Total body phosphorus is about 1 kg. Nearly 80% of this is present in bones and teeth. Phosphate is mainly an intracellular ion.

### Sources of Phosphorus

Milk and cheese are good sources. Egg and fish are moderate sources. Calcitriol increases phosphate absorption.

### Requirement of Phosphorus

- Adults: 500 mg/day
- Children: 1,000 mg/day

### Distribution of Phosphorus

The major part of phosphate ions in serum is complexed with cations and proteins, while the remaining part exists in free ionized form.

### Serum Level of Phosphorus

Serum level of phosphate is 3–4 mg/dL in normal adults and is 5–6 mg/dL in children. The phosphate level is regulated by excretion

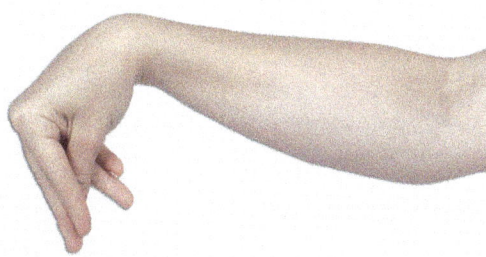

**Fig. 17.2:** Carpopedal spasm in tetany.

through urine. Renal threshold is 2 mg/dL. Usually 500 mg of phosphate is excreted through urine per day.

## Biological Role of Phosphorus

- Formation of bones and teeth.
- Production of high-energy phosphate compounds like ATP, GTP, and so on.
- Synthesis of certain coenzymes such as $NADP^+$.
- Synthesis of DNA and RNA, where phosphodiester linkages (between sugar and phosphate) form the backbone of the structure.
- Formation of phosphoproteins and phospholipids.
- Phosphate buffer system is an important buffer system in the blood, which helps to maintain the normal blood pH.
- Phosphate esters serve as intermediates in many metabolic pathways, for example, glucose-6-phosphate.

## Deficiency Manifestations of Phosphorus

Decrease in the absorption of phosphorus from intestine and vitamin D deficiency cause decrease in serum phosphate level. This leads to defects in the calcification of bones.

## Toxicity of Phosphorus

Toxicity is rare, but renal failure may lead to phosphate retention, thus causing toxicity. This leads to the inhibition of calcitriol production and so may lead to low serum calcium level.

## MAGNESIUM

Magnesium is mainly seen in intracellular fluid. Total body magnesium is about 25 g, 60% of which is complexed with calcium in bone.

## Requirement of Magnesium

The requirement is about 400 mg/day for men and 300 mg/day for women. More is required during lactation. Major sources are cereals, beans, leafy vegetables, and fish.

## Normal Serum Level of Magnesium

The normal serum level of $Mg^{++}$ is 1.8–2.2 mg/dL.

## Functions of Magnesium

- $Mg^{++}$ is the activator of many enzymes requiring ATP. Alkaline phosphatase, hexokinase, fructokinase, phosphofructokinase, and so on need magnesium.
- Neuromuscular irritability is lowered by magnesium.

## Hypomagnesemia

When serum magnesium level falls below 1.7 mg/dL, it is called hypomagnesemia. Vomiting, nasogastric suction, and diarrhea are the common causes. Deficiency of magnesium leads to neuromuscular hyperirritability and cardiac arrythmias.

## Hypermagnesemia

It is due to the excessive intake of magnesium orally (antacids), rectally (enema), or parenterally. Magnesium intoxication causes depression of the neuromuscular system, causing lethargy, hypotension, respiratory depression, and bradycardia.

## SODIUM

Sodium is the major cation present in extracellular fluid. It exists in association with anions like chloride, bicarbonate, and phosphate.

## Sources of Sodium

Table salt is the most usual source for sodium. All food stuffs of plant and animal origin contain sodium.

## Requirement of Sodium

Normal diet contains about 5–10 g of sodium per day. The requirement may be slightly high in tropical countries to compensate for the loss in perspiration.

## Normal Level of Sodium in Plasma

The normal level of sodium in plasma is 136–145 mEq/L.

## Biological Role of Sodium

- It is the major cation of extracellular fluid.
- Sodium regulates extracellular fluid volume and plasma volume.
- Sodium in the form of sodium bicarbonate is important in the regulation of acid–base balance.

## Sodium Deficiency (Hyponatremia)

Severe dehydration as seen in vomiting and diarrhea will lead to sodium deficiency. It also leads to fall in blood pressure. Loss of sodium due to severe sweating may cause muscle cramps and headache.

## Sodium Toxicity (Hypernatremia)

Sodium increase in the body causes retention of water. This causes an increase in blood pressure and may lead to cardiac failure.

# POTASSIUM

It is the major intracellular cation. It helps in maintaining osmotic pressure.

## Sources of Potassium

Tender coconut water, banana, oranges, pineapple, dates, and potato are good sources of potassium. Potassium is present in most of the food stuffs as it is an intracellular cation and so dietary deficiency is rare.

## Requirement of Potassium

Normal intake of potassium is about 4 g/day.

## Normal Level of Potassium in Plasma

The normal level of potassium in plasma is 3.5–4.5 mEq/L. Normal range of plasma potassium is very narrow so even minor imbalance can be life-threatening.

## Biological Functions of Potassium

- Potassium is necessary to maintain intracellular osmotic pressure.
- It is necessary for the transmission of nerve impulses.
- It is necessary for the contraction of heart and for muscle functions.

## Deficiency of Potassium (Hypokalemia)

Manifestations include muscular weakness, cardiac arrhythmias, and injury to myocardium.

## Toxicity of Potassium (Hyperkalemia)

When plasma potassium level is more than 5.5 mEq/L, the condition is called hyperkalemia. Manifestation of hyperkalemia includes decreased cardiac excitability, which can lead to bradycardia, peripheral vascular collapse, and cardiac arrest.

# CHLORIDE

## Sources of Chloride

Table salt (sodium chloride) is a good source of chloride.

## Normal Level of Chloride in Plasma

The normal level of chloride is 96–106 mEq/L. The concentration of chloride is highest in cerebrospinal fluid (CSF), where the chloride concentration is about 125 mEq/L.

## Biological Role (Functions) of Chloride

- Chloride is important for the formation of HCl, which is part of the gastric juice.
- It is involved in the acid–base balance of the body.

## Chloride Deficiency (Hypochloremia)

- Excessive vomiting may cause hypochloremia.

- In Addison's disease, renal reabsorption of chloride is decreased and more chloride is excreted.

## Chloride Toxicity (Hyperchloremia)

- In severe diarrhea, an increased loss of bicarbonate ions is accompanied by an increase in chloride ions. So chloride retention is seen.
- In Cushing's syndrome, the reabsorption of chloride from kidney tubules is increased.

# SULFUR

## Sources of Sulfur

Proteins rich in sulfur-containing amino acids like cysteine and methionine.

## Functions of Sulfur

- Sulfur forms part of sulfur-containing amino acids, which are important constituents of body proteins, for example, insulin, keratin, immunoglobulins.
- It is also part of mucopolysaccharides, for example, dermatan sulfate, heparin sulfate, chondroitin sulfate.
- Many enzymes contain –SH groups in their active sites, for example, glutathione.
- Sulfur is a constituent of certain B-complex vitamins like thiamine, biotin, and so on.

# TRACE ELEMENTS

They are essential for various functions in the body. If the requirement is less than 100 mg/day, then such minerals are classified as trace elements.

# IRON

Iron is present in almost all the cells of the body. Nearly 75% of iron is present in blood. Liver, bone marrow, and muscles also contain iron. Total iron content is about 3–5 g.

## Biological Role of Iron

Heme is the major iron-containing substance. Iron-containing compounds, especially proteins, are:
- Heme-containing proteins are hemoglobin, myoglobin, and enzymes like peroxidase, catalase, and cytochrome oxidase.
- Non-heme-containing proteins are transferrin, ferritin, and hemosiderin.

## Sources of Iron

Green leafy vegetables, jaggery, and dates are good sources. Liver and meat are moderate sources of iron. Milk is a poor source of iron.

## Requirement of Iron

Requirement of iron for men is 20 mg/day and women is 30 mg/day.

Only about 1–2 mg of dietary iron will be absorbed. Iron obtained from animal sources is easily absorbed than from vegetarian sources because vegetables contain substances that inhibit absorption.

## Absorption of Iron

Iron absorption takes place in the first and second part of the small intestine. Iron is absorbed in the ferrous form, and the absorption is helped by hydrochloric acid (HCl) of gastric juice, ascorbic acid, and so on. Phytate and oxalates of a vegetarian diet decrease iron absorption.

*Mucosal Block Theory*

Duodenum and jejunum are the sites of absorption. Iron metabolism is unique because homeostasis is maintained by regulation at the **level of absorption** and not by excretion. No other nutrient is regulated in this manner. In other words, iron is a "one-way element."

When iron stores in the body are depleted, absorption is enhanced. When adequate quantity of iron is stored, absorption is decreased. This is referred to as the **mucosal**

**block** of regulation of absorption of iron. Only ferrous (and not ferric) form of iron is absorbed. **Ferrous iron** in the intestinal lumen binds to mucosal cell protein called divalent metal transporter-1 (**DMT-1**). (Therefore, all other divalent ions, including calcium, copper, and lead will competitively inhibit the iron absorption). The bound iron is then transported into the mucosal cell. The rest of the unabsorbed iron is excreted.

Inside the mucosal cell, the ferric iron is formed and is complexed with apoferritin to form **ferritin**. It is kept temporarily in the mucosal cell. If there is anemia, the iron is further absorbed into the bloodstream. If transferrin is saturated with iron, any iron accumulated in the mucosal cell is lost when the cell is desquamated. The fraction of iron absorbed and retained is decided by the iron status. When iron is in excess, absorption is reduced; this is the basis of the "mucosal block."

Later, iron is released from the intestinal cell to the bloodstream. Iron in the ferritin is released and then crosses the mucosal cell. But this can happen only when there is free transferrin in the plasma to bind the iron. Iron crosses the cell membrane in ferrous form. In the blood it is reoxidized to ferric state and transported by **transferrin**.

## Transport of Iron

The transport form of iron in blood is **transferrin**. Normal plasma level of transferrin is 250 mg/dL. Apotransferrin takes up iron in the ferric state ($Fe^{3+}$) to form transferrin. The oxidation of iron from $Fe^{2+}$ to $Fe^{3+}$ state is by the enzyme ferroxidase. In the blood, **ceruloplasmin** (a copper-containing ferroxidase) oxidizes ferrous to ferric state.

Apotransferrin + 2$Fe^{2+}$ ⟶ Transferrin with 2$Fe^{3+}$

Each molecule of transferrin can take up two ferric ions. Except mature red blood cells (RBCs), all other cells take up the iron–transferrin complex. Iron is utilized by the cells, whereas transferrin is transported back to plasma.

## Storage of Iron

Iron is stored as **ferritin** in intestinal mucosal cells, liver, spleen, and bone marrow. In patients with iron deficiency anemia, the ferritin amount is reduced.

## Normal Level of Iron in Plasma

The normal level of iron in plasma is about 100 µg/dL.

## Conservation of Iron

Iron pool in the body is conserved. The mature RBCs are lysed to liberate hemoglobin. This hemoglobin is broken down to form heme and globin. The iron in the heme is reutilized or conserved (**Fig. 17.3**).

Homeostasis of iron takes place at the level of absorption. The unabsorbed iron is excreted through feces. Negligible amount of iron is excreted through urine. Any type of bleeding will cause loss of iron from the body. In women, menstrual flow is the major cause for loss of iron.

As almost all cells contain iron, the normal wear and tear of the upper layers of skin also forms a route for the excretion of iron from the body. A summary of iron metabolism is shown in **Figure 17.4**.

## Causes of Iron Deficiency

- Nutritional deficiency of iron.
- Hookworm infection
- Chronic blood loss as seen in peptic ulcers, uterine problems, piles, and so on
- Defect in absorption
- Repeated pregnancies at short intervals.

## Iron Deficiency Manifestations

Iron deficiency leads to **anemia**. Iron deficiency anemia is one of the most common nutritional deficiency diseases. About 85% of pregnant women suffer from anemia. Maternal anemia contributes to the increase in perinatal mortality. Anemia often leads to

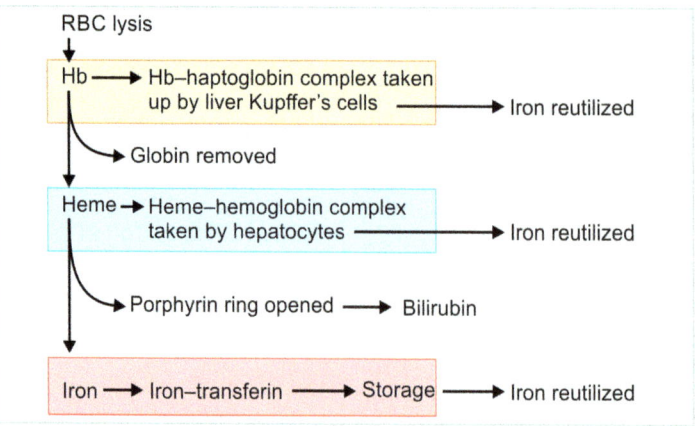

**Fig. 17.3:** Conservation of iron in the body.
(RBC: red blood cell)

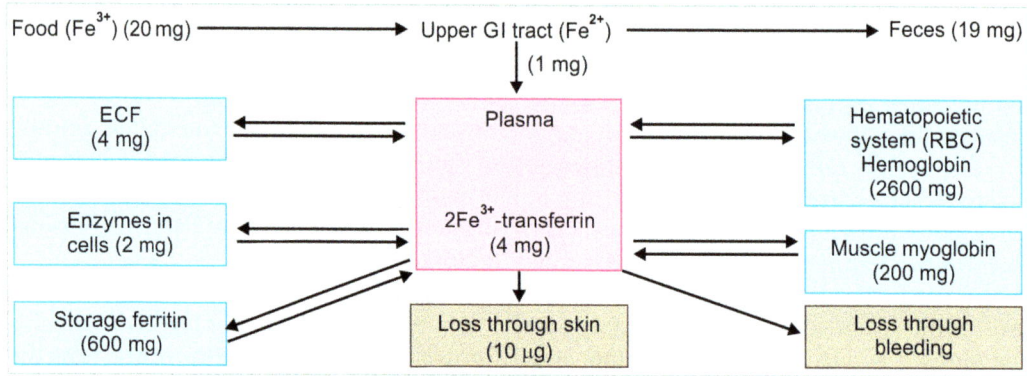

**Fig. 17.4:** Normal iron kinetics.
(ECF: extracellular fluid; GI: gastrointestinal; RBC: red blood cell)

irreversible impairment of the child's learning ability. In adults, anemia results in impaired work capacity. Iron deficiency anemia is characterized by **microcytic hypochromic anemia (Fig. 17.5)**. Anemia is diagnosed when **hemoglobin level is less than 10 g/dL.** In patients with anemia, as hemoglobin level reduces, body cells lack oxygen. All metabolic processes become sluggish, and patient becomes uninterested in the surroundings (apathy). There will be impaired attention, irritability, lowered memory, and reduced learning ability in children. A clinically oriented classification of anemia is shown in **Table 17.4**.

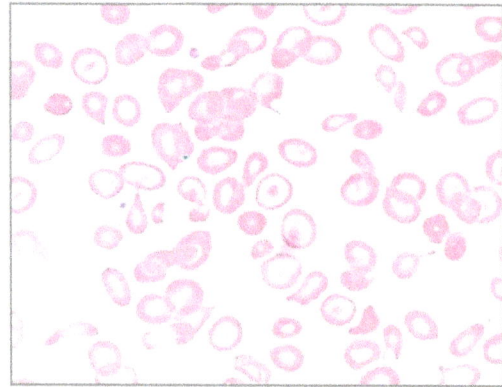

**Fig. 17.5:** Peripheral blood smear. Iron deficiency manifests as microcytic hypochromic anemia.

**Table 17.4:** Classification of anemias.

1. **Impaired production of RBCs:**
   - Defect in heme synthesis: Deficiency of iron, copper, pyridoxal phosphate, folic acid, vitamin $B_{12}$, or vitamin C. Lead will inhibit heme synthesis.
   - Defect in regulators: Lack of erythropoietin due to chronic renal failure.
   - Defect in stem cells: Aplastic anemia due to drugs (e.g., chloramphenicol), infections, and malignant infiltrations may lead to anemia
2. **Intracorpuscular defects:**
   - Hemoglobinopathies: HbS, HbC, HbM
   - Thalassemias: Major and minor
   - Abnormal shape: Spherocytosis
   - Deficiency of glucose-6-phosphate dehydrogenase
3. **Extracorpuscular causes:**
   - Infections: Malaria, streptococcal infection
   - Autoimmune hemolysis
4. **Hemorrhage:** Hematuria, hematemesis, hemoptysis, peptic ulcer, menorrhagia, hemorrhoids, hemophilia [absence of antihemophilic globulin (AHG)], thrombocytopenia

## Treatment for Iron Deficiency

The treatment for iron deficiency is to give oral iron supplementation. 100 mg of **iron** + 500 mg of **folic acid** are given to pregnant women, while 20 mg of iron + 100 mg of folic acid are given to children. Iron tablets are usually given along with **vitamin C** to convert it into ferrous form for easy absorption.

## Iron Toxicity

Iron excess is not very common. This condition is called **hemosiderosis**. Hemosiderin pigments are golden-brown granules formed from partially denatured ferritin and may contain up to 25% of iron. It is mainly seen in the spleen and liver. Hemosiderosis is usually seen in patients receiving repeated blood transfusions (as in hemophilia). In hereditary hemosiderosis, the iron absorption increases.

Due to iron overload, the increased hemosiderin deposition in liver cells leads to the death of hepatocytes and liver cirrhosis. Destruction of the cells of the pancreas leads to diabetes. Hemosiderin deposition under the skin causes yellow-brown discoloration called hemochromatosis. These three conditions of cirrhosis, hemochromatosis, and diabetes are collectively referred to as *bronze diabetes*.

## COPPER

The total body content of copper is about 100 mg. It is mainly seen in muscles, liver, bone marrow, kidney, brain, and hair. Copper is stored in the liver and bone marrow. It is mainly excreted through bile.

### Biological Role of Copper

Many enzymes contain copper. Copper-containing enzymes are ceruloplasmin, cytochrome oxidase, aminolevulinic acid (ALA) synthase, tyrosinase, and so forth. Copper also serves as a cofactor for vitamin C in hydroxylation reactions.

**Ceruloplasmin**, a copper-containing enzyme, promotes the oxidation of $Fe^{2+}$ to $Fe^{3+}$ (ferric) and its incorporation into transferrin. Normal serum level of ceruloplasmin is 25–50 mg/dL, which is nearly equivalent to 3–5 µg of copper.

### Requirement of Copper

The requirement of copper of adults is 1 mg/day.

### Dietary Sources of Copper

Green leafy vegetables, cereals, nuts, meat, and liver are good sources of copper. Milk is a poor source of copper.

### Metabolic Abnormalities of Copper

- **Wilson's disease:** It is an inherited disorder. Copper-containing ceruloplasmin level is greatly reduced. So, the copper is not

excreted and this leads to copper toxicity. Copper gets deposited in the liver leading to hepatocellular degeneration and cirrhosis. Copper deposition in the basal ganglia of brain leads to neurological symptoms. Copper is deposited as a golden-brown ring around the cornea also.
- **Copper deficiency anemia:** Copper is required for hemoglobin synthesis. So in patients with copper deficiency, microcytic anemia may be manifested.
- **Menkes kinky hair syndrome:** It is due to a genetic defect. In patients with copper deficiency, cross-linking of collagen fibers of connective tissue is defective. So it leads to wrinkled hair, fragile blood vessel walls, cerebral degeneration, and mental retardation.
- Copper is a part of tyrosinase enzyme, which is required for **melanin** synthesis. Due to the deficiency of copper, hypopigmentation of skin and hair may be seen.
- Excess copper intake causes copper toxicity. It is manifested as diarrhea, hemolysis, and renal failure.

## IODINE

Almost all cells of the body contain iodine. But 80% of the total iodine is stored in the thyroid glands. In the blood, the iodine level is normally between 5 and 10 µg/dL.

### Requirement of Iodine

The requirement of iodine is about 150–200 µg/day.

### Sources of Iodine

Drinking water, iodized table salt, fish, and vegetables contain iodine. The ingredients in food stuffs that prevent the absorption of iodine are called **goitrogenic substances**. These inhibitors are found in cabbage, tapioca, mustard seeds, and mustard oil. Upper regions of mountains contain less iodine because iodine-containing top soil is eroded. Such areas are called "goiterous belts."

### Biological Role/Functions of Iodine

The major function of iodine is for the synthesis of thyroid hormones **thyroxin** ($T_4$) and **triiodothyronine** ($T_3$).

### Deficiency Manifestations of Iodine

Iodine deficiency may lead to **goiter** (thyroid enlargement), which may or may not be associated with abnormal function. Iodine deficiency causes **cretinism** in infants, characterized by physical and mental retardation and **myxedema** in adults, characterized by physical and mental sluggishness.

## ZINC

Zinc is an intracellular mineral. Zinc is present in the skeletal muscles, bones, prostate gland, liver, pancreas, nail, and hair.

### Biological Role of Zinc

- Zinc serves as a cofactor for many enzymes. Example for zinc-containing enzymes (a type of metalloenzymes) are carboxypeptidase, alkaline phosphatase, lactate dehydrogenase, and RNA polymerase.
- Insulin when stored in the β-cells of the pancreas contains zinc. Insulin in the blood does not contain zinc.

### Requirement of Zinc

The requirement of zinc is 10 mg/day.

### Dietary Sources of Zinc

Beans, nuts, grains, and meat are good dietary sources of zinc.

### Normal Serum Level of Zinc

The normal serum level of zinc is 100 µg/dL.

## Zinc Deficiency Manifestations

Deficiency of zinc leads to poor wound healing, skin lesions, impaired spermatogenesis, and growth failure. Zinc deficiency can lead to central nervous system (CNS) manifestations like depression, dementia, and psychiatric disorders. Skin manifestation includes dermatitis and alopecia.

## Toxicity of Zinc

Zinc toxicity is usually seen in professional welders due to the inhalation of zinc oxide fumes. Symptoms include nausea, vomiting, gastric ulcers, anemia, and headache.

## FLUORIDE

### Requirement of Fluoride

The safe limit of fluoride is 1 ppm (parts per million) in drinking water.

### Dietary Source of Fluoride

The dietary source of fluoride is drinking water.

### Biological Role of Fluoride

Fluoride is known to prevent dental caries. Pits and fissures of molar and pre-molar teeth are sites of bacterial fermentation of food particles, which leads to acid production and erosion of teeth enamel causing inflammation and tooth ache. Application of fluoride results in the protective coating of fluoride on the enamel, thus protecting it from decay by acids.

### Toxicity of Fluoride

Fluoride levels more than 2 ppm in drinking water causes gastroenteritis, loss of appetite, and loss of weight. Levels more than 5 ppm in drinking water causes discoloration of teeth. Levels more than 20 ppm will cause osteoporosis and brittle bones. This condition is called **fluorosis.**

## SELENIUM

Selenium is present in soil and is a growth stimulant in agriculture.

### Requirement of Selenium

The requirement of selenium is 50–100 µg/day.

### Biological Role of Selenium

- It functions mainly as an antioxidant.
- It protects cell membrane from lipid peroxidation. Both vitamin E and selenium are antioxidants and are complementary to each other.
- Selenium is part of the enzyme glutathione peroxidase.

**Table 17.5** gives a summary of the minerals necessary for normal health.

## HEAVY METAL POISONING

### Lead Poisoning and its Consequences

Lead is the most common heavy metal poison in India. About 30% of the population are already affected by lead poisoning. It is not biodegradable. It is dispersed into air, food, soil, and water. Paint is the major source for exposure, especially in children, as they bite painted toys. Also, paint peels off as small flakes from the walls of living rooms. Increased content of lead is seen in the air, water, and vegetables in cities and near highways. This is due to the tetraethyl lead derived from the exhaust of vehicles. Statutory use of lead-free petrol has reduced this type of contamination. Lead pipes are important sources for contamination. Newspapers and xerox copies contain lead, which is adsorbed into fingertips, and later contaminate the foodstuff taken by hands. One pack of cigarette contains 15 mg of lead and chronic smokers have higher levels of lead in their blood. Battery repair, radiator repair, soldering, painting, and printing are occupations prone to get lead poisoning.

**Table 17.5:** Summary of mineral metabolism.

|  | Requirement of minerals for adult male per day | Blood level | Deficiency manifestations |
|---|---|---|---|
| Calcium | 500 mg | 9–11 mg/dL | Tremors |
| Phosphorus | 500 mg | 3–4 mg/dL | Defect in calcification of bone |
| Sodium | 5–10 g | 136–145 mEq/L | Muscle cramps |
| Potassium | 3–4 g | 3.5–5 mEq/L | Cardiac arrythmias |
| Chloride | 2–5 g | 96–106 mEq/L | Acid–base balance disturbance |
| Iron | 20 mg | 120 mg/dL (plasma) | Anemia |
| Copper | 1 mg | 100 mg/dL | Anemia |
| Iodine | 150 mg | 5–10 mg/dL | Hypothyroidism |
| Zinc | 15 mg | 100 mg/dL | Skin lesions |

Lead is a cumulative poison and is accumulated in tissues over the years. More than 10 mg/dL of lead in children and more than 25 mg/dL of lead in adults lead to toxic manifestations.

Miscarriage, stillbirth, and premature birth are reported due to lead poisoning in pregnant women. Developing brains are more susceptible to lead. Permanent neurological sequelae, cerebral palsy, and optic atrophy may be seen with lead poisoning. In children, mental retardation, learning disabilities, behavioral problems, hyperexcitability, and seizures are seen. Anemia, abdominal colic, and loss of appetite are very common. Lead inhibits heme synthesis.

If the level of lead in the blood is more than 70 mg/dL, acute toxicity is manifested as encephalopathy, convulsions, mania, neuropathy, abdominal colic, severe anemia, and kidney damage. Discoloration and blue line along the gums are characteristic features of acute lead poisoning.

Preventive measures include the following Lead pipes should be avoided. Adequate iron and calcium will reduce the absorption of lead. Plants and grass around the house will absorb lead and make the air cleaner. Lead-free petrol should be used. Strict laws on the lead content of paints should be enacted and implemented.

## SUMMARY OF THE CHAPTER

- Factors favorably influencing calcium absorption are calcitriol, parathyroid hormone (PTH), acidic pH, and arginine.
- Factors decreasing calcium absorption are phytic acid, high phosphate content, and malabsorption syndromes.
- Calcium activates enzymes such as protein kinases, glycogen synthase, pyruvate carboxylase, and pyruvate dehydrogenase.
- Calcium mediates the contraction of muscle fibers.
- Calcium mediates the secretion of insulin, PTH, calcitonin, and ADH and can act as a second messenger for some hormones, for example, glucagon.
- Calcium is known as factor IV in blood coagulation.
- Hypercalcemia may occur in hyperparathyroidism, multiple myeloma, bone cancer, and Paget's disease.
- Hypocalcemia leads to tetany as seen in renal tubular acidosis.
- Hypernatremia is seen in Cushing's syndrome and after prolonged cortisone therapy. Hyponatremia is seen in Addison's disease.
- Renal loss of potassium is seen in renal tubular acidosis, tubular necrosis, and metabolic alkalosis.

- Hyperchloremia is seen in dehydration, Cushing's syndrome, respiratory acidosis, and renal tubular acidosis.
- Iron homeostasis is maintained by the regulation of absorption and not by excretion.
- Transport form of iron is transferrin. Storage form of iron is ferritin.
- Iron deficiency may be caused by hookworm infection, nephrosis, and lead toxicity, and leads to microcytic hypochromic anemia.
- Iron toxicity leads to hemosiderosis.
- Copper in the plasma is bound to ceruloplasmin. It functions as a cofactor for vitamin C-mediated hydroxylations.
- Copper is essential for the formation of Hb, because copper deficiency leads to microcytic normochromic anemia.
- Copper level in the blood is lowered in Wilson's hepatocellular degeneration.
- Zinc is stored in combination with metallothionein. More than 300 enzymes are known to be zinc dependent.

## QUESTIONS FOR SELF-ASSESSMENT

1. Mention the examples of macronutrient and micronutrient.
2. Mention the functions of calcium. Add a schematic diagram regarding calcium homeostasis.
3. Mention the function of fluoride in our body. What is fluorosis?
4. Mention why iron is known as "one-way element."
5. Mention the mucosal block theory of iron absorption.
6. Mention the major functions of the metal copper.
7. What are the consequences of hypochloremia and hyperchloremia
8. Mention a list of enzymes that are zinc dependent.
9. Why restriction of common salt in diet is restricted in hypertension.
10. Mention the function of selenium in our body.

## MULTIPLE CHOICE QUESTIONS (MCQs)

17-1. The following favors iron absorption but exception is:
a. Vitamin C
b. Ferric iron
c. Pregnancy
d. Increased erythropoiesis

17-2. Regarding iron metabolism, one of the following statements is incorrect:
a. Ferritin is the storage form of iron.
b. Transferrin is the transport from of iron.
c. Iron is oxidized to $Fe^{3+}$ after absorption.
d. In the case of iron overload, more iron is transported as transferrin.

17-3. The following conditions results in iron deficiency, *except:*
a. Dermatitis
b. Malabsorption
c. Hookworm infestation
d. Hemorrhagic shock

17-4. Hypophosphatemia is observed in all the following conditions, *except:*
a. Rickets
b. Hyperparathyroidism
c. Fanconi syndrome
d. Renal failure

17-5. Calcium is required for the following, *except:*
a. Coagulation
b. Neuromuscular transmission
c. Absorption of iron
d. Intracellular messenger

17-6. All are major functions of sodium, *except:*
a. It maintains the osmotic pressure.
b. It regulates the electrolytes and pH balance of the extracellular compartment
c. It controls the movement of ions in muscle.
d. It helps in the active transport of glucose.

17-7. Concerning the functions of copper, all the statements are true, *except:*
a. Ceruloplasmin is a copper-containing protein

b. Glutaminase is a copper-dependent enzyme
c. Superoxide dismutase is a copper-dependent enzyme
d. Tyrosinase is a copper-dependent enzyme

17-8. **Normal level of potassium in plasma is:**
a. 3–4 mg/dL
b. 9–11 mg/dL
c. 4–5 mEq/L
b..96–106 mEq/L

17-9. **In a patient, the serum calcium level was 12 mg/dL and serum inorganic phosphate was 2 mg/dL. The above findings are characteristic of:**
a. Excess calcitonin secretion
b. Vitamin D deficiency
c. Hyperparathyroidism
d. Hypervitaminosis D

17-10. **The serum calcium level is:**
a. 10–15 g/dL
b. 11–15 mg/dL
c. 4–5 mg/dL
d. 9–11 mg/dL

## ANSWER KEYS TO MCQs

17-1. (b)  17-2. (d)  17-3. (a)  17-4. (d)  17-5. (c)  17-6. (c)  17-7. (b)  17-8. (b)
17-9. (c)  17-10. (d)

# CHAPTER 18

# Nutrition

At the completion of this chapter, the reader will be able to answer questions on the following topics:

- Respiratory quotient
- Specific dynamic action (SDA)
- Basic metabolic rate (BMR)
- Dietary carbohydrates
- Dietary fibers
- Dietary fats
- Dietary proteins
- Protein calorie malnutrition (PCM)
- Protein energy malnutrition (PEM)
- Balanced diet
- Recommended dietary allowance (RDA)

The food we take undergoes digestion and absorption. The absorbed food undergoes metabolic changes, so that it can provide energy for all the vital functions of the body as well as the building materials for the body tissues.

## CALORIFIC VALUE

The energy content of food materials is measured in calories. **One calorie** is the heat required to raise the temperature of 1 g of water by 1°C. Since it is a very small unit, in the medical practice, the energy content is usually expressed in **kilocalorie** (kcal), which is equal to 1,000 calories. The maximum available energy contained in a food can be measured by burning it in an atmosphere of oxygen in a calorimeter. The calorific value of nutrients, otherwise known as **energy density** (energy yield per unit weight of food), is given in **Table 18.1**.

Table 18.1: Calorific value of nutrients (energy yield from nutrients).

| Nutrient | Energy yield (calorific value in kcal/g) |
|---|---|
| Carbohydrates | 4 |
| Fats | 9 |
| Proteins | 4.2 |
| Alcohol | 7 |

## RESPIRATORY QUOTIENT

It is defined as the ratio of the volume of $CO_2$ produced to the volume of oxygen consumed.

$$RQ = \frac{\text{Volume of } CO_2 \text{ produced}}{\text{Volume of } O_2 \text{ consumed}}$$

The RQ of 1 is for carbohydrates. Fats and proteins have values less than 1. For a mixed diet, RQ is taken as 0.8.

## ENERGY REQUIREMENTS OF A NORMAL PERSON

While calculating the energy requirements, we have to consider the energy required for the (1) maintenance of basal metabolic rate (BMR), (2) specific dynamic action or thermogenic effect of food; and (3) extra energy expenditure for physical activities.

## Specific Dynamic Action

Specific dynamic action (SDA) is defined as the increased heat production over and above the energy of the food. This may be due to the energy expenditure for the digestion and absorption of food. SDA is maximum for proteins (30%). For lipids and carbohydrates, SDA is 15% and 5%, respectively. This energy is taken from previously available energy, so that the actual energy from the food is lesser than that of theoretical calculation. This energy is to be supplied initially.

## Basal Metabolic Rate

Basal metabolic rate (BMR) is defined as the energy requirement of a person when the person is physically, mentally, emotionally, and at digestive rest when he or she is awake. It is the minimum amount of energy required to maintain life or sustain vital functions like the working of the heart, circulation, brain function, respiration, and so on. This is not the minimum energy need of the body. The minimum energy requirement is during sleep. The metabolic rate during sleep is less than BMR.

Basal metabolic rate can be estimated directly by measuring the heat evolved. But it is a difficult procedure. Therefore, BMR is generally estimated indirectly by measuring the oxygen consumed and $CO_2$ given out. BMR is expressed usually in kcal/hr/sq.m of the body. For easier calculations, BMR for an adult is fixed as **24 kcal**/kg body weight per day. It can also be expressed in terms of percentage increase or decrease from the normal BMR value.

### Factors Affecting the Basal Metabolic Rate

- **Surface area:** The most important is the surface area, which depends on the height and weight of the individual.
- **Age:** During the period of active growth, BMR is high. It reaches a maximum by five years of age. During old age, BMR is lowered.
- **Sex:** Males have higher BMR than females.
- **Physical activity:** The increase in BMR during exercise is due to increased cardiac output.
- **Climatic conditions:** BMR increases in cold climate as a compensatory mechanism to maintain body temperature.
- **Thyroid hormones:** Since thyroid hormones have a general stimulant effect on the rate of metabolism and heat production, BMR is raised in hyperthyroidism and lowered in hypothyroidism.
- **Physical activity:** The energy requirements would depend on the occupation, physical activity, and lifestyle of the individual. The activity level may be divided into three groups—sedentary, moderate, and heavy. Additional calories are to be added for each category:

For sedentary work, +30% of BMR; for moderate work, +40% of BMR; and for heavy work, +50% of BMR should be added (**Table 18.2**). The energy requirement of a 55-kg male doing moderate work may be calculated as shown in **Table 18.3**.

| Table 18.2: Energy requirement and occupation. | |
|---|---|
| Type of activity | Occupation |
| Light | Office workers, lawyers, accountants, doctors, teachers |
| Moderate | Students, industry workers, farm workers, housewives without mechanical appliances |
| Very active | Agricultural workers, miners, unskilled laborers, athletes |
| Heavy work | Lumberjacks, blacksmiths, and construction workers |

| Table 18.3: Calculation for the energy requirement of a 55-kg person doing moderate work. | | |
|---|---|---|
| For BMR | = 24 × 55 kg | = 1,320 kcal |
| + For activity | = 40% of BMR | = 528 kcal |
| Subtotal | = 1320 + 528 | = 1,848 kcal |
| + Need for SDA | = 1848 × 10% | = 184 kcal |
| Total | = 1848 + 184 | = 2,032 kcal |
| Rounded to the nearest multiple of 50 | | = 2,050 kcal |

## PROXIMATE PRINCIPLES

The proximate (major) principles of diet are carbohydrates, fats (lipids), and proteins. Other important constituents are minerals, vitamins, and water.

### Dietary Carbohydrates

The major energy requirement of the body is met by carbohydrate. Generally speaking, 60–70% of the energy needs should be met from carbohydrates. Available carbohydrates (such as starch) give energy. Unavailable carbohydrates are otherwise called **dietary fibers**. They are so called because they cannot be digested easily. But, they are necessary to maintain the normal motility of the gastrointestinal tract and can reduce the incidence of colon cancer. They also have a hypolipidemic effect.

### Dietary Fats

The dietary fat is essential for the body and is composed of saturated fatty acids (SFAs), monounsaturated fatty acids (MUFAs), and polyunsaturated fatty acids (PUFAs). PUFAs are the source of essential fatty acids like linoleic, linolenic, and arachidonic acids, which are needed for the synthesis of prostaglandins. PUFA and omega-3 fatty acids present in fish oil can reduce the serum level of LDL cholesterol.

**Visible fats** are fats that are easily seen and identified such as butter, ghee, and oils. **Invisible fats** are fats present as part of other food items, for example, egg, fish, meat, cereals, nuts, and oilseeds. More than half of the essential fatty acids in the Indian diet is in the form of invisible fat.

The atherogenic effect of **cholesterol** and the risk of coronary artery disease in people with hypercholesterolemia are described in Chapter 9. Food items known to be rich in cholesterol (egg yolk, animal liver, brain, and kidney) are to be consumed in limited amounts. Vegetables, cereals, and pulses do not contain any cholesterol. Further, vegetable sterols will inhibit cholesterol absorption.

Fats provide a concentrated source of energy. A minimum intake of lipids is essential since the requirements of fat-soluble vitamins and essential fatty acids are to be met. Fats increase the taste and palatability of food. They are the favored cooking medium.

The ideal fat intake is about 25% of the total calories, out of which about 25% may be PUFA. This will be a total of about 25 g of oils and about 3 g of PUFA for a normal person. PUFA should not be more than 30% of total fat. Moreover, the fat content should be such that SFA:MUFA:PUFA may be in 1:1:1 ratio. Further, cholesterol intake should be less than 250 mg/day.

### Dietary Proteins

Proteins are the sources of essential amino acids to the body, which are called the building blocks. A good quality protein should contain all the essential amino acids in the correct proportion needed for the body. In a normal diet, the energy from proteins should be limited to a maximum of 15%.

### Nitrogen Balance

A normal healthy adult is said to be in nitrogen balance, because the dietary intake (I) equals the daily loss through urine (U), feces (F), and skin (S). In other words, I = U + F + S **(Fig. 18.1)**.

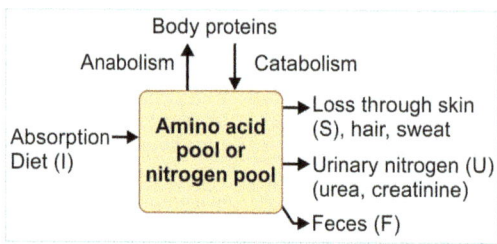

**Fig. 18.1:** Nitrogen balance.

When the excretion exceeds intake, it is **negative** nitrogen balance. When the intake exceeds excretion, it is a state of **positive** nitrogen balance.

### Factors Affecting Nitrogen Balance

- **Growth:** During the period of active growth, a state of positive nitrogen balance exists. On an average, when a person gains 5 kg, about 1 kg of proteins (160 g nitrogen) are added to the body.
- **Hormones:** Growth hormone, insulin, and androgens promote positive nitrogen balance, while corticosteroids cause a negative nitrogen balance.
- **Pregnancy:** A pregnant woman will be in a state of positive nitrogen balance due to the growth of fetus.
- **Convalescence:** A person convalescing after an illness or surgery will be in positive nitrogen balance due to active regeneration of tissues.
- **Acute illness:** Negative nitrogen balance is seen in subjects immediately after surgery, trauma, and burns.
- **Protein deficiency:** The deficiency of even a single amino acid can cause negative nitrogen balance. Prolonged starvation is another important cause.

## Protein Requirement

The recommended daily allowance of protein is 1 g/kg body weight. During infancy and childhood as well as during pregnancy and lactation, the requirement may go from 1.5–2 g/kg body weight per day. Requirement will be more in the case of infants, children, adolescents, pregnancy, lactation, and convalescence. As growth stops, protein requirement also decreases.

### Limiting Amino Acids

Limiting amino acid limits the weight gain when a protein is supplied to an animal. This essential amino acid that is lacking in that protein is said to be the **limiting amino acid** (Table 18.4).

### Supplementation

This problem may be overcome by taking a mixture of proteins in the diet. **Mutual supplementation of proteins** is thus achieved (Table 18.4). For example, pulses are deficient in methionine but rich in lysine. On the other hand, cereals are deficient in lysine but rich in methionine. Therefore, a combination of pulses plus cereal (e.g., chapati + dal) will cancel each other's deficiency and become equivalent to first-class protein.

### Protein Calorie Malnutrition or Protein–Energy Malnutrition

This is characterized by a severe deficiency of nutrients, especially in quantity and quality. **Marasmus** and **kwashiorkor** are the two different forms of protein calorie malnutrition

**Table 18.4:** Limiting amino acids in proteins.

| Protein | Limiting amino acid | Protein supplemented to cancel the deficiency |
| --- | --- | --- |
| Rice | Lysine, threonine | Pulse proteins |
| Wheat | Lysine, threonine | Pulse proteins |
| Tapioca | Phenyl alanine, tyrosine | Fish proteins |
| Bengal gram | Cysteine, methionine | Cereals |

(PCM). Marasmus is due to the combined prolonged deficiency of protein and calorie, while kwashiorkor is predominantly due to the deficiency of protein. **Table 18.5** shows the comparison between marasmus and kwashiorkor. A severe continued PCM may lead to mild to moderate mental retardation.

## OBESITY

This is due to the increased energy intake and decreased energy expenditure. The **obesity index** is calculated as $W/H^2$ (where $W$ = weight in kilograms and $H$ = height in meters); it is used to assess the obesity. "Obesity index" is an old terminology; the modern expression is **body mass index** (BMI). A person is obese when BMI exceeds 27.8 kg/m$^2$ if the person is a male and 27.3 kg/m$^2$ if the person is a female (excess of 120% of desirable body weight).

Obesity can occur only as a result of ingestion of food in excess of the body's needs. The major causes are food habits (intake of calorie-rich food in excess amounts) and lack of exercise. There is an increase in the number (hyperplasia) and size (hypertrophy) of the adipocytes.

### Diseases Related to Obesity

Sensitivity of peripheral tissues to insulin is decreased. The major ill effects of obesity are increased risk of **coronary artery disease, diabetes mellitus, hypertension, metabolic syndrome,** and a reduced life span. Obesity is a lifestyle disease. Treatment is to reduce the intake of calories and fat. Controlled exercise is very useful.

## BALANCED DIET

A balanced diet contains all the essential nutrients in optimum quantity required for the growth and development of the body as well as to maintain health. The food should be sufficient not only in quantity but also in quality. The protein in the diet should be a good quality protein that contains all the essential amino acids. The lipids should provide all the essential fatty acids required for the body. While formulating a diet one should take into account the physiological status of the persons as well as the physical activity in which the individual is involved. Milk and egg are considered to be complete diet as they contain all the essential nutrients required for the body. (However, iron is deficient in milk).

### Prescription of a Diet

When prescribing a diet, the following points are to be kept in mind:
- It should be a balanced diet containing all essential nutrients.
- The diet should be simple, locally available, palatable, and digestible.
- Adequate protein content with essential amino acids should be supplied. This is achieved by a cereal–pulse mixture with additional animal proteins, if necessary.
- Calorie intake should be correct and should balance the energy expenditure.

**Table 18.5:** Salient features of kwashiorkor and marasmus.

|  | Marasmus | Kwashiorkor |
| --- | --- | --- |
| Deficiency of | Calorie | Protein |
| Cause | Early weaning and repeated infection | Starchy diet after weaning, precipitated by an acute infection |
| Growth retardation | Marked | Present |
| Attitude | Irritable and fretful | Lethargic and apathetic |
| Appetite | Normal | Anorexia |
| Hair | No characteristic change | Sparse, soft, and thin hair; curls may be lost |
| Serum albumin | 2–3 g/dL | <2 g/dL |

- Special care should be taken to see that adequate quantity of calcium and iron are obtained from the diet. The absorption of these minerals is reduced by other factors in Indian diet.

*Steps to be Taken When Prescribing a Diet*

1. What is the requirement of the person with regard to calorie and other essential nutrients?
2. What is the quantity of proximate principles required?
3. Which composition of food will give the above requirement?
4. How can a palatable diet that contains these compositions be prescribed?
5. The total quantity may be divided into three or four meals.

**First Step: Calorie Requirement**

For a 60-kg sedentary man, the energy requirement is 60 × 30 = 1,800 kcal plus additional allowance for specific dynamic action (1,800 × 10% = 180). Therefore, the total requirement is roughly 2,000 kcal. Balanced diet should contain calories from carbohydrate, proteins, and fat in the ratio of 60:20:20.

**Second Step: Proximate Principles**

He requires 60 g of proteins. This will give 60 × 4 = 240 kcal of energy. His total requirement is 2,000 kcal. Therefore carbohydrates plus fats should produce (2,000–240) = 1,760 kcal.

As a general rule, about 20% of total calories are supplied by fat. Therefore, fats should supply 1,760 × 20% = 350 kcal, which is provided by 350/9 = about 35 g of fats. (About 30% of the total fat may be supplied as PUFAs).

The rest 1,400 kcal are supplied by 350 g of carbohydrates. These calculations are based on the fact that 1 g of carbohydrate provides 4 kcal, 1 g of fat supplies 9 kcal, and 1 g of protein gives rise to 4 kcal.

**Third Step: General Composition of Food**

The third step is to calculate how these proximate principles are supplied as common foodstuffs.

**Mutual supplementation** of cereals and pulses: Although the protein content of pulses is more than cereals, the average Indian diet contains more cereals, and hence proteins are mainly supplied by cereals. But pulses give good quality proteins. A judicious combination of cereals and pulses provides all the essential amino acids (pulses are deficient in methionine, while cereals lack in lysine). An accepted formula is that the food should contain pulses and cereals in the ratio of 1:5 to provide good quality proteins.

**Fourth Step: Determine the Items of Food**

The prescribed diet should satisfy the requirements regarding protein (60 g) fats (45 g), calories (2,000 kcal), calcium (400 mg), and iron (25 mg). When calories alone are to be increased, as in the case of a person having severe exercise, tubers and roots will serve this purpose.

As he requires 60 g of proteins, this will give 60 × 4 = 240 kcal of energy. As a general rule, about 25% of total calories are supplied by fat, or about 25 g of fat. (About 30% of the total fat may be supplied as PUFAs). The rest 1,400 kcal are supplied by 350 g of carbohydrates.

Table 18.6 gives a balanced diet for a 60-kg sedentary man. As the activity of the person is increased (manual labor), the calorie requirement will be more.

**Table 18.6:** A balanced diet for a 60-kg sedentary man.

| Item | Quantity of vegetarian food | Quantity of nonvegetarian food |
|---|---|---|
| Cereals | 350 g | 350 g |
| Pulses | 75 g | 60 g |
| Vegetable oil | 40 mL | 25 mL |
| Milk | 250 mL | 150 mL |
| Leafy vegetables | 200 g | 200 g |
| Sugar | 25 g | 25 g |
| Fish/meat | – | 60 g |

## Recommended Dietary Allowance

The dietary requirements for an individual is recommended on the basis of the need of the individual, taking into consideration all the physical and physiological variables. Sufficient amount of overages are allowed to compensate the loss of nutrients during cooking and processing. The RDA will also be based on the special needs of the person like physical activity and physiological conditions, such as sex, age, state of growth, pregnancy, lactation, and so on. The RDAs for Indians are given by the Indian Council of Medical Research (ICMR; *see* Appendix 2). Nutritional values of certain food items are shown in Appendix 3.

## Glycemic Index

It is assessed by the glucose tolerance test and comparing it with a reference meal. The reference meal is always taken as 50 g of glucose (*see* **Fig. 31.9**).

$$\text{Index} = \frac{\text{Incremental area under glucose tolerance curve after 50 g of test meal}}{\text{Incremental area under curve after 50 g of reference meal (glucose)}} \times 100$$

Simple carbohydrates such as glucose or sugar will have a high glycemic index. But the same quantity of complex carbohydrates (such as starch) will not increase the blood sugar as much. Because the digestion and absorption are slow with minimal increment in blood sugar. The glycemic index of complex carbohydrate is lesser than cane sugar. As a general rule, the glycemic index of carbohydrate is lowered if it is combined with protein, fat, or fiber, preferably at least two of the three.

## SUMMARY OF THE CHAPTER

- One calorie is the heat required to raise the temperature of 1 g of water by 1°C.
- Respiratory quotient (RQ) is defined as the ratio of the volume of $CO_2$ produced to the $O_2$ consumed. Carbohydrates, fats, and proteins have RQs of 1, 0.7, and 0.8, respectively.
- BMR is defined as energy required by an awake individual during physical, emotional, and digestive rest.
- Increased heat production following the intake of food is referred to as *specific dynamic action* (SDA). Values of SDA for proteins, lipids, and carbohydrates are 30%, 15%, and 5%, respectively.
- Dietary fibers are essential to maintain the normal motility of GI tract, prevent constipation, decrease cholesterol levels, and to improve glucose tolerance.
- Marasmus and kwashiorkor are two conditions of protein energy malnutrition.

## QUESTIONS FOR SELF-ASSESSMENT

1. Define BMR. Mention the factors affecting BMR.
2. Define balanced diet.
3. Define specific dynamic action (SDA).
4. Write a note on respiratory quotient (RQ).
5. Mention the clinical role of dietary fibers.
6. Define the term *PEM*.
7. What are the characteristic features of kwashiorkor?
8. Define the features of marasmus.
9. Write a note on obesity. Mention the diseases related to obesity.
10. Define glycemic index and mention its importance.

## MULTIPLE CHOICE QUESTIONS (MCQs)

18-1. The caloric value for 1 g of fat is:
   a. 9
   b. 6
   c. 3
   d. 4

18-2. The respiratory quotient for carbohydrate is:
   a. 0.8
   b. 1.0
   c. 2.0
   d. 0.7

18-3. **All the following foods give complete proteins, *except*:**
 a. Meat
 b. Fish
 c. Legumes
 d. Cheese

18-4. **Concerning fat, one of the following statements is incorrect:**
 a. It provides 90% of our energy needs
 b. It helps to maintain the cell membrane structure.
 c. It helps in the absorption of vitamins A and D
 d. Protein-rich foods are deficient in fat

18-5. **Which has the highest calorific value?**
 a. Glucose
 b. Palmitic acid
 c. Albumin
 d. Ethanol

18-6. **Regarding dietary fiber, all the following statements are true, *except*:**
 a. They are resistant to breakdown by the human digestive enzymes
 b. Insoluble fiber promotes normal elimination by providing bulk for stool formation
 c. Increased amounts of fiber in the diet is the reason for irritable bowel syndrome
 d. Fruits and vegetables are rich in soluble fiber

18-7. **Negative nitrogen balance is seen in:**
 a. Unconsciousness
 b. Muscle loss and weakness
 c. Hepatomegaly
 d. Overeating

18-8. **The minimum daily requirement of proteins in food for a normal healthy adult is:**
 a. 0.2 g/kg body weight
 b. 0.4 g/kg body weight
 c. 0.7 g/kg body weight
 d. 0.9 g/kg body weight

18-9. **Kwashiorkor is characterized by the following features, *except*:**
 a. Edema
 b. Diarrhea
 c. Night blindness
 d. Weight gain

18-10. **A person with 60-kg weight was given food containing 65 g of proteins per day for 2 weeks. After the experiment, his weight was found to be 60 kg itself. His nitrogen excretion is said to be:**
 a. Minus 5 g
 b. Minus 2.5 g
 c. 0 g
 d. Plus 2.5 g

## ANSWER KEYS TO MCQs

18-1. (a)  18-2. (b)  18-3. (d)  18-4. (b)  18-5. (b)  18-6. (c)  18-7. (b)  18-8. (c)
18-9. (a)  18-10. (c)

# CHAPTER 19

# Acid–Base and Electrolyte Balance

At the completion of this chapter, the reader will be able to answer questions on the following topics:

- Buffer system in blood
- Bicarbonate buffer system
- Phosphate buffer system
- Respiratory mechanism
- Renal mechanism
- Acid–base imbalance
- Acidosis and anion gap
- Metabolic acidosis
- Respiratory acidosis
- Metabolic alkalosis
- Respiratory alkalosis
- Assessment of acid–base status
- Electrolyte composition of body fluid compartments
- Regulation of sodium and water balance
- Antidiuretic hormone
- Renin–angiotensin system
- Sodium metabolism
- Hypernatremia and hyponatremia
- Potassium metabolism
- Hyperkalemia and hypokalemia
- Chloride metabolism
- Hyperchloremia and hypochloremia
- Magnesium
- Hypermagnesemia and hypomagnesemia

The oxidation of various metabolites in the living organism results in the formation of a variety of acids and bases. But under normal condition, the pH of blood does not vary beyond the range of 7.35–7.45, which is maintained approximately at 7.4. Maintenance of this constant blood pH is one of the prime requisites of life. Variation in the pH disturbs the vital process and may lead to death.

Acids produced in the body are carbonic acid, sulfuric acid, phosphoric acid, pyruvic acid, lactic acid, acetoacetic acid, β-hydroxybutyric acid, etc. Bases produced in the body are bicarbonate, phosphate, acetate, ammonia, etc.

## ACIDS AND BASES

### Definition

According to the definition proposed by Bronsted, acids are substances that are capable of donating protons and bases are those that accept protons. Acids are proton donors and bases are proton acceptors, for example:

$HCl \rightarrow H^+ + Cl^-$
$HCO_3^- + H^+ \rightarrow H_2CO_3$

### Weak and Strong Acids

The extent of dissociation decides whether they are strong acids or weak acids. Strong acids dissociate completely in solution, while weak acids ionize incompletely, for example:

HCl → H⁺ + Cl (Complete)
$H_2CO_3 \leftrightarrow H^+ + HCO_3^-$ (Partial)

In a solution of HCl, almost all the molecules dissociate and exist as H⁺ and Cl⁻ ions. Hence, the concentration of H⁺ is very high and it is a strong acid. But in the case of a weak acid (e.g., acetic acid), it will ionize only partially. So the number of acid molecules existing in the ionized state is much less and may be only 50%.

## Acidity of a Solution and pH

The H⁺ concentration is the negative of the logarithm of hydrogen ion concentration and is designated as the pH. Therefore,

$$pH = -\log[H^+] = \log \frac{1}{[H^+]}$$

Thus, the **pH value is inversely proportional to the acidity**. Lower the pH, higher the acidity or hydrogen ion concentration; while higher the pH, the acidity is lower **(Table 19.1)**. At a pH of 1, the hydrogen ion concentration is 10 times that of a solution with a pH 2 and 100 times that of a solution with a pH of 3 and so on. The **pH 7 indicates the neutral pH**.

## Buffers

A buffer is defined as a solution that resists the change in pH while adding small amounts of acid or base. Buffer consists of (a) mixture of weak acid and their corresponding salt of strong base or (b) weak base and their corresponding salt of strong acid.

**Table 19.1:** Relation between hydrogen ions, hydroxyl ions, and pH of aqueous solutions. Ionic product of water = [H⁺][OH⁻] = 10⁻¹⁴.

| −log[H⁺] = pH | pOH | Inference |
|---|---|---|
| 1 | 13 | Strong acid |
| 4 | 10 | Acid |
| 7 | 7 | Neutral |
| 10 | 4 | Alkali |
| 13 | 1 | Strong alkali |

## Henderson–Hasselbalch Equation

The relationship between pH, pKa, concentration of acid, and conjugate base (or salt) is expressed by the Henderson-Hasselbalch equation,

$$pH = pKa + \log \frac{[base]}{[acid]} \text{ or } pH = pKa + \log \frac{[salt]}{[acid]}$$

*Composition of a Buffer*

Buffers are of two types:
1. Mixtures of weak acids with their salt with a strong base or
2. Mixtures of weak bases with their salt with a strong acid. A few examples are given below:
   a. Carbonic acid and sodium bicarbonate ($H_2CO_3$/$NaHCO_3$) (bicarbonate buffer)
   b. $Na_2HPO_4$/$NaH_2PO_4$ (phosphate buffer)

*Factors Affecting the pH of a Buffer*

The pH of a buffer solution is determined by two factors:
1. **The value of pK:** The lower the value of pK, the lower is the pH of the solution.
2. **The ratio of salt to acid concentrations:** Actual concentrations of salt and acid in a buffer solution may be varied widely, with no change in pH, so long as the ratio of the concentrations remains the same.

*Factors Affecting Buffer Capacity*

On the other hand, the buffer capacity is determined by the actual **concentrations of salt and acid** present as well as by their ratio. Buffering capacity is the number of grams of strong acid or alkali which is necessary for a change in pH of one unit of 1 L of buffer solution. The buffering capacity of a buffer is defined as **the ability of the buffer to resist changes in pH when an acid or base is added.**

*How Do Buffers Act?*

Buffer solutions consist of mixtures of a weak acid or base and its salt. For example, when hydrochloric acid is added to the acetate buffer, the salt reacts with the acid forming the weak acid, acetic acid and its salt. Similarly when a base is added, the acid reacts with it forming salt and water. Thus, changes in the pH are minimized.

$CH_3-COOH + NaOH \rightarrow CH_3-COONa + H_2O$
$CH_3-COONa + HCl \rightarrow CH_3-COOH + NaCl$

*Effective Range of a Buffer*

A buffer is most effective when the concentrations of **salt and acid are equal** or when the pH = pKa. The effective range of a buffer is **1 pH unit higher or lower** than pKa.

## Normal pH

The **pH of plasma is 7.4** (average hydrogen ion concentration of 40 nmol/L). The pH of plasma is maintained within a **narrow range of 7.38–7.42**. The pH of the interstitial fluid is generally 0.5 units below that of the plasma.

## Acidosis

**If the pH is below 7.38, it is called acidosis.** Life is threatened when the pH is lowered below 7.25. Acidosis leads to central nervous system (CNS) depression and coma. Death occurs when the pH is below 7.0.

## Alkalosis

**When the pH is more than 7.42, it is alkalosis.** It is very dangerous if the pH increases above 7.55. Alkalosis induces neuromuscular hyperexcitability and tetany. Death occurs when the pH is above 7.6.

## MECHANISM OF REGULATION OF pH IN THE BODY

To accomplish the normal blood pH, the body has three lines of defense mechanisms.
1. Buffer system in the blood
2. Respiratory mechanism
3. Renal mechanism.

## Buffer Systems in the Blood

- Bicarbonate buffer system ($NaHCO_3/H_2CO_3$)
- Phosphate buffer system ($Na_2HPO_4/NaH_2PO_4$)
- Protein buffer system (NaPr/HPr)
- Hemoglobin buffer system.

*Bicarbonate Buffer System*

It is the main buffer system in the blood plasma. It consists of weak acid ($H_2CO_3$) and its corresponding salt with strong base ($NaHCO_3$). Normal ratio of $NaHCo_3/H_2CO_3$ in blood is 20:1 (**Table 19.2**).

The bicarbonate buffer system is called the alkali reserve. This buffer system neutralizes strong and nonvolatile acids entering the blood to weak and volatile acid ($H_2CO_3$). Carbonic acid thus formed is eliminated through the diffusion of carbon dioxide ($CO_2$) through alveoli of lungs. The normal ratio of bicarbonate buffer system in blood is maintained with the help of lungs; hence, the bicarbonate buffer system is directly linked with respiration. It also neutralizes strong bases.

*Phosphate Buffer System*

The concentration of phosphate buffer in the blood plasma is about 8% of that of the bicarbonate buffer. So its buffering capacity is much lower than the bicarbonate buffer system. The phosphate buffer system consists of dibasic phosphate ($Na_2HPO_4$) and monobasic phosphate ($NaH_2PO_4$). The normal ratio of the dibasic phosphate and the monobasic phosphate is 4:1. This ratio is kept constant with the help of kidneys. Thus, the phosphate buffer system is directly linked with kidneys (**Table 19.2**).

*Protein Buffer System*

The buffering capacity of plasma protein is much less than hemoglobin buffer system. In acidic medium, protein acts as a base; and in alkaline medium, it acts as an acid (**Table 19.2**).

Table 19.2: Buffer systems of the body.

| | Extracellular fluid | Intracellular fluid | Erythrocyte fluid |
|---|---|---|---|
| 1. | $\dfrac{NaHCO_3}{H_2CO_3}$ (bicarbonate) | $\dfrac{K_2HPO_4}{KH_2PO_4}$ (phosphate) | $\dfrac{K^+ Hb}{H^+ Hb}$ (hemoglobin) |
| 2. | $\dfrac{Na_2HPO_4}{NaH_2PO_4}$ (phosphate) | $\dfrac{K^+ \text{Protein}}{H^+ \text{Protein}}$ (protein buffer) | $\dfrac{K_2HPO_4}{KH_2PO_4}$ (phosphate) |
| 3. | $\dfrac{Na^+ \text{Albumin}}{H^+ \text{Albumin}}$ | $\dfrac{KHCO_3}{H_2CO_3}$ | $\dfrac{KHCO_3}{H_2CO_3}$ |

## Hemoglobin Buffer System

They are involved in buffering $CO_2$ inside erythrocytes. The buffering capacity of hemoglobin depends on its oxygenation and deoxygenation. The buffering capacity of hemoglobin is due to the presence of imidazole nitrogen group (**Table 19.2**).

## Respiratory Mechanism

The respiratory system serves as a mechanism for rapid adjustment of acid–base balance by regulating the concentration of $H_2CO_3$ in blood and other body fluids. The respiratory system controls $H_2CO_3$ concentration in plasma by the removal of $CO_2$ through expired air.

$$H_2CO_3 \longrightarrow CO_2 + H_2O$$

In acidic state, the respiratory center located in the medulla of the brain is stimulated and the respiratory rate increases (hyperventilation). Thus, increased amount of $CO_2$ is blown off in the expired air with the residual hydrogen ion being left as water. In the alkaline state, the reverse occurs (hypoventilation), and more $CO_2$ is retained; hence the concentration of $H_2CO_3$ is increased. This mechanism helps to maintain the normal ratio of bicarbonate buffer system.

## Renal Mechanism

Kidneys excrete urine (pH around 6) with a pH lower than that of extracellular fluid (pH = 7.4). This is called **acidification of urine**. The pH of the urine may vary from as low as 4.5 to as high as 9, depending on the amount of acid excreted. The major kidney mechanisms for regulation of pH are the following:

- Excretion of $H^+$ and generation of bicarbonate
- Excretion of titratable acid and excretion of $NH4^+$ (ammonium ions)

### Excretion of $H^+$ and Generation of Bicarbonate

The $CO_2$ combines with water to form $H_2CO_3$, with the help of carbonic anhydrase. The $H_2CO_3$ then ionizes to $H^+$ and bicarbonate. The hydrogen ions are secreted into the tubular lumen; in exchange for $Na^+$ reabsorbed. These $Na^+$ ions along with $HCO_3^-$ will be reabsorbed into the blood. There is net excretion of hydrogen ions and net generation of bicarbonate. So this mechanism serves to increase the alkali reserve (**Fig. 19.1**).

### Excretion of $H^+$ as Titratable Acid

The hydrogen ions are generated in the tubular cell by a reaction catalyzed by **carbonic anhydrase**. The term titratable acidity of urine refers to the number of milliliters of N/10 NaOH required to titrate 1 L of urine to pH 7.4. This is a measure of net acid excretion by the kidney. Due to the $Na^+$-to-$H^+$ exchange occurring at the renal tubular cell border, the $Na_2HPO_4$ (basic phosphate) is converted to $NaH_2PO_4$ (acid phosphate). As a result, the pH of the tubular fluid drops. The phosphate buffer is considered as the urinary buffer (**Fig. 19.2**).

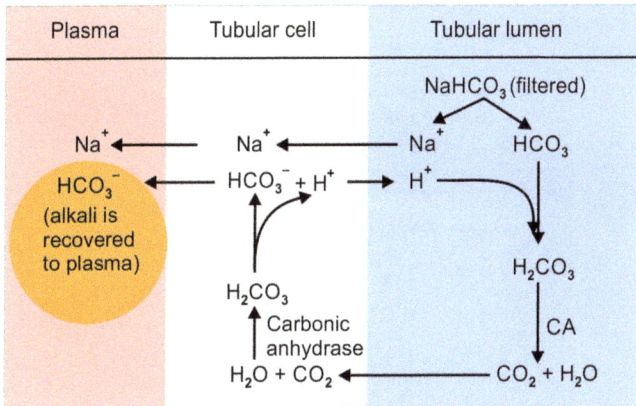

**Fig. 19.1:** Excretion of hydrogen ions in the proximal tubules. Reabsorption of bicarbonate from the tubular fluid. (CA: carbonic anhydrase)

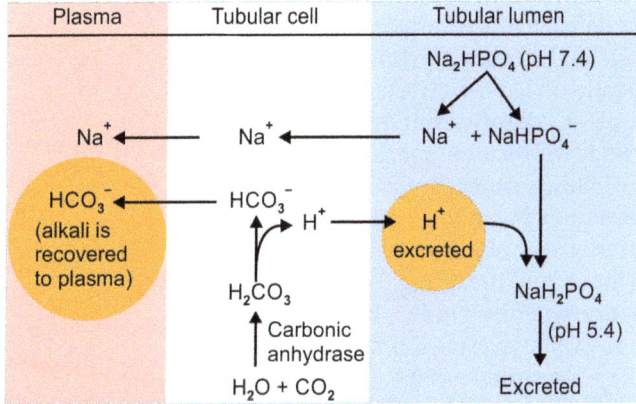

**Fig. 19.2:** Phosphate mechanism in tubules.

## Excretion of Ammonium Ions

This mechanism also helps to trap hydrogen ions in the urine, so that large quantity of acid can be excreted with minor changes in pH. The excretion of ammonia helps in the elimination of hydrogen ions without appreciable change in the pH of the urine. The **glutaminase** present in the tubular cells can hydrolyze glutamine to ammonia and glutamic acid. The ammonia diffuses into the luminal fluid and combines with hydrogen ion to form ammonium ion **(Fig. 19.3)**.

## ACID–BASE IMBALANCE

Acid-base imbalance can manifest as acidosis and alkalosis.
Acidemia → increased concentration of $H^+$
Alkalemia → decreased concentration of $H^+$
Euphemia → state where the pH is within the normal limit. The clinical state of acid accumulation which is referred to as acidosis (fall in pH) and accumulation of base is called alkalosis (rise in pH).
- Acidosis can be (1) metabolic acidosis and (2) respiratory acidosis
- Alkalosis can be (1) metabolic alkalosis and (2) respiratory alkalosis.

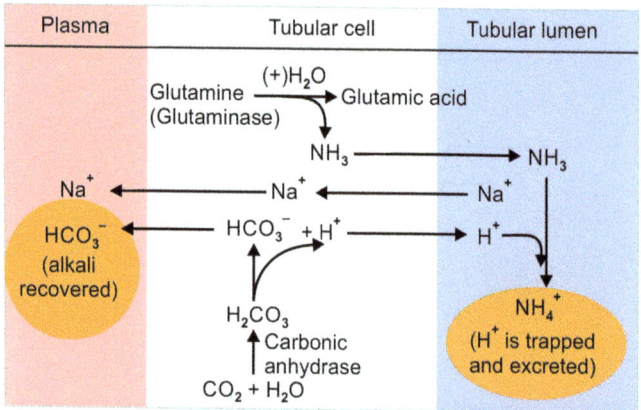

**Fig. 19.3:** Ammonia mechanism.

## Metabolic Acidosis

It is due to a primary deficit in the bicarbonate (primary alkali deficit). In case of primary deficit of $HCO_3^-$, the ratio of $(HCO_3^-)/(H_2CO_3)$ = 20:1 decreases, i.e., pH decreases resulting in metabolic acidosis. Body will try to compensate the condition by the following mechanisms:

a. **Primary compensatory mechanism:** The respiratory center is stimulated causing deep and rapid (Kussmaul) breathing. This increased ventilation will result in $CO_2$ loss and reduction in $H_2CO_3$. As a result, the ratio of $(HCO_3^-)/(H_2CO_3)$ is restored as 20:1. But low $PCO_2$ depressing the respiratory center is set against each other and respiratory compensation is only partial.

b. **Secondary compensatory mechanism:** It is the renal mechanism to correct the disturbances. Details are shown above.

### Causes of Metabolic Acidosis

- Excessive production of acid ion in:
  - Diabetic acidosis
  - Lactic acidosis
  - Starvation
  - Increased fever
  - Shock
  - Anoxia.
- Ingestion of acidifying salt, dietary or iatrogenic, e.g., phosphoric acid, acetylsalicylic acid, HCl, etc.
- Renal insufficiency
- Abnormal loss of $HCO_3^-$ in severe diarrhea, vomiting, etc.

### Anion Gap

The anion gap is a mathematical approximation of the difference between the anions and cations routinely measured in serum. Routine electrolyte measurement include $Na^+$ and $K^+$ (95% cations) and $Cl^-$ and $HCO_3^-$ (86% anions). The unmeasured cations, i.e., $Ca^{2+}$ and $Mg^{2+}$ together will be average 7 mmol/L; and the unmeasured anions, i.e., $PO_4^-$ and $SO_4^-$ together average 24 mmol/L.

The sum of cations and anions in extracellular fluid (ECF) is always equal, so that the electrical neutrality is maintained. The **unmeasured anions constitute the anion gap.** The anion gap is calculated as the difference between $[Na^+ + K^+]$ and $[HCO_3^- + Cl^-]$. Normally this is about $12 \pm 5$ mmol/L.

The alteration in the anion gap is extremely useful in the clinical assessment of patients with acid-base disorders. The increase in anion gap occurs in acidosis resulting from renal failure, diabetic ketoacidosis, and lactic acidosis. Anion gap increases due to the accumulation of other buffer anions.

### Normal Anion Gap Acidosis

When there is a loss of both anions and cations, the anion gap is normal but acidosis may prevail, e.g., in diarrhea.

**Decreased Anion Gap Acidosis**

This occurs due to either (a) increased unmeasured cations or (b) decreased unmeasured anions.

## Respiratory Acidosis

A primary excess of $H_2CO_3$ is a feature of respiratory acidosis. If excretion of $CO_2$ through lung is impaired, the $CO_2$ will accumulate in blood, resulting in excess $H_2CO_3$ formation. This results in lowering the ratio of $(HCO_3^-)/(H_2CO_3)$, lowering the pH.

*Compensatory Mechanism in Respiratory Acidosis*

**Respiratory mechanism:** Increased $CO_2$ concentration stimulates the respiratory center and leads to hyperventilation. This compensatory mechanism becomes less effective if the acidosis is due to depression in respiratory center.

**Renal mechanism:** More $HCO_3^-$ (bicarbonate) is reabsorbed from tubule to restore the ratio of $(HCO_3^-)/(H_2CO_3)$ at 20:1.

*Causes of Respiratory Acidosis*

- Brain damage
- Drug poisoning, e.g., morphine, barbiturates, etc.
- Loss of ventilator function
- Effect of pain
- Emphysema
- Pulmonary edema
- Asthma
- Congenital heart diseases
- Pneumonia

## Metabolic Alkalosis

Primary excess of $HCO_3^-$ (bicarbonate) is the characteristic feature of metabolic alkalosis. Excess accumulation of $HCO_3^-$ (alkali reserve) causes an increase in the ratio of $(HCO_3^-)/(H_2CO_3)$, i.e., pH increases.

*Compensatory Mechanism of Metabolic Alkalosis*

Respiratory center is inhibited, causing shallow irregular breathing. This results in $CO_2$ retention and increase in $H_2CO_3$ level to make up the ratio. The following renal mechanisms will come into play:

- Increased excretion of $HCO_3^-$
- Increased excretion of $K^+$
- Reduced $NH_3$ formation and reduced excretion of nonvolatile acids.

*Causes of Metabolic Alkalosis*

- This results from either loss of acid or the gain of base. Loss of acid may result from severe vomiting or gastric aspiration leading to the loss of chloride and acid; therefore, hypochloremic alkalosis occurs.
- Alkali ingestion and alkali administration.
- Hyperaldosteronism causes retention of sodium and loss of potassium, leading to hypokalemia in metabolic alkalosis.

## Respiratory Alkalosis

A primary deficit of $H_2CO_3$ is described as respiratory alkalosis. It involves a decrease in $H_2CO_3$ fraction, with no corresponding change in bicarbonate ($HCO_3^-$) in plasma. Excessive quantities of $CO_2$ may be washed out of the blood by hyperventilation, resulting in the decreased $H_2CO_3$. Thus, the ratio of $(HCO_3^-)/(H_2CO_3)$ increases, i.e., the pH increases.

*Compensatory Mechanism of Respiratory Alkalosis*

- Increased excretion of bicarbonate ($HCO_3^-$)
- Decreased excretion of acid
- Decreased excretion of ammonia ($NH_3$)
- Retention of chloride in blood.

*Causes of Respiratory Alkalosis*

- Stimulation of respiratory center in
  - CNS diseases such as meningitis, encephalitis, etc.
  - Salicylate poisoning
  - Hyperpyrexia

**Table 19.3:** Types of acid–base disturbances.

| Disturbance | pH | Primary change | Secondary change |
|---|---|---|---|
| Metabolic acidosis | Decreased | Deficit of bicarbonate | Decrease in $PCO_2$ |
| Metabolic alkalosis | Increased | Excess of bicarbonate | Increase in $PCO_2$ |
| Respiratory acidosis | Decreased | Excess of carbonic acid | Increase in bicarbonate |
| Respiratory alkalosis | Increased | Deficit of carbonic acid | Decrease in bicarbonate |

($PCO_2$: partial pressure of carbon dioxide)

- High-altitude effect—hyperpnea
- Hepatic coma
- Nonjudicious use of respirator.

**Table 19.3** gives a summary of the different acid–base disturbances.

### Relationship of pH with K⁺ Ion Balance

In general, **acidosis is associated with hyperkalemia** and alkalosis with hypokalemia. Sudden hypokalemia may develop during the correction of acidosis. K⁺ may go back into the cells, suddenly lowering the plasma K⁺. Hence, it is important to maintain the K⁺ balance during correction of alkalosis. Similarly alkalosis can be corrected only if the potassium balance is normalized.

## Assessment of Acid–Base Status

- Usually by blood gas analyzer (ABG).
- Measure pH, $PCO_2$, and $PO_2$ directly by means of electrode.
- $PO_2$ value helps to get an idea about the functional status of the respective system
- Heparinized arterial blood is used.
- Analyzed within ½ hour after collection.
- There should not be any contact with air during the collection or analysis; $HCO_3^-$ can be calculated from pH and $PCO_2$.

### Reference Values in Blood

- pH = 7.3–7.4
- $PO_2$ = 85–100 mm Hg
- $PCO_2$ = 35–45 mm Hg
- $O_2$sat. = 95–98%
- $HCO_3^-$ = 22–29 mmol/L
- $H_2CO_3$ = 1.35 mEq/L

## ELECTROLYTE BALANCE

Important electrolytes in the body are sodium, potassium, chloride, and magnesium. Electrolytes especially sodium is directly related with the water balance in the body.

## Regulation of Sodium and Water Balance

The major regulatory factors are the hormones [aldosterone, antidiuretic hormone (ADH)] and the renin–angiotensin system.

### Aldosterone

It is secreted by the zona glomerulosa of the adrenal cortex regulates the Na⁺ → K⁺ exchange and Na⁺ → H⁺ exchange at the renal tubules. The net effect is the sodium retention.

### Antidiuretic Hormone

When osmolality of the plasma rises, the osmoreceptors of hypothalamus are stimulated, resulting in ADH secretion. The ADH will increase the water reabsorption by the renal tubules. Therefore, proportionate amounts of sodium and water are retained to maintain the osmolality. When the osmolality decreases, ADH secretion is inhibited. When the ECF volume expands, the aldosterone secretion is cutoff.

### Renin–Angiotensin System

When there is a fall in the ECF volume, the renal plasma flow decreases and this would result in the release of renin from the kidney. The factors that stimulate renin release are

(a) decreased blood pressure and (b) salt depletion. The inhibitors of renin release are (a) increased blood pressure, (b) salt intake, and (c) angiotensin-II. Renin will activate angiotensin, which increases the blood pressure by causing vasoconstriction of the arterioles.

*Atrial Natriuretic Peptides*

They are secreted in response to the stimulation of atrial stretch receptors. They inhibit renin and aldosterone secretion and eliminate sodium. **Table 19.4** gives the electrolyte concentration in body fluids.

## Disturbances in Fluid and Electrolyte Balance

When the effective osmolality increases, the body fluid is called **hypertonic** and when osmolality decreases, the body fluid is called **hypotonic**.

Clinical effects of **increased** effective osmolality are due to dehydration of cells. A patient may be comatose when the serum sodium increases up to 170 mmol/L rapidly. A sudden **reduction** in effective osmolality may cause brain cells to swell, leading to headache, vomiting, and medullary herniation.

**Table 19.4:** Electrolyte concentration of body fluid compartments.

| Solutes | Plasma (mEq/L) | Interstitial fluid (mEq/L) | Intracellular fluid (mEq/L) |
|---|---|---|---|
| **Cations** | | | |
| Sodium | 140 | 146 | 12 |
| Potassium | 4 | 5 | 160 |
| Calcium | 5 | 3 | --- |
| Magnesium | 1.5 | 1 | 34 |
| **Anions** | | | |
| Chloride | 105 | 117 | 2 |
| Bicarbonate | 24 | 27 | 10 |
| Phosphate | 2 | 2 | 140 |
| Protein | 15 | 7 | 54 |

*Isotonic Contraction of Extracellular Fluid*

This results from the loss of fluid that is isotonic with plasma. The most common cause is loss of gastrointestinal fluid, due to small intestinal obstruction where fluid accumulates in the lumen. Since equivalent amounts of sodium and water are lost, the plasma sodium is often normal. Hemoconcentration is seen. In severe cases, hypotension may occur.

*Hypotonic Contraction*

There is predominant sodium depletion. The causes are (a) **infusion of fluids** with low sodium content like dextrose. Therefore, in **postoperative cases**, care should be taken to adequately replace sodium by giving sufficient quantity of normal saline. (b) Deficiency of aldosterone in **Addison's disease**.

*Hypertonic Contraction*

It is predominantly water depletion. The commonest cause is **diarrhea**, where the fluid lost has only half of the sodium concentration of the plasma. Another important cause is **vomiting**. The increase in osmolality will stimulate thirst and increase the water intake.

*Isotonic Expansion*

Water and sodium retention are often manifested as **edema** and occur secondary to **hypertension or cardiac failure**. Hemo-dilution is the characteristic finding. Edema is also seen in hypoalbuminemia (nephrotic syndrome, protein malnutrition).

## SODIUM

Sodium level is intimately associated with water balance in the body. Sodium regulates the extracellular fluid volume. Total body sodium is about 4000 mEq. About 50% of it is in bones, 40% in extracellular fluid, and 10% in soft tissues. Sodium is the major cation of **extracellular** fluid. **Sodium pump** is operating in all the cells, so as to keep sodium extracellular. This mechanism is ATP dependent (*see* Chapter 2). Sodium (as $NaHCO_3$) is also important in the regulation of acid–base balance.

Normal level of $Na^+$ in plasma is **136–145 mEq/L** and in cells 12 mEq/L. Normal diet contains about 5–10 g of sodium, mainly as sodium chloride. The same amount of sodium is daily excreted through urine. However, body can conserve sodium to such an extent that on a sodium-free diet, the urine does not contain sodium. Ideally dietary sodium intake should be lower than potassium, but processed food have increased sodium intake.

Sodium excretion is regulated at the distal tubules of the kidney. Aldosterone increases sodium reabsorption in distal tubules. The ADH increases reabsorption of water from tubules. The amount of sodium reabsorbed is under the control of aldosterone.

In **edema**, along with water, sodium content of the body is also increased. When diuretic drugs are administered, they increase sodium excretion. Along with sodium, water is also eliminated. **Sodium restriction** in diet is, therefore, advised in congestive cardiac failure and hypertension.

**Causes of hypernatremia** (increased sodium in blood) are as follows:
- Prolonged cortisone therapy
- In pregnancy, steroid hormones cause sodium retention in the body.
- In dehydration, when water is predominantly lost, blood volume decreases with apparent increased concentration of sodium.
- Elderly patients with poor water intake and inability to express thirst.
- Cushing's disease (adrenal overactivity)
- Drugs: Osmotic diuretics

**Causes of hyponatremia** (decreased sodium in blood) are the following:
- Vomiting
- Diarrhea
- Addison's disease (adrenal insufficiency)
- Chronic renal failure, nephrotic syndrome
- Congestive cardiac failure
- Hyperglycemia and ketoacidosis
- Excess nonelectrolyte (glucose) intravenous infusion.

**Treatment** of hyponatremia is to give sodium chloride infusion intravenously or to increase oral salt intake.

The correction of hypernatremia is to give water orally or intravenously. Care should be taken to prevent sudden overhydration and water intoxication. Rapid correction can also cause brain herniation and permanent neurologic deficit.

## POTASSIUM

Total body potassium is about 3500 mEq, out of which 75% is in skeletal muscle. Potassium is the major **intracellular** cation and maintains intracellular osmotic pressure.

The depolarization and contraction of heart require potassium. During transmission of nerve impulses, there is sodium and potassium efflux, with depolarization. After the nerve transmission, these changes are reversed. The intracellular concentration gradient is maintained by the $Na^+$-$K^+$ ATPase pump.

### Requirement

Potassium requirement is 3–4 g/day.

### Sources

Sources rich in potassium but low in sodium are banana, orange, apple, pineapple, almond, dates, beans, yam, and potato. Tender coconut water is a very good source of potassium.

### Normal Level

Plasma potassium level is **3.5–5 mEq/L**. The cells contain 160 mEq/L, so precautions should be taken to prevent hemolysis when taking blood for potassium estimation.

### Potassium Excretion

Excretion of potassium is mainly through urine. Aldosterone and corticosteroids increase the excretion of K⁺. On the other hand, K⁺ depletion will inhibit aldosterone secretion.

## Hypokalemia

This term denotes that plasma potassium level is below 3 mmol/L, when mortality will be very high. Hypokalemia is seen in:
- Cushing's syndrome
- Hyperaldosteronism
- Alkalosis
- Diarrhea
- Vomiting
- Deficient intake or low potassium diet
- Excess saline infusion in excess

**Drugs:** Insulin, osmotic diuresis, corticosteroids.

**Signs and symptoms:** Hypokalemia is manifested as muscular weakness, fatigue, muscle cramps, hypotension, decreased reflexes, palpitation, cardiac arrhythmias, and cardiac arrest. Electrocardiograph (ECG) waves are flattened, **T wave is inverted**, and ST segment is lowered. This may be corrected by oral feeding of orange juice.

**Redistribution** of potassium can occur following insulin therapy. For diabetic coma, the standard treatment is to give **glucose and insulin**. This causes entry of glucose and potassium into the cell and hypokalemia may be induced. K⁺ should be supplemented in such cases. Redistribution is also seen in **alkalosis**, where the potassium moves into the cell in exchange for H⁺.

## Hyperkalemia

Plasma potassium level above 5.5 mmol/L is known as hyperkalemia. Since the normal level of K⁺ is kept at a very narrow margin, even minor increase is life-threatening. In hyperkalemia, there is increased membrane excitability, which leads to ventricular arrhythmia and ventricular fibrillation. Hyperkalemia is characterized by flaccid paralysis, bradycardia, and **cardiac arrest**. The ECG shows **elevated T wave**, widening of QRS complex, and lengthening of PR interval.

Causes of hyperkalemia are the following:
- Renal failure
- Deficient aldosterone (Addison's)
- Increased hemolysis
- Excess potassium supplementation
- Metabolic acidosis.

## CHLORIDE

Intake, output, and metabolism of sodium and chloride run in parallel. The homeostasis of Na⁺, K⁺, and Cl⁻ are interrelated. Chloride is important in the formation of hydrochloric acid in gastric juice. Chloride concentration in plasma is **96–106 mEq/L**; and in CSF, it is about 125 mEq/L. Chloride concentration in CSF is higher than any other body fluids. Since CSF protein content is low, Cl⁻ is increased to maintain Donnan membrane equilibrium. Excretion of Cl⁻ is through urine and is parallel to Na⁺. Daily excretion of Cl⁻ is about 5–8 g/day.

## Hyperchloremia

Increased chloride in blood is seen in (a) dehydration, (b) Cushing's syndrome, where mineralocorticoids cause increased reabsorption from kidney tubules, (c) severe diarrhea which leads to loss of bicarbonate and compensatory retention of chloride, and (d) renal tubular acidosis.

### Causes for Hypochloremia

Excessive vomiting is the most common cause. Through vomit, the HCl is lost, so plasma Cl⁻ is lowered. There will be compensatory increase in plasma bicarbonate. This is called **hypochloremic alkalosis**. In Addison's disease, aldosterone diminishes, renal tubular reabsorption of Cl⁻ decreases, and more Cl⁻ is excreted.

## MAGNESIUM

Magnesium is the fourth most abundant cation in the body and second most prevalent intracellular cation. Magnesium is mainly seen in intracellular fluid. Total body magnesium is about 25 g, 60% of which is complexed with calcium in bone. One third of skeletal magnesium is exchangeable with serum. Magnesium orally produces diarrhea; but intravenously it produces CNS depression.

## Requirement

The requirement is about 400 mg/day for men and 300 mg/day for women. Doses above 600 mg may cause diarrhea. More is required during lactation. Major sources are cereals, beans, leafy vegetables, and fish.

## Normal Serum Level of Magnesium

Normal serum level $Mg^{++}$ is 1.8–2.2 mg/dL. Serum must be separated from the clot as soon as possible or the level of magnesium will increase because of its elution from the red blood cells. Hemolyzed samples as well as blood collected with citrate, oxalate or EDTA are unacceptable for analysis. Homeostasis is maintained by intestinal absorption as well as by excretion by kidney.

## Functions of Magnesium

- Magnesium is the activator of many enzymes requiring ATP. Alkaline phosphatase, hexokinase, fructokinase, phosphofructokinase, adenyl cyclase, etc., need magnesium.
- Neuromuscular irritability is lowered by magnesium.
- Insulin-dependent uptake of glucose is reduced in magnesium deficiency. Magnesium supplementation improves glucose tolerance.

## Hypomagnesemia

It is commonly seen in hospital patients. When the serum magnesium level falls below 1.7 mg/dL, it is called hypomagnesemia. Vomiting, nasogastric suction, diarrhea, liver cirrhosis, protein-calorie malnutrition, and diuretic therapy are the common causes. Deficiency of magnesium leads to neuromuscular hyperirritability and cardiac arrhythmias. Deficiency is treated by giving parenteral magnesium. Oral therapy may lead to diarrhea, hence intravenous magnesium sulfate is given.

## Hypermagnesemia

It is uncommon and always due to excessive intake orally (antacids), rectally (enema), or parenterally. Magnesium intoxication causes depression of neuromuscular system, causing lethargy, hypotension, respiratory depression, bradycardia, and weak tendon reflexes.

## SUMMARY OF THE CHAPTER

- Acids are capable of donating protons.
- Strong acids dissociate completely, while weak acids ionize incompletely.
- Acidity of a solution is measured by noting the hydrogen ion concentration.
- The pH is inversely proportional to acidity.
- Neutral pH is 7.
- Buffers are solutions which can resist changes in pH when acid or alkali is added.
- Buffers can be made by mixtures of weak acids with their salt with a strong base.
- The pH of buffer is calculated by the Henderson–Hasselbalch equation.
- Most important buffer system in plasma is the bicarbonate–carbonic acid system.
- Bicarbonate represents the alkali reserve.
- Normal bicarbonate level in plasma is 24 mmol/L.
- Main intracellular buffer is phosphate buffer.
- First defense against acid entry into blood is by bicarbonate buffer system.
- Second defense against acid is by respiratory regulation of pH.
- Third defense system is the renal regulation.
- Renal regulation has three components: (1) excretion of hydrogen ion, (2) $Na^+/H^+$ exchange, and (3) excretion of ammonium ions.
- Respiratory acidosis means primary excess of carbonic acid.

- Metabolic acidosis means primary deficiency of bicarbonate.
- Respiratory alkalosis is primary deficiency of alkali.
- Metabolic alkalosis means primary excess of bicarbonate.
- The major factors controlling the water intake are thirst and the rate of metabolism.
- Major determinant factor of osmolality is sodium.
- Major regulatory factors of sodium and water balance are aldosterone, ADH, and rennin–angiotensin system.
- Osmotic balance is maintained between the extra- and intracellular fluid compartments even though there is a difference in solute content and composition.
- Kidney is the major organ regulating water and electrolyte balance.
- Regulation of sodium and water balance is by rennin angiotensin system.
- Any fall in ECF volume stimulates the juxtaglomerular cells to secrete renin. Renin in turn cleaves angiotensinogen to angiotensin, which will stimulate the adrenal cortex to secret aldosterone. Retention of sodium and water leads to restoration of ECF volume.
- The ADH secretion and renin production are under control by changes in osmolality.
- Gain or loss of water and sodium and relative amounts of each decides the tonicity of ECF.

## QUESTIONS FOR SELF-ASSESSMENT

1. Define buffers. Mention the various buffers present in our body.
2. Explain the mechanism of buffer action.
3. Mention the importance of pH in buffer regulation.
4. Mention the significance of Henderson–Hasselbalch equation.
5. Add a note on renal regulation with a schematic diagram.
6. What are the consequences of acid–base imbalance?
7. What is metabolic acidosis and mention its probable causes?
8. Define metabolic alkalosis and mention one example.
9. What are the causes of respiratory acidosis?
10. Add a note on respiratory alkalosis.
11. Add a note on potassium balance.
12. Add a note on sodium balance.
13. Add a note on hyperkalemia.
14. Add a note on hyponatremia.
15. Add a note on hyperchloremic acidosis.
16. Add a note on hypochloremic alkalosis.

## MULTIPLE CHOICE QUESTIONS (MCQs)

19-1. **The buffering action of protein depends mainly on the following:**
   a. pKa value of its ionizable side chains of histidine
   b. pKa value of tyrosine
   c. Basic amino acid content
   d. Neutral amino acid content

19-2. **Metabolic acidosis occurs in all the following conditions, *except*:**
   a. Vomiting
   b. Uncontrolled diabetes mellitus
   c. Prolonged fasting
   d. Strenuous exercises

19-3. **Deficit of $HCO_3^-$ leads to:**
   a. Metabolic alkalosis
   b. Metabolic acidosis
   c. Respiratory acidosis
   d. Respiratory alkalosis

19-4. **Respiratory acidosis occurs due to:**
   a. Excess $CO_2$
   b. Excess $HCO_3^-$
   c. Excess $H^+$
   d. Excess $H_2CO_3$

19-5. **Respiratory alkalosis results from:**
   a. Occurs in hyperventilation
   b. Occurs in normal pregnancy
   c. Does not occur in type I respiratory failure
   d. May occur in type II respiratory failure

19-6. **The recognized causes of metabolic acidosis with a normal anion gap is:**
   a. Salicylate poisoning
   b. Starvation

c. Diarrhea
d. Pancreatic secretion

**19-7. Which of the following is the major intracellular cation?**
a. Magnesium
b. Potassium
c. Sodium
d. Calcium

**19-8. Hyponatremia is seen in:**
a. Addison's disease
b. Conn's syndrome
c. Diabetes insipidus
d. Compulsive water drinking

**19-9. Administration of diuretics causes loss of potassium, which may lead to:**
a. Metabolic acidosis
b. Respiratory acidosis
c. Respiratory alkalosis
d. Metabolic alkalosis

**19-10. Patients with diarrhoea may develop the following complications, *except*:**
a. Metabolic alkalosis
b. Hypertonic contraction
c. Dehydration
d. Metabolic acidosis

## ANSWER KEYS TO MCQs

19-1. (a)   19-2. (a)   19-3. (b)   19.4. (a)   19.5. (a)   19-6. (c)   19-7. (b)   19-8. (a)
19.9. (d)   19-10. (a)

# CHAPTER 20

# Liver and Renal Function Tests

At the completion of this chapter, the reader will be able to answer questions on the following topics:

- Functions of liver
- Liver function tests, classification
- Tests based on bilirubin metabolism
- Plasma proteins
- Detoxification
- Serum enzymes
- Normal constituents of urine
- Abnormal constituents of urine
- Benedict's test
- Ketone bodies
- Bile pigments
- Albumin
- Renal function tests
- Clearance tests
- Gastric function tests
- Augmented histamine test
- Cerebrospinal fluid

## LIVER FUNCTION TESTS

The biochemical tests of liver function are useful in detecting several treatable causes of liver dysfunction, where the patients may have vague symptoms or are asymptomatic. When liver disease is evident, the nature, duration, and severity can also be assessed by the biochemical tests. Major functions of liver are shown in **Table 20.1**.

## CLINICAL MANIFESTATIONS OF LIVER DYSFUNCTION

### Jaundice

Jaundice is the yellowish discoloration of sclera, skin, and mucus membrane **(Fig. 20.1)**. It is characteristic of liver disease. Jaundice is seen when the rate of hemolysis increases leading to elevation of serum bilirubin. Normal serum bilirubin value is less than 1 mg/dL. When it exceeds to more than 2 mg/dL, bilirubin deposits in tissues.

### Portal Hypertension

The entire venous drainage of gastrointestinal tract, the spleen, the pancreas, and the gallbladder constitutes portal circulation with a pressure of 5 mm Hg. Any obstruction in the course of portal circulation will cause portal hypertension. **Effects** of portal hypertension are as follows:

- Due to increase in pressure, veins of the portal system get dilated (varices).
- Portosystemic shunting leads to deterioration of the metabolic functions of the liver.
- Failure of detoxification of ammonia by urea synthesis leads to hyperammonemia and hepatic encephalopathy.
- Decrease in albumin synthesis leads to hypoalbuminemia which predisposes to oozing of fluid into the peritoneal cavity causing ascites. It is a common presenting feature of cirrhosis.

### Table 20.1: Major functions of liver.

1. **Synthetic function**
   a. Synthesis of plasma proteins (albumin, coagulation factors, many globulins)
   b. Synthesis of cholesterol
   c. Synthesis of triacylglycerol
   d. Lipoprotein synthesis

2. **Metabolic function**
   a. Carbohydrates: glycolysis, glycogen synthesis, glycogen breakdown, gluconeogenesis
   b. Ketogenesis, fatty acid synthesis and breakdown
   c. Protein catabolism
   d. Citric acid cycle, production of adenosine triphosphate (ATP)

3. **Detoxification and excretion**
   a. Ammonia to urea
   b. Bilirubin (bile pigment)
   c. Cholesterol
   d. Drug metabolites

4. **Homeostasis**: Blood glucose regulation

5. **Storage function**: Vitamins A, D, K, $B_{12}$

6. **Secretory function:** Production of bile salts that help in digestion

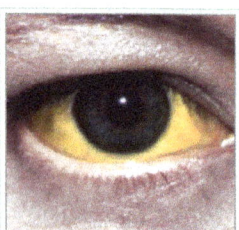

Yellow color of sclera is seen in jaundice. Normal serum bilirubin value is 1 mg/dL. When it exceeds 2 mg/dL, bilirubin deposits in tissues

**Fig. 20.1:** Jaundice.

- Diminished synthesis of clotting factors predisposes to bleeding.
- Loss of thrombolytic factors (e.g., antithrombin III) predisposes to hypercoagulability and venous thrombosis.
- A detailed classification of the liver function tests (LFTs) is shown in **Table 20.2**. Important LFTs are described below:

### Table 20.2: Classification of liver function tests.

**Group I (tests of hepatic excretory function)**
- Serum: Bilirubin; total, conjugated, and unconjugated
- Urine: Bile pigments, bile salts, and urobilinogen

**Group II: Liver enzyme panel**
- Alanine aminotransferase (ALT; marker of liver injury)
- Aspartate aminotransferase (AST; marker of liver injury)
- Alkaline phosphatase (ALP; marker of cholestasis)
- Gamma glutamyltransferase (GGT; marker of cholestasis)

**Group III: Plasma proteins (tests for synthetic function of liver; marker of chronic liver diseases)**
- Total proteins
- Serum albumin, globulins, albumin/globulin (A/G) ratio
- Prothrombin time
- Blood ammonia

**Group IV: Special tests (tests for metabolic liver disease)**
- Ceruloplasmin
- Ferritin and iron
- Alpha fetoprotein (AFP)

## Indications for Liver Function Tests

- Jaundice
- Suspected liver metastasis
- Alcoholic liver disease
- Any undiagnosed chronic illness
- Annual checkup of diabetic patients
- Coagulation disorders
- Therapy with statins to check hepatotoxicity.

## Liver Function Test: The Routine Markers of Liver Disease

- Serum bilirubin—total and conjugated
- Hepatic enzyme panel—alanine aminotransferase (ALT), aspartate aminotransferase (AST), alkaline phosphatase (ALP), and gamma glutamyltransferase (GGT)

- Serum proteins and albumin/globulin (A/G) ratio
- Prothrombin time (PT)
- Urine analysis for bile pigments and salts.

## TESTS BASED ON BILIRUBIN METABOLISM

Bilirubin is the bile pigment. It is the catabolic end product of hemoglobin. Details are shown in Chapter 15.

Hb → Globin → Heme → Biliverdin → Bilirubin

- Serum bilirubin (conjugated and unconjugated) estimation
- Detection of urine bilirubin
- Detection of urine bile salt.

## Conjugated Bilirubin (Direct Bilirubin)

In the liver, bilirubin is conjugated with glucuronic acid to produce bilirubin diglucuronide, so as to make it water soluble for easy excretion.

## Unconjugated Bilirubin (Indirect Bilirubin)

Bilirubin formed in the reticuloendothelial cell is insoluble in water. The lipophilic bilirubin is therefore transported in plasma bound to albumin called unconjugated bilirubin.

### Normal Range of Bilirubin

- Total bilirubin: 0.2–1.0 mg/dL.
- Conjugated bilirubin (direct): 0.1–0.4 mg/dL.
- Unconjugated bilirubin (indirect): 0.2–0.6 mg/dL.

### Estimation of Bilirubin (Van den Bergh Reaction)

Diazo reagent (diazotized sulfanilic acid, i.e., sulfanilic acid, hydrochloric acid, and sodium nitrite) reacts with bilirubin to form a purple azobilirubin. The intensity of purple is directly proportional to the concentration of bilirubin in serum.

## Jaundice

Chapter 15 describes jaundice and its clinical significance. Jaundice is characterized by the yellow coloration of conjunctiva, mucous membrane, and skin **(Fig. 20.1)**. In jaundice, the serum total bilirubin is above 2 mg/dL. Urinary findings are shown in **Table 20.3**.

### Hemolytic Jaundice

It is due to the increased breakdown of hemoglobin, so liver cells are unable to conjugate all the bilirubin produced. Its causes are hemolytic disease of newborn (HDN), malaria, glucose-6-phosphate dehydrogenase (G6PD) deficiency or autoimmune hemolytic anemia. These cases demonstrate increased **unconjugated bilirubin** level.

### Hepatic Jaundice or Hepatocellular Jaundice

It is the disease of parenchymal cells of liver caused due to viral hepatitis or drug-induced cholestasis. Defective bilirubin conjugation is caused by liver cells. In these cases, either or both types of bilirubin may be increased. The most common cause for hepatocellular jaundice is infection with hepatitis viruses (viral hepatitis). It may be due to hepatitis A virus (HAV), which is transmitted by the intake of contaminated food and water. Type A disease is usually self-limiting.

**Table 20.3:** Urinary findings in jaundice.

| Types of jaundice | Bile pigment | Bile salt | Urobilinogen |
| --- | --- | --- | --- |
| Prehepatic (hemolytic) | Nil | Nil | ++ |
| Hepatocellular | ++ | + | Normal or– |
| Posthepatic (obstructive) | +++ | ++ | Nil or– |

Infection by **hepatitis B virus** (HBV) is transmitted mainly through parenteral contamination by infected blood or blood products. The virus is highly contagious. It is a DNA virus, which destroys the hepatic cells. The **surface antigen** (HBs) is seen in patient's circulation. About 5% of the world population is carriers of HBV. In most cases of hepatitis B infection, complete recovery is possible, but about 1% case progresses to cirrhosis and eventual hepatic failure. In another 2–5% cases, the disease goes to a chronic carrier state. In about 1% case, it leads to chronic **cirrhosis** and eventually hepatic failure. In fact, the most common cause for cirrhosis in developing countries is HBV. In a small fraction of such cases, the development of hepatocellular carcinoma (HCC) is also noticed. Thus, HBV is an **oncogenic virus**. Medical personnel, including medical students, doctors, nurses, and technicians, are advised to take the hepatitis B vaccination. Hepatitis viruses type A, B, C, D, E, and G are identified. While A and E are transmitted by oral route; B, C, D and G are transmitted through parenteral route.

*Obstructive Jaundice*

Obstructive jaundice involves obstruction to the flow of bile due to gallstone or carcinoma of the head of pancreas. In this case, **conjugated bilirubin** increases. The tests to distinguish the three different forms of jaundices are shown in **Table 20.4**.

## Urine Bilirubin

Detection of bile pigment in urine is based on the oxidation of bilirubin to different colored compounds (blue or green) by reacting with **Fouchet's reagent** (ferric chloride in trichloroacetic acid). In obstructive jaundice, bilirubin is found in urine (**Tables 20.3 and 20.4**).

## Urine Bile Salt

Bile salts (taurocholate and glycocholate) are observed in urine in obstructive jaundice. It is detected by **Hays sulfur powder test**. (Bile salts and bile pigments are different. Bile salts are produced from cholesterol and are necessary for fat absorption. Bile pigments are produced from hemoglobin breakdown and are unnecessary excretory products.)

**Table 20.4:** Tests useful to distinguish different types of jaundice.

| Specimen | Test | Prehepatic or hemolytic jaundice | Hepatocellular jaundice | Posthepatic or obstructive jaundice |
|---|---|---|---|---|
| Blood | Unconjugated bilirubin (van den Bergh indirect test) | ++ | ++ | Normal |
| Blood | Conjugated bilirubin (van den Bergh direct test) | Normal | It is the first activity to be impaired activity to be impaired | ++ |
| Urine | Alkaline phosphatase 2–3 times increased 10–12 times | Normal | 2–3 times increased | 10 times increased |
| Urine | Bile salt (Hays test) | Absent | Absent | Present |
| Urine | Conjugated bilirubin (Fouchet's test) | Absent | Present | Present |
| Urine | Urobilinogens (Ehrlich test) | +++ | + ve or -ve | Absent |
| Feces | Urobilin | ++ | Normal or decreased | Clay colored |

## TESTS BASED ON PLASMA PROTEIN ESTIMATION

### Normal Range of Plasma Proteins

- Total protein: 6.0–8.0 g/dL
- Albumin: 3.5–5.5 g/dL
- Globulin: 1.8–3.6 g/dL

Albumin is synthesized only by liver. So in hepatic diseases, albumin level decreases. Total protein level may also decrease. However, a corresponding increase in globulin can be observed in hepatic diseases.

### Prothrombin Time

The term prothrombin time is the time required for clotting to takes place in citrated plasma to which the optimum amounts of thromboplastin and calcium have been added. Normal range 12–14 seconds. Prothrombin is synthesized only in the liver.

### TESTS BASED ON SERUM ENZYME ACTIVITY

Normal ranges are as follows:
- Aspartate transaminase [serum glutamic-oxaloacetic transaminase (SGOT)]: 0–41 IU/L
- Alanine transaminase [serum glutamic pyruvic transaminase (SGPT)]: 0–45 IU/L
- Alkaline phosphatase: 23–92 IU/L
- Gamma glutamyltransferase: 10–47 IU/L

### Enzymes Indicating Hepatocellular Damage

In all liver diseases, the levels of transferases (ALT and AST) in serum are found to be elevated. Very high levels (more than 1000 units) are seen in acute hepatitis (viral and toxic drugs) and in ischemia due to cardiac failure. Moderate elevation in aminotransferases (between 100 and 300 U/L) is seen in alcoholic and nonalcoholic chronic hepatitis. Minor elevation of less than 100 U/L is seen in chronic viral hepatitis **(Table 20.5)**.

**Table 20.5:** Clinical significance of AST/ALT ratio.

**Normal AST/ALT ratio is 0.8:1**
*AST more than ALT is seen in:*
- Alcoholic hepatitis
- Hepatitis with cirrhosis
- Nonalcoholic steatohepatitis (NASH)
- Liver metastases
- Myocardial infarction

*Higher ALT levels than AST is seen in:*
- Acute hepatocellular injury
- Toxic exposure
- Extrahepatic obstruction (cholestasis)

(AST: aspartate aminotransferase; ALT: alkaline phosphatase)

## MARKERS OF OBSTRUCTIVE LIVER DISEASE

### Alkaline Phosphatase

Very high levels of ALP are noticed in patients with cholestasis or hepatic carcinoma. In parenchymal diseases of the liver, mild elevation of ALP is noticed. Even in bone diseases, an increased level of ALP is observed **(Table 20.4)**.

### Gamma Glutamyltransferase

Elevated levels of GGT are observed in chronic alcoholism. In liver diseases, GGT elevation parallels to that of ALP.

## RENAL FUNCTION TESTS

The functional unit of kidney is called **nephron**. Kidney is composed of approximately 1 million nephrons. About 1200 mL of blood (650 mL plasma) pass through the kidneys every minute. From this, about 120–125 mL is filtered per minute by the kidneys and this is referred to as glomerular filtration rate (GFR). The glomerular filtrate formed in adult is about 175–180 L/day, out of which only 1.5L are excreted as urine. Thus, more than 99% of the glomerular filtrate is reabsorbed by the kidneys. Important **functions of kidney** are the following:

- **Maintenance of homeostasis:** The kidneys are largely responsible for the regulation of water, electrolyte (sodium and potassium), and acid–base balance in the body.
- **Excretion of metabolic waste products:** The end products of protein and nucleic acid metabolism are eliminated from the body, e.g., urea, creatinine, uric acid, sulfate, phosphate, acids, bases, toxins, and drug metabolites.
- **Hormonal functions:** Kidney produces rennin, erythropoietin, and calcitriol.
- **Filtration and reabsorption:** Urine formed per day is about 180 L/day, out of which 178.5 L are reabsorbed. Further all glucose and amino acids are reabsorbed; most of the sodium and chloride are reabsorbed. Thus, kidneys retain several substances of biochemical importance in the body, such as glucose, amino acids, etc. Renal function tests are summarized in **Table 20.6**.

## NORMAL CONSTITUENTS OF URINE

### Normal Inorganic Constituents of Urine

- Chloride
- Sulfate
- Phosphate
- Calcium
- Sodium
- Potassium
- Ammonia

### Normal Organic Constituents of Urine

- Urea
- Uric acid
- Creatinine
- Urobilinogen

### Observations of Urine Sample

- Volume: 1.5–2 L/day
- Appearance: Clear and transparent
- Odor: Faint aromatic
- Color: Straw colored
- Specific gravity: 1.017–1.025

## ABNORMAL CONSTITUENTS OF URINE

### Glucose

Benedict's Test: To 5 mL of Benedict's reagent add 8 drops of urine and heated to boil for 2 minutes (or place in a boiling water bath for 5 minutes). Cool and note the color. Benedict's reagent contains copper sulfate, sodium citrate, and sodium carbonate. On heating in alkaline medium, the sugar reduces the cupric ions to cuprous ions. Depending on the extent of reduction, the color of the solution may vary.

- Green color: 0.5% of glucose is present in urine.
- Yellow: 1% of glucose is present in urine.
- Orange: 1.5% of glucose is present in urine.
- Red: 2% of glucose is present in urine.

Glycosuria (excretion of glucose through urine) is observed in **diabetes mellitus**, excessive secretion of cortisol, and hyperthyroidism.

### Ketone Bodies

They are acetoacetic acid, β-hydroxybutyric acid, and acetone. Ketonuria (excretion of ketone bodies through urine) is observed in **diabetes mellitus**, starvation, persistent vomiting, etc.

---

**Table 20.6:** Classification of renal function tests.

**A. To screen for kidney disease**
- Complete urine analysis, including volume, pH, specific gravity, osmolarity, and presence of abnormal constituents such as proteins, blood, ketone bodies, and glucose
- Plasma urea and creatinine
- Plasma electrolytes

**B. To assess renal glomerular function**
- Glomerular filtration rate
- Clearance tests (urea, creatinine, inulin)
- Glomerular permeability (proteinuria)

**C. To assess renal glomerular function**
- Reabsorption studies
- Secretion tests
- Concentration and dilution tests
- Renal acidification

**Rothera's test:** Saturate 5 mL of urine with solid ammonium sulfate. Add 3 drops of freshly prepared sodium nitroprusside and gently shake. Carefully add a few milliliters of strong ammonia through the sides of the tube, so as to form a layer at the top. Let it stand for a few minutes. Appearance of a violet ring at the junction of the two fluids indicates the presence of ketone bodies. Specificity of this test is more toward acetoacetic acid. β-Hydroxybutyric acid will not answer this test. Sodium nitroprusside reacts with both acetone and acetoacetic acid in the presence of alkali to produce a violet-colored compound.

## Bile Pigments

Bilirubin appears in the urine during obstructive jaundice and hepatocellular jaundice **(Tables 20.3 and 20.4)**.

**Fouchet's test:** To 10 mL of urine add a few crystals of magnesium sulfate and boil it. While boiling add 10% barium chloride solution drop by drop till maximum precipitate is attained. This precipitate of barium sulfate adsorbs bile pigment and thus turns yellow. Filter and discard the filtrate. Spread the filter paper with the precipitate on a dry filter paper and press lightly to remove the moisture. To the precipitate add two drops of Fouchet's reagent (ferric chloride in trichloroacetic acid). A green or blue indicates the presence of a bile pigment. Ferric chloride oxidizes bilirubin to biliverdin (green) and bilicyanin (blue).

## Bile Salts (Sodium or Potassium Taurocholate and Glycocholate)

Bile salts are found in urine during the early phase of obstructive jaundice.

**Hays test:** To 5 mL of urine in the test tube, fine powder of sulfur powder is sprinkled carefully. Sulfur will float in normal urine, but it will sink to the bottom when the bile salts are found in the urine. This is due to the lowering of surface tension by the bile salts.

## Albumin

**Albuminuria** is an important index of renal disease. In normal urine, protein concentration cannot be detected by the usual tests.

**Heat and acetic acid test:** Fill two third of a test tube with the urine sample and heat the top of the column to boiling. Add 3 drops of 1% acetic acid. Compare the top portion with the lower. Cloudiness on the top shows the presence of albumin. Albumin coagulates when heated, which precipitates at the isoelectric point, when acetic acid is added.

**Microalbuminuria** or minimal albuminuria or paucialbuminuria is identified when small quantities of albumin (30–300 mg/day) are seen in urine. It is an early indication of nephropathy in patients with diabetes mellitus and hypertension. It is an indicator of future renal failure.

**Glomerular proteinuria:** The glomeruli of kidney are not permeable to big molecular weight substances and so plasma proteins are not found in normal urine. When glomeruli are damaged or diseased, they become more permeable and plasma proteins may appear in urine. The smaller molecules of albumin pass through the damaged glomeruli more readily compared to the heavier globulins. Albuminuria is always pathological. Large quantities (a few grams per day) of albumin are lost in urine in nephrosis. Small quantities are seen in urine in acute nephritis.

**Overflow proteinuria:** When there is an increase in small-molecular-weight proteins in the blood, they overflow into the urine. For example, hemoglobin can pass through normal glomeruli; and therefore, if it exists in free form (as in hemolytic conditions), hemoglobin can appear in urine (**hemoglobinuria**). Yet another example is the **Bence Jones proteins** (monoclonal light chains produced by plasmacytomas). **Table 20.7** gives the varied appearance of urine in different clinical conditions. **Table 20.8** gives a summary of the abnormal constituents of urine.

**Table 20.7:** Appearance of urine in different conditions.

| Appearance | Significance |
|---|---|
| Clear | Normal urine is straw colored |
| Cloudy/opalescent | Urine turns cloudy on standing due to the precipitation of phosphates on refrigeration. Presence of pus also causes cloudiness. |
| High color | Concentrated urine, oxidation of urobilinogen to urobilin |
| Yellow | Bilirubinuria in jaundice; B-complex intake |
| Smoky red | Presence of blood |
| Brownish red | Hemoglobinuria |
| Orange | High levels of bilirubin; rifampicin intake |
| Red | Porphyria; ingestion of red beetroot |
| Black urine | Alkaptonuria, formic acid poisoning |
| Milky urine | Chyluria |

**Table 20.8:** Abnormalities detected in urine by dipstick method (important for bedside clinics).

| Test and normal range | Interpretations |
|---|---|
| 1. Specific gravity (SG) 1.005–1.025 | • Low SG in renal tubular dysfunction<br>• High SG in inadequate water intake; volume depletion |
| 2. pH 5.5–6.5 | • Low pH in high protein diet and acidosis<br>• Recent meal (alkaline tide)<br>• High pH in low protein diet |
| 3. Blood | • Menstruation, traumatic catheterization, glomerulonephritis, stones, tumor, and trauma of urinary tract<br>• Hemoglobinuria, hemolysis |
| 4. Protein | Fever, exercise, orthostatic proteinuria; glomerulonephritis, urinary tract infection, tubular diseases |
| 5. Glucose | Diabetes mellitus, renal glycosuria; Fanconi's syndrome |
| 6. Ketone bodies | Diabetes mellitus, starvation |
| 7. Bilirubin | Hepatitis, obstructive jaundice |
| 8. Urobilinogen | Concentrated urine, hepatitis, intravascular hemolysis low in obstructive jaundice |
| 9. Bile salts | Obstructive jaundice |
| 10. Nitrite | Urinary tract infection |
| 11. Leukocyte esterase | Urinary tract infection, fever |

## CLEARANCE TESTS (MARKERS FOR GLOMERULAR FILTRATION RATE)

Measurement of the clearance is predominantly a test of **GFR**. The GFR provides the most useful index for the severity of renal damage. The relation between clearance value and GFR is shown in **Figure 20.2** and **Table 20.9**.

**Clearance is defined** as the quantity of blood or plasma completely cleared off a substance per unit time and is expressed as milliliter per minute. It is the milliliter of plasma that contains the amount of substance excreted by the kidney within a minute.

$$C = \frac{U \times V}{P}$$

C = Clearance
U = Concentration of the substance in urine
P = Concentration of the substance in plasma
V = Volume of urine in milliliter excreted per minute.

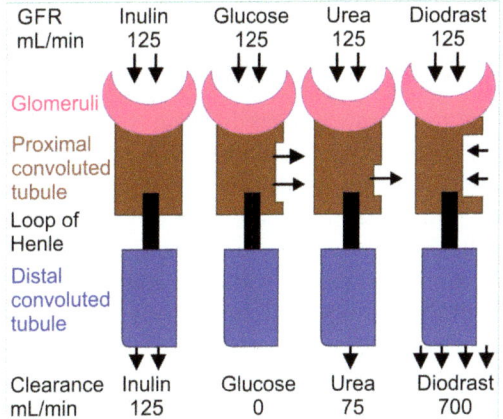

**Fig. 20.2:** Tubules handle substances differently. Inulin is neither absorbed nor secreted. Glucose is absorbed completely. Urea is partially absorbed. Diodrast is actively secreted. (GFR: glomerular filtration rate)

**Table 20.9:** Relationship of GFR with clearance.

| Mechanism | Result | Example |
|---|---|---|
| Substances filtered; neither reabsorbed nor excreted | GFR = clearance | Inulin creatinine |
| Substances filtered; partially reabsorbed | Clearance < GFR | Urea |
| Substances filtered; secreted but not reabsorbed | Clearance > GFR | Diodrast, para aminohippuric acid (PAH) |

(GFR: glomerular filtration rate)

## Creatinine Clearance Test

**Definition**: Creatinine clearance test is the milliliter of plasma that contains the amount of creatinine excreted by the kidney within a minute.

Creatinine is an excretory product derived from creatine phosphate. The excretion of creatinine is rather constant and is not influenced by body metabolism or dietary factors. The value of creatinine clearance is close to GFR, hence its measurement is a sensitive and good approach to assess the glomerular function.

### Procedure for Creatinine Clearance Test

Give 500 mL of water to the patient to promote good urine flow. After about 30 minutes, ask to the patient to empty the bladder and discard the urine. Exactly after 60 minutes, again void the bladder and collect the urine and note the volume. Take one blood sample. Creatinine level in blood and urine are tested and calculated.

Uncorrected clearance = $(U/P) \times V$

where U is the urine creatinine concentration, P is the plasma creatinine concentration, and V is the urine flow in milliliter/minute (the 24-hour urine collection is not necessary for the creatinine clearance test).

Reference value of creatinine clearance is shown in **Table 20.10**. The value is slightly lower in women due to decreased muscle mass. Clearance value up to 75% of the average normal value may indicate adequate renal function.

### Advantages of Creatinine Clearance

- Extrarenal factors will rarely interfere.
- Conversion of creatine phosphate to creatinine is spontaneous and nonenzymatic. As the production is continuous, the blood level will not fluctuate. Blood may be collected at any time.
- It is not affected by diet or exercise.

**Table 20.10:** Normal reference values of creatinine.

|  | Serum creatinine | Creatinine clearance (GFR) |
|---|---|---|
| Adult male | 0.7–1.4 mg/dL | 95–115 mL/minute |
| Adult female | 0.6–1.3 mg/dL | 85–110 mL/minute |
| Children | 0.5–1.2 mg/dL | — |

(GFR: glomerular filtration test)

*Disadvantages of Creatinine Clearance*

- Of the total excretion, about 10% is tubular component. When the GFR is reduced, the secretion component increases and vitiates the results.
- Very early stages of decrease in GFR may not be identified by creatinine clearance (creatinine blind area).

## Urea Clearance Test

**Definition**: Urea clearance test is the milliliter of plasma that contains the amount of urea excreted by the kidney within a minute.

Urea is the end product of protein metabolism. After being filtered by the glomeruli, it is partially reabsorbed by the renal tubules. Hence, urea clearance is less than that of the GFR. It is further influenced by the protein content of the diet. Thus, urea clearance is not as sensitive as creatinine clearance. Reference value is 75 mL/min.

Estimation of urea, creatinine, uric acid, electrolyte, calcium, phosphorus, proteins, etc., in blood or serum will help in accessing the renal function. Reference values of important blood parameters are shown in **Table 20.11**.

**Table 20.11:** Reference values (normal values) of important blood parameters.

| Substance | Reference range |
|---|---|
| Urea: | 20–40 mg/dL |
| Creatinine, adult male | 0.7–1.4 mg/dL |
| Uric acid, adult male | 3.5–7 mg/dL |
| Sodium ($Na^+$) | 135–145 mEq/L |
| Potassium ($K^+$) | 3.5–5 mEq/L |
| Chloride ($Cl^-$) | 96–106 mEq/L |
| Calcium | 9–11 mg/dL |
| Phosphorous | 3–4 mg/dL |
| Total proteins | 6–8 g/dL |
| Albumin | 3.5–5 g/dL |
| Globulins | 2.5–3.5 g/dL |

## GASTRIC FUNCTION TESTS

### Mechanism of Hydrochloric Acid Secretion

The gastric mucosa has different types of cells: (a) the mucous secreting surface epithelial cells, (b) the oxyntic or parietal cells that secrete acid, and (c) the chief cells or peptic cells that secrete enzymes. The daily volume of gastric secretion is around 2,000 mL. The hydrochloric acid secreted by the parietal cells is of high concentration (0.15 M) with a pH as low as 0.8.

The parietal cells transport protons against a concentration gradient at the extracellular fluid pH of 7.4. It is an energy-requiring process. The $K^+$-activated ATPase is necessary for the production of HCl (**Fig. 20.3**). The $H^+$ ions are generated within the cell by ionization of carbonic acid by the enzyme **carbonic anhydrase**. One molecule of ATP is hydrolyzed for every molecule of $H^+$ secreted (**Fig. 20.3**). The ATP hydrolysis is coupled with

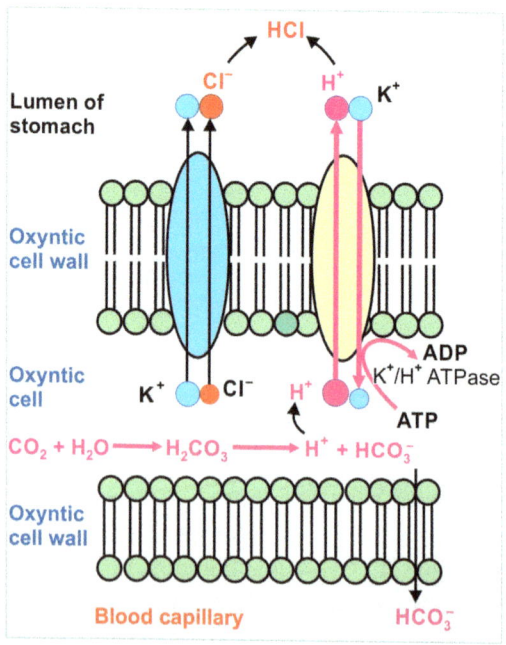

**Fig. 20.3:** Mechanism of hydrochloric acid secretion. (ADP: adenosine diphosphate; ATP: adenosine triphosphate)

an exchange of K⁺ for H⁺. The hydrogen ions are then secreted into gastric lumen. At the same time of H⁺ to K⁺ exchange, a bicarbonate to chloride exchange also takes place. When the bicarbonate level within the cell increases (formed from $H_2CO_3$), it is reabsorbed into the bloodstream, in exchange for $Cl^-$. The chloride is then secreted into the lumen to form HCl. This would account for the **alkaline tide** of plasma and urine, following hydrochloric acid secretion, immediately after meals.

## Other Constituents of Gastric Secretions

The major enzyme present in gastric juice is **pepsin**. It digests the proteins in the food. One of the functions of HCl is to activate the zymogen pepsinogen to pepsin by partial proteolysis. In addition, the HCl helps in the **absorption of iron** and calcium. The gastric juice also contains a glycoprotein required for the absorption of $B_{12}$, the Castle's **intrinsic factor**.

## FRACTIONAL TEST MEAL

### Steps

- Introduction of Ryle's tube in stomach of patient fasting overnight fasting.
- Removal of residual gastric contents and its analysis.
- Ingestion of the test meal.
- Removal of gastric contents at specific interval after meal by aspiration and its analysis.

### Test Meal

**Ewald's test meal**: Two pieces (35 g) of toast with approximately 250 mL light tea.

**Oat meal porridge**: Prepared by adding 2 tablespoon full of oat meal to one quarter of boiling water and strained the porridge through fine mesh cloth.

### Collection of Sample

At intervals of exactly 15 minutes, about 10 mL of gastric contents are removed by means of syringe attached to the tube. If the stomach is not empty at the end of 3 hours, the remaining stomach contents are removed and the volume is noted.

### Analysis

Each sample is strained through a fine mesh cloth. The residue on the cloth is examined for mucus, bile, and starch. The strained sample are analyzed for free and total acidity.

### Result and Interpretations

*Normal Response*

- Free acid: 0–30 mEq/L
- Total acid: 10–40 mEq/L
- Volume: 20–50 mL
- Bile: Found occasionally but not usually
- Blood: Not present
- Mucus: Present in small amount
- Starch: Nil

**Starch content** is present up to 1¾ hours and trace obtained up to 2¼ hours and nil at 2½ hours. Initially the **free acid** level will be low and then rises steadily to reach a maximum of about 1 hour, $1^1/_4$ hours, and $1^1/_2$ hours after which the concentration of free acid begins to decrease. The free acid ranges from 15 to 45 mEq/L at the maximum, with the total of about 1 unit higher. About 8% of normal people fall within these limits. **Blood** should not be present and there should not be any appreciable amount of bile.

*Abnormal Response*

**Hyperchlorhydria**: Free acid reaches a higher concentration in duodenal ulcer, gastric ulcer (blood), gastric carcinoma, and appendicitis.

**Hypochlorhydria**: Free acid concentration is below the normal range in gastric carcinoma and pernicious anemia.

**Achlorhydria**: Free acid is absent in gastritis, carcinoma, and microcytic hypochromic anemia. Normal value of acid secretion is

**Table 20.12:** Normal hydrochloric acid secretion.

| | Acid output in mmol/hour | | | |
| --- | --- | --- | --- | --- |
| | Men | | Women | |
| | Lower Limit | Upper Limit | Lower limit | Upper limit |
| Basal acid output | — | 10 | — | 5.5 |
| Maximal acid output | 7 | 45 | 5 | 30 |
| Peak acid output | 12 | 60 | 8 | 40 |

**Table 20.13:** Causes for hyperacidity and hypoacidity.

| Hyperacidity is seen in | Hypoacidity is seen in |
| --- | --- |
| Duodenal ulcer | Gastritis |
| Gastric cell hyperplasia | Gastric carcinoma |
| Carcinoid tumors | Partial gastrectomy |
| Zollinger–Ellison syndrome | Pernicious anemia |
| Multiple endocrine neoplasia (MEN) | Chronic iron-deficiency anemia |

shown in **Table 20.12**. Important causes for hypoacidity and hyperacidity are shown in **Table 20.13**.

## Histamine Stimulation Test

Histamine is a powerful stimulant for the secretion of HCl in the normal stomach. It acts on receptors on oxyntic cell, increasing the cyclic AMP (cAMP) level, which causes secretion of highly acidic gastric juice.

## Augmented Histamine Test

This test has been used for two purposes: (1) to show the inability to secrete acid and (2) to assess the maximum possible acid secretion.

*Procedure*

- After an overnight fast, pass a Ryle's tube and remove the residual content for analysis.
- Collect the resting content every 20 minutes for an hour.
- At the end of 1 hour give histamine (0.04 mg histamine acid phosphate per kg body weight)

*Clinical Significance*

- In pernicious anemia, no free HCl is secreted after augmented histamine stimulation.
- In duodenal ulcers, increased secretions are observed.

## CEREBROSPINAL FLUID

The surface of the central nervous system is covered by the meninges made of three layers called outer most dura mater, middle arachnoid mater, and inner pia mater. Cerebrospinal fluid (CSF) is the area between the pia and arachnoids, that is, subarachnoid space. The CSF is formed by active secretion from the cells of the choroids plexus, the vascular structures lying within the ventricles of the brain. The CSF is also found around the spinal cord. In normal individuals, the total volume of CSF is around 100 to 200 mL.

*Collection of CSF sample:* The CSF is obtained by passing a lumbar puncture needle aseptically between third and fourth lumbar vertebrae into the subarachnoid space. A sample of CSF submitted to chemical analysis should be fresh and free from blood. The fluid should be collected in sterile containers and sent to the laboratory at the earliest.

In addition to the cell count, microbiological and serological tests are also carried out. The biochemical analyses that are commonly and routinely carried out on CSF are protein and glucose estimations. Estimation of CSF chloride is rarely done nowadays.

## Composition of Normal Cerebrospinal Fluid

- Color and appearance: Clear, colorless and no coagulum and deposit.
- pH: 7.3
- Protein: 15–45 mg/dL
- Glucose: 40–80 mg/dL
- Chloride: 120–130 mmol/L
- Urea: 20–40 mg/dL
- Calcium: 5.5–6 mg/dL

## Cerebrospinal Fluid Protein

Protein concentration in CSF can rise in a variety of pathological conditions. Mild elevation up to 300 mg/dL will be seen in viral meningitis, neurosyphilis, brain tumor, and subdural hematoma. Moderate increase in protein, with disproportionate increase in globulin is seen in multiple sclerosis and neurosyphilis. Pronounced increase up to 1 g/dL is observed in acute bacterial meningitis, tubercular meningitis, tumor of spinal cord, cerebral hemorrhage, intracranial tumor and Guillain-Barré syndrome (postinfective polyneuritis).

In most diseases, cell count and protein concentration tend to change parallelly. However, protein concentration rises, with relatively few cells in CSF observed in degenerative diseases of CNS, e.g., Guillain-Barré syndrome, spinal tumor, and multiple sclerosis.

## Cerebrospinal Fluid Glucose

Decrease in glucose level is of diagnostic significance. Almost complete disappearance is seen in bacterial and fungal meningitis. Moderate reduction can occur in tubercular meningitis and leukocyte infiltration. No significant change is observed in viral meningitis, multiple sclerosis, polyneuritis, brain tumor, etc.

## Cerebrospinal Fluid Chloride

Decrease in CSF chloride is seen in tubercular meningitis.

## SUMMARY OF THE CHAPTER

- Bilirubin is estimated by van den Bergh reaction. Normal serum does not give a positive van den Bergh reaction.
- When bilirubin is conjugated, the purple color is produced immediately on mixing with the reagent, the response is said to be van den Bergh direct positive. When the bilirubin is unconjugated, the color is obtained only when alcohol is added, and this response is known as indirect positive.
- The most common cause for hepatocellular jaundice is the infection with hepatitis viruses (viral hepatitis).
- Elevated levels of GGT are observed in chronic alcoholism.
- High levels of ALP are noticed in patients with cholestasis or hepatic carcinoma.
- The GFR of a person with 70 kg body weight is 120–125 mL/minute.
- The glomeruli of kidney are not permeable to substances with molecular weight of more than 69,000.
- Microalbuminuria is seen as a complication of diabetes mellitus and hypertension.
- Ketonuria may be detected by Rothera's test.
- Clearance is defined as the quantity of blood or plasma completely cleared of a substance per unit time and is expressed as milliliter per minute.
- Maximum urea clearance is found to be 75 mL/min in normal.

## QUESTIONS FOR SELF-ASSESSMENT

1. Enumerate liver function tests and describe with clinical significance.

2. Classify jaundice. Give an account of the biochemical tests that will help in differentiating the types of jaundice.
3. Mention the role of bilirubin in normal and abnormal conditions.
4. Mention the enzymes used as liver function tests.
5. Discuss the biochemical alterations seen in blood and urine in different types of jaundice.
6. Add a note on fractional meal test.
7. Name the renal clearance tests. Give details of any one of them.
8. How creatinine clearance test is done?
9. Mention the significance of GFR.
10. Mention the biochemical investigations in CSF.

## MULTIPLE CHOICE QUESTIONS (MCQs)

**20-1.** An increase in serum unconjugated bilirubin occurs in:
   a. Hemolytic jaundice
   b. Obstructive jaundice
   c. Defect in intestinal absorption
   d. Glomerulonephritis

**20-2.** Conjugated hyperbilirubinemia and raised alkaline phosphatase levels are characteristic of:
   a. Obstructive jaundice
   b. Hemolytic jaundice
   c. Viral hepatitis
   d. Physiological jaundice

**20-3.** All are features of obstructive jaundice, *except:*
   a. Increased level of conjugated bilirubin in blood
   b. Clay-colored stools
   c. Presence of bile salts in urine
   d. Increased excretion of urobilinogen in urine

**20-4.** Hypoacidity is found in all the following conditions, *except:*
   a. Pernicious anemia
   b. Carcinoma of stomach
   c. Insulinoma
   d. Atrophic gastritis

**20-5.** All the following biochemical parameters are indices of kidney function, *except:*
   a. Bilirubin
   b. Urea
   c. Albumin
   d. Creatinine

**20-6.** All the following biochemical parameters are indices of liver function, *except:*
   a. Bilirubin
   b. Cholesterol
   c. Albumin
   d. Creatinine

**20-7.** Normal specific gravity of urine is:
   a. 1.003–1.010
   b. 1.010–1.015
   c. 1.015–1.025
   d. 1.025–1.035

**20-8.** Polyuria is a characteristic feature of all the following, *except:*
   a. Diabetes mellitus
   b. Glomerulonephritis
   c. Diabetes insipidus
   d. Chronic renal failure

**20-9.** Specific gravity of urine increases in:
   a. Chronic glomerulonephritis
   b. Diabetes mellitus
   c. Liver diseases
   d. Intake of vegetables

**20-10.** A patient with infective hepatitis is likely to have all the following findings, *except:*
   a. Hyperbilirubinemia
   b. Bilirubinuria
   c. Absence of bile salts in urine
   d. Elevated AST

## ANSWER KEYS TO MCQs

20-1. (a)   20-2. (a)   20-3. (d)   20-4. (c)   20-5. (c)   20-6. (d)   20-7. (c)   20-8. (b)
20-9. (b)   20-10. (b)

# CHAPTER 21

# Molecular Biology

At the completion of this chapter, the reader will be able to answer questions on the following topics:

- Purines and pyrimidines
- Nucleosides and nucleotides
- Uric acid and gout
- Purine and pyrimidine analogues as anticancer drugs
- Watson–Crick model of DNA structure
- Replication of DNA
- Ribonucleic acid
- Messenger RNA
- Transcription process
- Transfer RNA
- Ribosomes
- Genetic code
- Protein biosynthesis
- Inhibitors of protein synthesis

## COMPOSITION OF NUCLEOTIDES

Nucleic acids are deoxyribonucleic acid (DNA) and ribonucleic acid (RNA). These are made up of nucleotides. A nucleotide is made up of three components:
1. Nitrogenous base (a purine or a pyrimidine).
2. Pentose sugar, either ribose or deoxyribose.
3. Phosphate groups esterified to the sugar.

When a base combines with a pentose sugar, a **nucleoside** is formed. When the nucleoside is further esterified to another phosphate group, it is called a **nucleotide** or nucleoside monophosphate. When a second phosphate gets esterified to the existing phosphate group, a nucleoside diphosphate is generated. The attachment of a third phosphate group results in the formation of a nucleoside triphosphate (NTP). The nucleic acids (DNA and RNA) are the polymers of nucleoside *monophosphates*.

## Bases Present in the Nucleic Acids

Two types of nitrogenous bases: the **purines** and **pyrimidines** are present in nucleic acids.

*Purine Bases*

The purine bases present in RNA and DNA are the same: **adenine** and **guanine**. The numbering of the purine ring with the structure of adenine and guanine is shown in **Figure 21.1**.

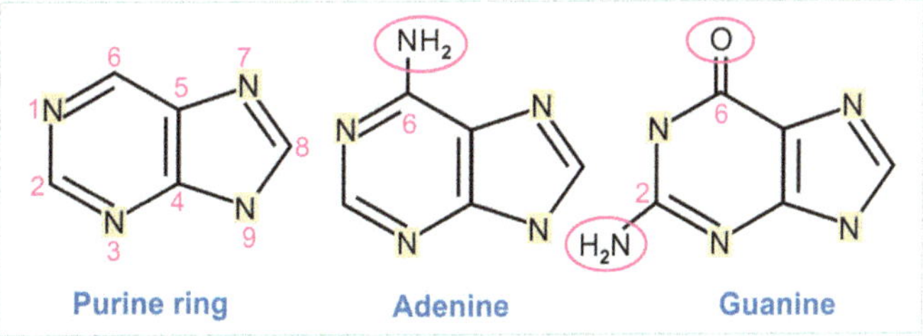

**Fig. 21.1:** Structure of purines.

*Pyrimidine Bases*

The pyrimidine bases present in nucleic acids are cytosine, thymine, and uracil. **Cytosine** is present in both DNA and RNA. **Thymine** *is present in DNA and* **uracil** *in RNA*. Structures are shown in **Figure 21.2**. Nucleosides with purine bases have the suffix **"-sine"**, while pyrimidine nucleosides end with **"-dine"**. *Uracil combines with ribose only; and thymine with deoxyribose only* (**Table 21.1**).

## Nucleotides

These are phosphate esters of nucleosides. Base plus pentose sugar plus phosphoric acid is a nucleotide (**Table 21.2**). Most of the nucleoside phosphates involved in biological function are 5'-phosphates (**Table 21.3**). Since 5'-nucleotides are more often seen, they are simply written without any prefix. For example, 5'-AMP is abbreviated as AMP. Many coenzymes are derivatives of adenosine monophosphate. Examples are nicotinamide adenine dinucleotide (NAD$^+$), nicotinamide adenine dinucleotide, nicotinamide adenine dinucleotide phosphate (NADP), flavin adenine dinucleotide (FAD), and coenzyme A.

**Table 21.1:** Base + sugar are nucleosides.

| Ribonucleosides | | |
|---|---|---|
| Adenine | + Ribose | → Adenosine |
| Guanine | + Ribose | → Guanosine |
| Uracil | + Ribose | → Uridine |
| Cytosine | + Ribose | → Cytidine |
| Hypoxanthine | + Ribose | → Inosine |
| Xanthine | + Ribose | → Xanthosine |
| Deoxyribonucleosides | | |
| Adenine | + Deoxyribose | → Deoxy adenosine (d-adenosine) |
| Guanine | + Deoxyribose | → d-guanosine |
| Cytosine | + Deoxyribose | → d-cytidine |
| Thymine | + Deoxyribose | → d-thymidine |

**Table 21.2:** Base + sugar + phosphate = nucleotide.

| Ribonucleotides | | |
|---|---|---|
| Adenosine | + Pi | Adenosine monophosphate (AMP) (adenylic acid) |
| Guanosine | + Pi | Guanosine monophosphate (GMP) (guanylic acid) |
| Cytidine | + Pi | Cytidine monophosphate (CMP) (cytidylic acid) |
| Uridine | + Pi | Uridine monophosphate (UMP) (uridylic acid) |
| Deoxyribonucleotides | | |
| d-adenosine | + Pi | d-AMP (d-adenylic acid) |
| d-guanosine | + Pi | d-GMP (d-guanylic acid) |
| d-cytidine | + Pi | d-CMP (d-cytidylic acid) |
| d-thymidine | + Pi | d-TMP (d-thymidylic acid) |

## Nucleoside Triphosphates

Corresponding nucleoside di- and triphosphates are formed by esterification of further phosphate groups to the existing ones. In general, any nucleoside triphosphate is abbreviated as NTP or d-NTP (**Table 21.4**).

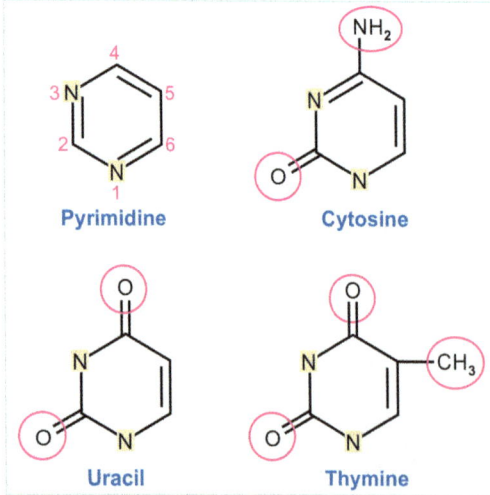

**Fig. 21.2:** Common pyrimidines.

**Table 21.3:** Nucleosides and nucleotides.

| Base | Sugar | Nucleoside | Phosphoric acid at | Nucleotide |
|---|---|---|---|---|
| Adenine | Ribose | adenosine | 5′ Position | AMP |
| Do | Do | Do | 3′ Position | 3′-AMP |
| Do | Deoxyribose | d-Adenosine | 5′ Position | d-AMP |
| Do | Do | Do | 3′ Position | d-3′-AMP |
| Cytosine | Ribose | Cytidine | 5′ Position | CMP |
| Do | Do | Do | 3′ Position | 3′-CMP |
| Do | Deoxyribose | d-Cytidine | 5′ Position | d-CMP |
| Do | Do | Do | 3′ Position | d-3′-CMP |

(CMP: cyclic monophosphate; AMP: adenosine monophosphate)

**Table 21.4:** Nucleoside triphosphates.

| Nucleoside | Nucleoside monophosphate (NMP) | Nucleoside diphosphate (NDP) | Nucleoside triphosphate (NTP) |
|---|---|---|---|
| Ribonucleoside phosphates | | | |
| Adenosine | AMP | ADP | ATP |
| Guanosine | GMP | GDP | GTP |
| Cytidine | CMP | CDP | CTP |
| Uridine | UMP | UDP | UTP |
| Deoxy ribonucleoside phosphates | | | |
| d-Adenosine | d-AMP | d-ADP | d-ATP |
| d-Guanosine | d-GMP | d-GDP | d-GTP |
| d-Cytidine | d-CMP | d-CDP | d-CTP |
| d-Thymidine | d-TMP | d-TDP | d-TTP |

(GMP: guanosine 5′-monophosphate; CMP: cytidine monophosphate; AMP: adenosine monophosphate; UMP: uridine 5′-monophosphate; ADP adenosine diphosphate; GDP: guanosine diphosphate; CDP: cytosine diphosphate; UDP: uridine diphosphate; ATP: adenosine triphospahate; GTP: guanosine triphospahate; CTP: cytosine triphospahate; UTP: uridine triphospahate; TMP: thymidine monophosphate; TDP: thymidine diphosphate; TTP: thymidine triphosphate)

Nucleoside diphosphate contains one high-energy bond and triphosphate has two high-energy bonds. ATP is the **universal energy currency (Fig. 21.3)**. It is formed during oxidative processes by trapping the released energy in the high-energy phosphate bond. More details on high-energy bonds are given in Chapter 13.

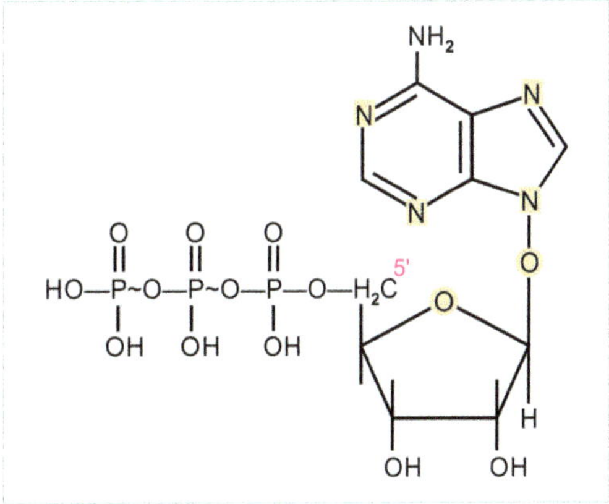

## BIOSYNTHESIS OF PURINE NUCLEOTIDES

The purine nucleotides are synthesized by most of the tissues. However, the major site is the liver. This pathway operates in the cytoplasm. The major pathway is denoted as

**Fig. 21.3:** Adenosine triphosphate (ATP).

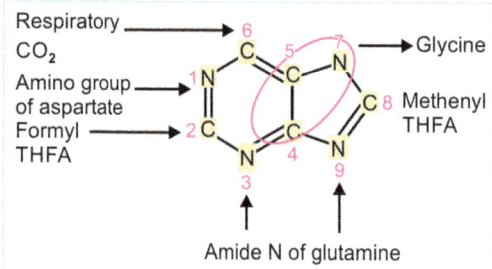

**Fig. 21.4:** The assembly of purine ring from various sources. [THFA (FH4): tetrahydrofolic acid].

de novo **synthesis**, because the purine ring is synthesized from different small components. The contribution of different atoms from different sources for the formation of the purine ring is shown in **Figure 21.4**.

During the de novo synthesis, *purine ring is built on a ribose-5-phosphate molecule.* Hence, the nucleotides are the products of the de novo synthesis. The de novo synthesis pathway involves 10 steps **(Table 21.5)**. The enzymes catalyzing these reactions exist as a multienzyme complex in eukaryotic cells; this arrangement increases the efficiency of the pathway.

**Table 21.5:** Summary of steps involved in purine synthesis.

| Step | Donor | Added atom |
|---|---|---|
| 1 | Glutamine | N9 (Rate limiting) |
| 2 | Glycine | C4, 5, N7 (ATP required) |
| 3 | Methenyl-THFA | C 8 |
| 4 | Glutamine | N3 (ATP required) |
| 5 | — | Ring closure (ATP required) |
| 6 | Carbon dioxide | C 6 |
| 7 | Aspartic acid | N1 (ATP required) |
| 8 | — | Fumarate removed |
| 9 | Formyl-THFA | C 2 |
| 10 | — | Ring closure |

(THFA: tetrahydrofolic acid; ATP: adenosine triphosphate)

## Salvage Pathway

This pathway ensures the recycling of purines formed by degradation of nucleotides. The free purines are salvaged by two different enzymes: adenine phosphoribosyltransferase (APRTase) and hypoxanthine guanine phosphoribosyltransferase (HGPRTase). The pathway is of special importance in tissues such as red blood cells (RBCs) and brain where the de novo pathway is not operating. Salvage pathway is summarized below:

$$\text{Adenine + PRPP} \xrightarrow{\text{APRTase}} \text{AMP + PPi}$$

$$\text{Guanine + PRPP} \xrightarrow{\text{HGPRTase}} \text{GMP + PPi}$$

## Regulation of Purine Synthesis

The committed step in de novo synthesis is the reaction catalyzed by amidotransferase (step 1). It is inhibited by AMP and GMP. Both AMP and GMP inhibit their own formation by feedback inhibition of adenylosuccinate synthetase and inosine monophosphate (IMP) dehydrogenase.

*Analogs as Purine Synthesis Inhibitors*

They act as competitive inhibitors of the naturally occurring nucleotides. They are utilized to synthesize DNA. Such DNA becomes functionally inactive. Thereby cell division is arrested. So they are useful as anticancer drugs. Examples are:
- Mercaptopurine inhibits the production of GMP and AMP.
- Folate antagonists (methotrexate) would affect the reactions involving one-carbon group transfers.

## Degradation of Purine Nucleotides

The end product of purine nucleotide catabolism is **uric acid** (urate). The structure is shown in **Figure 21.5**. This degradation is taking place mainly in the liver. The main

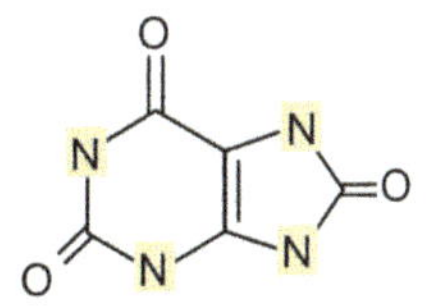

**Fig. 21.5:** Structure of uric acid.

enzyme, **xanthine oxidase**, contains FAD, molybdenum, and iron. The reaction produces hydrogen peroxide (reactive oxygen species).

## URIC ACID

In the body, purines are degraded into uric acid. Normal blood level of uric acid ranges from 3 to 7 mg/dL in males and 2 to 5 mg/dL in females. The daily excretion varies from 500 to 700 mg. Nucleic acid content is more in nonvegetarian diet. Uric acid is sparingly soluble in water.

## Disorders of Purine Metabolism

The most common abnormality is elevation of uric acid level in blood, which is referred to as **hyperuricemia**. It is defined as serum uric acid concentration exceeding 7 mg/dL in male and 6 mg/dL in female. It may or may not be associated with increased excretion of uric acid in urine, which is called **uricosuria**. The manifestations are due to the low solubility of uric acid in water.

## GOUT

- It is due to accumulation of uric acid crystals in the synovial fluid resulting in inflammation leading to acute arthritis.
- At 30°C, the solubility of uric acid is lowered to 4.5 mg/dL. Therefore, uric acid is deposited in cooler areas of the body to cause **tophi**. Thus, tophi are seen in distal joints of foot.
- Increased excretion of uric acid may cause deposition of uric acid crystals in the urinary tract, leading to **calculi** or stone formation with renal damage. Gout may be either primary or secondary.

## Primary Gout

About 10% of cases of primary gout are idiopathic. Primary gout may show a familial incidence. Incidence of primary gout is about 1:500 in total population. The enzyme, 5-amido phosphoribosyl transferase (old name, 5-phosphoribosyl amidotransferase), is active but not sensitive to feedback regulation by the inhibitory nucleotides. This would lead to overproduction of purine nucleotides.

## Secondary Hyperuricemia

It may be due to enhanced turnover rate of nucleic acids as seen in rapidly growing malignant tissues, e.g., leukemia, lymphoma, polycythemia, etc. Hyperuricemia is also seen in cancer patients on radiotherapy or chemotherapy (tumor lysis) due to increased cellular turnover. Hence, these patients are given allopurinol also to decrease the uric acid levels. Uric acid also increases when there is tissue damage due to trauma and raised rate of catabolism as in starvation.

### Clinical Finding of Gout

Many geniuses, including Isaac Newton, Gibbon, and Johnson, were suffering from gout. Gouty attacks may be precipitated by high purine diet and increased intake of **alcohol**. Often the patients have a few drinks, go to sleep symptomless but are awakened during the early hours of morning by excruciating joint pains. Alcohol leads to accumulation of lactic acid. The typical gouty **arthritis** affects the first metatarsophalangeal joint (big toe), but other joints may also be affected. The joints are extremely painful. Synovial fluid will show urate crystals.

In **chronic cases**, uric acid may get deposited around joints causing swelling (**tophi**), which is composed of sodium urate. In chronic gout, urate crystals are deposited in renal medulla, which leads to urolithiasis and renal damage.

## Treatment Policies in Gout

- Reduce dietary purine intake and restrict alcohol.
- Increase the renal excretion of urate by **uricosuric drugs**, which decrease the reabsorption of uric acid from kidney tubules, e.g., probenecid.
- Reduce urate production by **allopurinol**, an analog of hypoxanthine. Allopurinol is a **competitive** inhibitor of xanthine oxidase, thereby decreasing the formation of uric acid. Xanthine and hypoxanthine are more soluble and are excreted more easily. Xanthine oxidase converts allopurinol to alloxanthine. It is a more effective inhibitor of xanthine oxidase. This is a good example of '**suicide inhibition** (*see* Chapter 3).

## DE NOVO SYNTHESIS OF PYRIMIDINE

The pyrimidine ring (unlike the purine) is synthesized as free pyrimidine and then it is incorporated into the nucleotide. The derivation of atoms of pyrimidine nucleus is indicated in **Figure 21.6**.

In the first step, carbamoyl phosphate is produced in the cytoplasm (during urea synthesis, the reaction happens in mitochondria). The nitrogen of glutamine and bicarbonate react to form carbamoyl phosphate (step 1). The enzyme is carbamoyl phosphate synthetase II (CPS II). (CPS-I is described under urea, (*see* Chapter 11). Uridine monophosphate (UMP) is the first purine that is synthesized in the pathway.

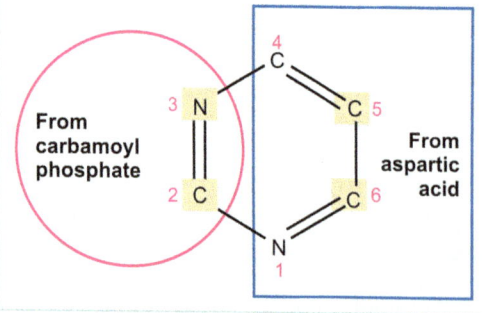

**Fig. 21.6:** Sources of C and N atoms of pyrimidine.

## Disorders of Pyrimidine Metabolism

### Orotic Aciduria

The condition results from the absence of certain enzymes of the pyrimidine synthesis pathway [orotate phosphoribosyltransferase (OPRTase) and orotidine 5'-monophosphate (OMP). It is an autosomal recessive disease. It causes retarded growth and megaloblastic **anemia**. The rapidly growing cells are more affected and hence anemia. Crystals are excreted in urine which may cause urinary tract obstruction. This condition can be successfully treated by feeding cytidine or uridine. They may be converted to UTP which can act as feedback inhibitor.

### Analogs as Purine Synthesis Inhibitors

They act as competitive inhibitors of the naturally occurring nucleotides. They are utilized to synthesize DNA, such DNA becomes functionally inactive. Thereby cell division is arrested. So they are useful as anticancer drugs. A few examples are mercaptopurine, methotrexate, azaserine, and 8-aza guanine.

### Anticancer Agents Acting on Pyrimidines

Anticancer drugs 5-fluorouracil, 5-iodouracil, 3-deoxy uridine, 6-azauridine, 6-azacytidine, and 5-iodo-2-deoxyuridine competitively inhibit pyrimidine synthesis. Two antimetabolites used in the treatment of viral infections, acyclovir (acycloguanosine) and zidovudine (azido deoxythymidine), are derivatives of nucleotides.

## STRUCTURE OF DNA

DNA is composed of four deoxyribonucleotides, i.e., deoxyadenylate (A), deoxyguanylate (G), deoxycytidylate (C), and deoxythymidylate (T). These units are combined through 3' to 5' **phosphodiester bonds** to polymerize into a long chain. The nucleotide is formed by a combination of base + sugar + phosphoric acid. The 3'-hydroxyl of one sugar is combined to the 5'-hydroxyl of another sugar through a phosphate group **(Fig. 21.7)**. In this particular example, the thymidine is attached to cytidine and then cytidine to adenosine through phosphodiester linkages **(Fig. 21.7)**. In the DNA, the **base sequence** is of paramount importance. The genetic information is coded in the specific sequence of bases; if the base is altered, the information is also altered.

## Watson–Crick Model of DNA Structure

The salient features of Watson–Crick model of DNA are given in **Figure 21.5**.

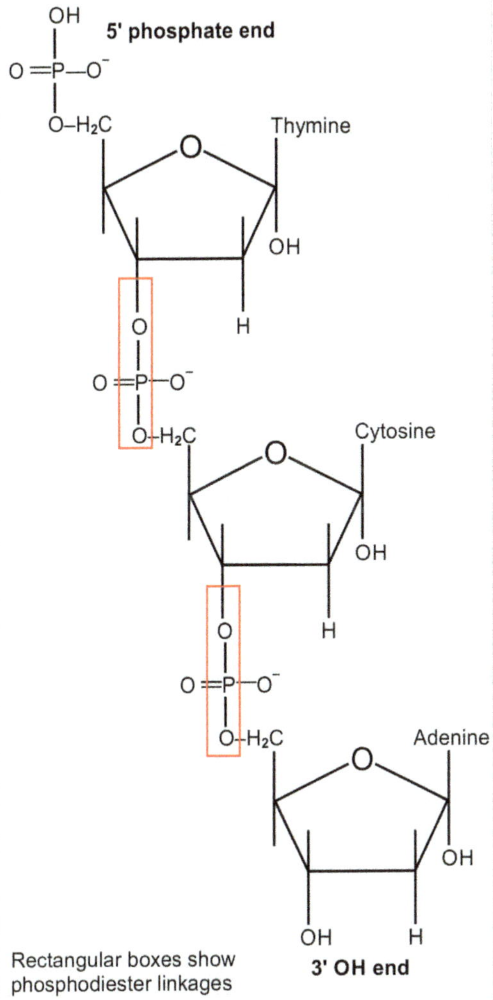

**Fig. 21.7:** Polynucleotide.

Rectangular boxes show phosphodiester linkages

### Right-Handed Double Helix

DNA consists of two polydeoxyribonucleotide chains twisted around one another in a right-handed double helix similar to a spiral staircase.

### Base-Pairing Rule

Always the two strands are **complementary** to each other. So the adenine of one strand will pair with the thymine of the opposite strand, while guanine will pair with cytosine. The base-pairing (**A with T; G with C**) is called **Chargaff's rule**, which states that the number of purines is equal to the number of pyrimidines **(Fig. 21.9)**.

### Hydrogen Bonding

The DNA strands are held together mainly by hydrogen bonds between the purine and pyrimidine bases. There are two hydrogen

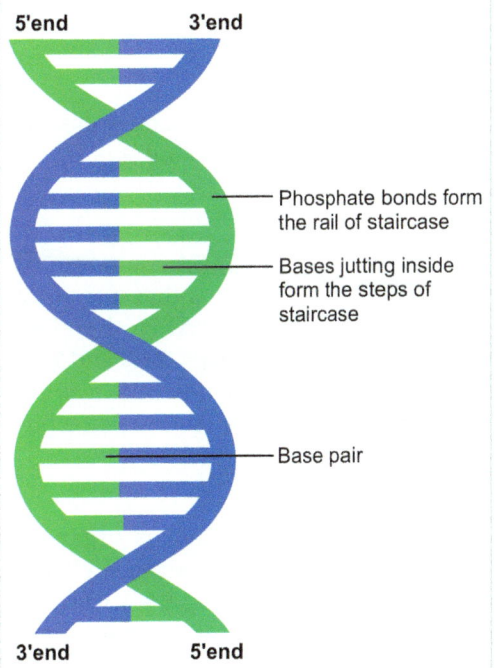

Fig. 21.8: Watson–Crick model of double-helical structure of DNA. Adjacent bases are separated by 0.34 nm. The diameter or width of the helix is 2 nm.

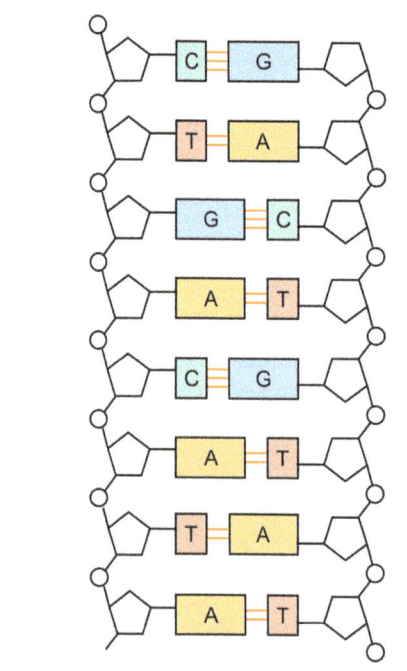

**Fig. 21.9:** Base-pairing rule. Base-pairing of A with T and G with C. Hydrogen bonds keep the bases in position.

bonds between A and T, while there are three hydrogen bonds between C and G.

*Antiparallel*

The two strands in a DNA molecule run antiparallel, which means that one strand runs in the 5' to 3' direction, while the other is in the 3' to 5' direction **(Figs. 21.8 and 21.9)**. This is similar to a road divided into two, each half carrying traffic in the opposite direction.

## Higher Organization of DNA

In higher organisms, DNA is organized inside the nucleus. Double-stranded DNA is wound round histones to form nucleosomes. **Chromatin** is a loose term employed for a long stretch of DNA in association with histones. Chromatin is then further and further condensed to form **chromosomes** (Fig. 21.10).

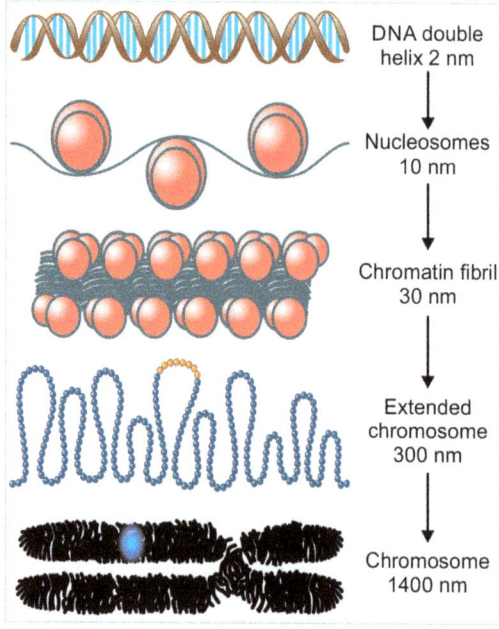

**Fig. 21.10:** DNA condenses repeatedly to form chromosomes.

## Size of DNA

Human diploid genome consists of about $7 \times 10^9$ base-pairs. So when placed end to end, it will be about 2 m long. If one nucleotide is added per second, it will take 250 years to synthesize the whole DNA of a human cell. The length of a DNA molecule is compressed to 8,000- to 10,000-fold to generate the chromosomes.

## Exons, Introns, and Cistrons

Only about 10% of the human DNA contain genes and the rest are silent areas. The segments of the gene coding for proteins are called **exons** (expressed regions). They are interspaced in the DNA with stretches of silent areas called **introns** (intervening areas). The primary transcripts contain intron sequences, which are later removed to produce mature messenger RNA (mRNA). Introns are not translated. A **cistron** is the unit of genetic expression. It is the biochemical counterpart of a "gene" of classical genetics. One cistron will code for one polypeptide chain. If a protein contains four subunits, these are produced under the direction of four cistrons ("one cistron–one polypeptide" concept).

## Repeat Sequences of DNA

About 30% of the genome consists of repetitive sequences, 5–500 base-pairs repeated many times. One such sequence, the Alu family is repeated about 500,000 times and accounts for about 5% of the total human DNA.

## REPLICATION OF DNA

During cell division, each daughter cell gets an exact copy of the genetic information of the mother cell. This process of copying the DNA is known as **DNA replication.** In the daughter cell, one strand is derived from the mother cell, while the other strand is newly synthesized. This is called **semiconservative** type of DNA replication.

## Steps of Replication

1. Unwinding of parental DNA to form a replication fork.
2. Each strand serves as a **template** or mold, over which a new **complementary** strand is synthesized **(Fig. 21.11)**. The enzyme, **DNA polymerase**, synthesizes a new complementary strand of DNA, by incorporating deoxynucleoside monophosphate (dNMP) sequentially in 5' to 3' direction, making use of single-stranded DNA as template.
3. The **base-pairing** rule is always maintained. The new strand is joined to the old strand by hydrogen bonds between base-pairs (A with T and G with C).
4. DNA synthesis is continuous in the leading strand (toward replication fork) by DNA polymerase.
5. DNA synthesis is discontinuous in the lagging strand (away from the fork), as Okazaki fragments.
6. In both strands, the synthesis is from 5' to 3' direction.

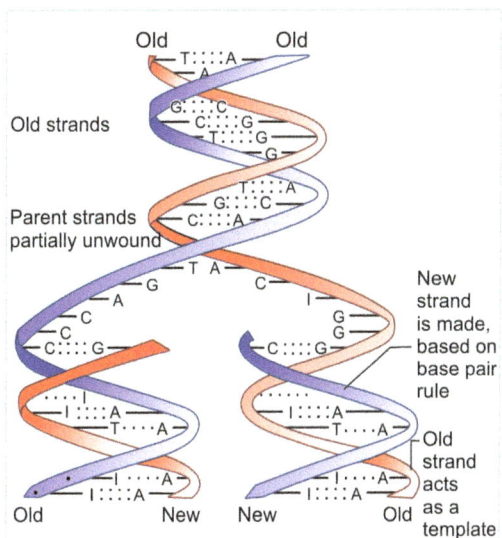

**Fig. 21.11:** Replication of DNA. Both strands are replicated.

7. Proofreading is done by DNA polymerase.
8. Thus, two double strands are produced. One goes to one daughter cell nucleus, while the other to the second daughter cell. But each daughter cell gets only one strand of the parent DNA molecule. Old DNA strand is not degraded but is conserved for the daughter cell, hence this is referred to as semiconservative synthesis.
9. Finally organized into chromatin.

## Inhibitors of DNA Replication

Certain compounds will inhibit bacterial enzymes but will not affect human cells. Such compounds or drugs are useful as **antibacterial agents**. Examples are ciprofloxacin, nalidixic acid, and novobiocin. Some other components will inhibit human enzymes, and they will arrest new DNA synthesis and also arrest the cell division. Those drugs are therefore useful as **anticancer agents**. Examples are etoposide, Adriamycin, and doxorubicin.

## RNA

RNA is also a polymer of purine and pyrimidine nucleotides linked by phosphodiester bonds.

**Table 21.6:** Differences between RNA and DNA.

|  | RNA | DNA |
|---|---|---|
| 1. | Mainly seen in cytoplasm | Mostly inside nucleus |
| 2. | Usually 100–5000 bases | Millions of base-pairs |
| 3. | Generally single stranded | Double stranded |
| 4. | Sugar is ribose | Sugar is deoxyribose |
| 5. | Purines: adenine and guanine | Purines: adenine and guanine |
| 6. | Pyrimidines: cytosine and uracil | Pyrimidines: cytosine and thymine |

However, RNA differs from DNA as shown in **Table 21.6**.

## Five Types of Cellular RNAs

1. *Messenger RNA:* The genetic information present in DNA is transcribed into mRNA, which is then used for the synthesis of a particular protein. They have a short half-life and generally degrade quickly.
2. *Ribosomal RNA (rRNA):* They are involved in protein biosynthesis and are very stable.
3. *Transfer RNA (tRNA):* The tRNAs are about 60 different species. They are very stable and carry amino acids.
4. *Small RNA:* The small RNAs are of about 30 different varieties. They are very stable. They are involved in mRNA splicing.
5. *Micro RNA (miRNA):* They regulate genetic expression by altering the mRNA function.

## Central Dogma of Molecular Biology

The information available in the DNA is passed to mRNA, which is then used for the synthesis of a particular protein (**Fig. 21.12**). DNA **replication** is like printing a copy of all the pages of a book. The replication process occurs only at the time of cell division. But **transcription** is taking place all the time. Only certain areas of the DNA are copied

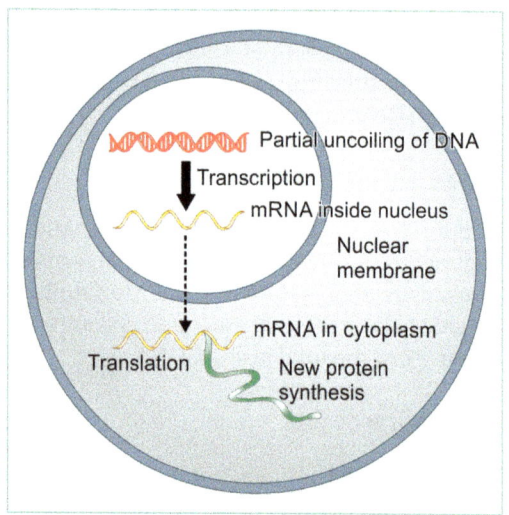

**Fig. 21.12:** Central dogma of molecular biology. (mRNA: messenger RNA).

(select regions on the sense strand). This is like taking photo copy of a particular page of the book. So the genetic information of the DNA is transcribed (copied) to the **mRNA**. During transcription, the message from the DNA is copied in the language of nucleotides (four-letter language). During **translation,** the nucleotide sequence is translated to the language of amino acid sequence (20-letter language).

## TRANSCRIPTION PROCESS

The template strand of DNA is transcribed to give rise to mRNA. Transcription is catalyzed by **RNA polymerase (RNAP).** First, the DNA helix partially unwinds, and the RNAP binds with the DNA and moves forward. Along with that the new mRNA strand is being synthesized. The mRNA will be complementary to the base present in the DNA. The RNAP moves along the DNA template. New nucleotides are incorporated in the nascent mRNA, one by one, according to the base-pairing rule. However, thymine is not present in RNA; instead uracil will be incorporated.

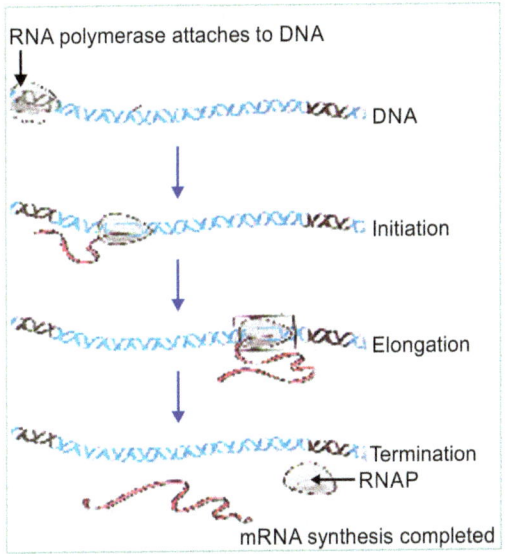

**Fig. 21.13:** Transcription process. (DNA: deoxyribonucleic acid; RNAP = ribonucleic acid polymerase)

## Steps of Transcription

1. Initiation of transcription
2. Elongation process
3. Termination of transcription
4. Posttranslational processing.

## Initiation of Transcription

The DNA helix partially unwinds and the RNAP binds with the promoter site on the DNA and moves forward **(Fig. 21.13)**. When it reaches the appropriate site on the gene, the first nucleotide of the mRNA attaches to the initiation site on the beta subunit of RNAP. This becomes the 5' end of the mRNA. It will be complementary to the base present in the DNA at that site. The next nucleotide attaches to the RNAP. A phosphodiester bond is formed. Then the enzyme moves to the next base on the template DNA.

## Signals for Initiation of Transcription

**Promoters:** Certain consensus sequences on DNA act as start signals and may be located upstream or downstream from the start site. The RNAP attaches at the promoter site on the template DNA strand.

**TATA box:** In the case of bacteria, about 10 base-pair upstream, there is the sequence 5'-TATAAT-3'. This is referred to as *TATA box*. In mammals, the exact sequence in T box is slightly different (TATAAA) and is known as *Goldberg–Hogness* box. This signal sequence located at -25 to -30 position indicates the start site.

**Enhancers** increase the rate of transcription and **silencers** decrease the rate.

## Elongation Process

The RNAP moves along the DNA template. New nucleotides are incorporated in the nascent mRNA, one by one, according to the base-pairing rule **(Fig. 21.13)**. Thus, A in DNA is transcribed to U in mRNA; similarly T to A, G to C, and C to G. The synthesis of mRNA is from 5' to 3' end. That means the reading of template DNA is from 3' to 5'. This is analogs to the polarity in DNA synthesis. As the RNAP moves on the DNA template, the DNA helix unwinds downstream and winds at the upstream areas. A transcription bubble containing RNAP, DNA, and nascent RNA is formed **(Fig. 21.13)**.

## Termination of Transcription

The specific signals are recognized by a termination protein, the rho factor (abbreviated with Greek letter, "ρ"). When it attaches to the DNA, the RNAP cannot move further. So the enzyme dissociates from DNA and consequently releases the newly formed mRNA.

## Post-transcriptional Processing

The mRNA formed and released from the DNA template is known as the primary transcript. It is also known as heteronuclear mRNA or hnRNA. In mammalian system, it undergoes extensive editing to become the mature mRNA. Its modifications are the following:

- **Poly(A) tailing**: The 3' terminus is polyadenylated in the nucleoplasm. This tail protects mRNA from attack of 3' exonuclease.
- **Removal of introns**: The primary transcript contains coding regions (exons) interspersed with noncoding regions (introns). These intron sequences are cleaved and the exons are spliced (combined together) to form the mature mRNA molecule. This processing is done in nucleus.

## Inhibitors of RNA Synthesis

Actinomycin D and mitomycin intercalate with DNA strands, thus blocking transcription. They are used as anticancer drugs. Rifampicin is widely used in the treatment of tuberculosis and leprosy.

## PROTEIN BIOSYNTHESIS

During protein biosynthesis, the DNA is **transcribed** to mRNA which is **translated** into protein with the help of ribosomes and tRNAs.

## Transfer RNA

The tRNAs transfer amino acids from cytoplasm to the ribosomal protein synthesizing machinery; hence, the name tRNA. Since they are easily soluble, they are also referred to as **soluble RNA** or **sRNA**. The tRNAs have **clover leaf**-like structure **(Fig. 21.14)**. It recognizes the message present in mRNA (triplet nucleotide codon present in mRNA). The tRNA molecule will show specificity in both aspects, i.e., in recognizing the mRNA codon and in accepting the specific amino acid coded by that codon. In this way, the tRNA molecules play a pivotal role in translation.

## Ribosomal RNA

Ribosomes provide necessary infrastructure for the mRNA, tRNA, and amino acids to interact with each other for the translation process. Thus, *ribosomal assembly is the protein synthesizing machinery.* They contain different rRNAs and specific proteins. The mammalian ribosome has a sedimentation velocity of **80S unit**. Bacteria has 70S ribosomes. So many antibiotics will inhibit bacterial protein synthesis but will do no harm to human cells.

## Genetic Code

A triplet sequence of nucleotides on the mRNA is the codon for each amino acid. Since there are four different bases, they can generate 64 ($4^3$) different codons or code words. The codes (triplet codons) are on the mRNA. Each codon is a consecutive sequence of three bases on the mRNA, e.g., UUU codes for phenylalanine. **One codon codes for only one amino acid (Fig. 21.14)**. The codons are universal, they are the same for the same amino acid in all species; they are the same for elephant and *Escherichia coli*. The genetic code has been highly preserved during evolution.

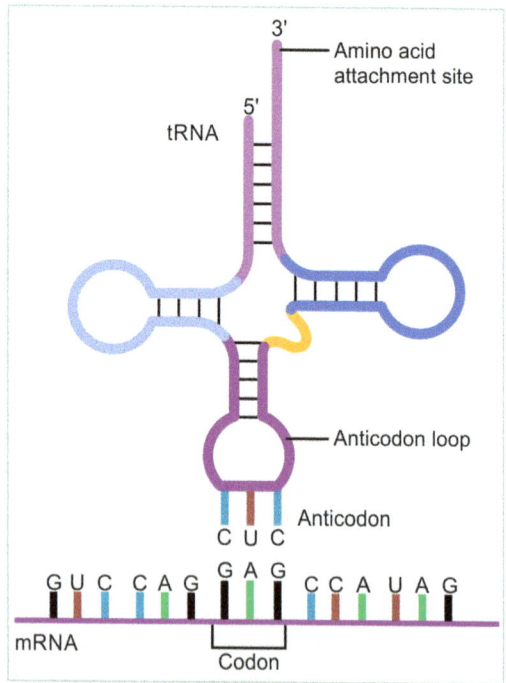

**Fig. 21.14:** Transfer RNA (tRNA) general structure. (mRNA: messenger RNA)

# TRANSLATION PROCESS

The translation is a cytoplasmic process. The enzymes **aminoacyl tRNA synthetases** add the specific amino acid to the corresponding tRNA. The tRNA carrying the particular amino acid binds with ribosome, which in turn binds with mRNA. The ribosome gives the necessary structure for protein synthesis. One by one, the amino acids are added to the new protein. Which particular amino acid is to be added is determined by the genetic code present in the amino acid. That particular amino acid is carried by the specific tRNA. Thus, genetic code is translated to protein, with the help of mRNA, ribosomes, and tRNAs.

## Steps of Translation

The translation is a cytoplasmic process. The mRNA is translated from 5' to 3' end. In the polypeptide chain synthesized, the first amino acid is the amino terminal one. The chain growth is from amino terminal to carboxyl terminal. The process of translation can be conveniently divided into the following five phases:

### 1. Activation of Amino Acid

The enzymes aminoacyl tRNA synthetases activate the amino acids. The enzyme is highly selective in the recognition of both the amino acid and the tRNA acceptor. The CCA 3' terminus of the acceptor arm carries the amino acid **(Fig. 21.14)**. Amino acid is first activated with the help of ATP. Then the carboxyl group of the amino acid is esterified with 3' hydroxyl group of tRNA.

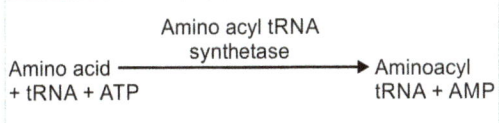

In this reaction, ATP is hydrolyzed to AMP, and so two high-energy phosphate bonds are consumed.

### 2. Initiation of Protein Synthesis

The first AUG triplet after the marker sequence is identified by the ribosome as the start codon. For the process, **initiation factors** (IFs) are required. In eukaryotes, the first amino acid incorporated is methionine (AUG codon).

**Formation of initiation complex**: The met-tRNA (tRNA carrying methionine) and **40S ribosomal** subunit are combined and then mRNA binds to form 48 S initiation complex.

**Formation of 80S ribosomal assembly**: The 48S initiation complex now binds with **60S** ribosomal unit to form the full assembly of 80S ribosome. This needs hydrolysis of **GTP**.

**The P and A sites of ribosomal assembly**: The whole ribosome contains two receptor sites for tRNA molecules. The "P" site or **peptidyl site** carries the peptidyl tRNA. It carries the growing peptide chain. The "A" site or **aminoacyl site** carries the new incoming tRNA with the amino acid to be added next. The tRNA-Met is now at the P site.

### 3. Elongation Process of Translation

**Binding of new amino acyl tRNA:** A new aminoacyl tRNA comes to the "A" site. The next codon in mRNA determines the incoming amino acid. GTP is required. The tRNA binds to the A site **(Fig. 21.15)**.

**Peptide bond formation**: The a-amino group of the incoming amino acid in the A site forms a peptide bond (CO-NH) with carboxyl group of the peptidyl tRNA occupying the P site. This reaction is catalyzed by the enzyme **peptidyl transferase**. Now the growing peptide chain occupies the A site **(Fig. 21.15)**.

**Translocation process**: At this time, the tRNA fixed at the P site does not carry any amino acid and is therefore released from the ribosome. Then the whole ribosome moves over the mRNA through the distance of one codon (three bases). The peptidyl tRNA is translocated to the P site. Now, the A site is ready to receive another aminoacyl

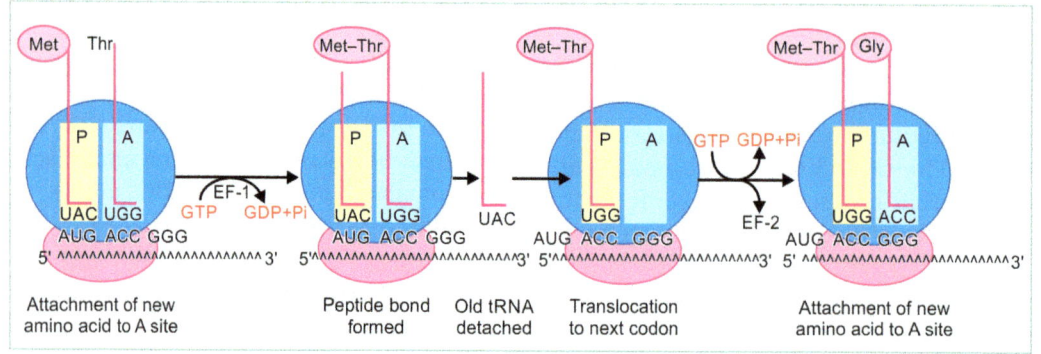

**Fig. 21.15:** Elongation phase in translation.
(tRNA: transfer ribonucleic acid; P: peptidyl site; A: aminoacyl site; AUG, ACC, GGG are codons on mRNA; UAC, UGG and ACC are anticodons on tRNA; Met: methione; Thr: threonine; EF: elongation factor; GTP: guanosine triphosphate; GDP: guanosine diphosphate)

tRNA bearing the appropriate anticodon. The new aminoacyl tRNA is fixed to the A site by base-pairing with the mRNA codon (**Fig. 21.15**).

The whole process is repeated. Translocation requires hydrolysis of GTP to GDP. The elongation reactions (steps 1, 2, and 3 above) are repeated till the polypeptide chain synthesis is completed.

**Energy requirements:** For each peptide bond formation, four high-energy phosphate bonds are used. The actual peptide bond formation (peptidyl transferase step) does not require any energy, because the amino acids are already activated.

### 4. Termination Process of Translation

After successive addition of amino acids, ribosome reaches the terminator codon sequence (UAA, UAG, or UGA) on the mRNA. Since there is no tRNA bearing the corresponding anticodon sequence, the A site remains free. The **releasing factor** (RF) enters this site along with the hydrolysis of GTP to GDP. The RF hydrolyzes the peptide chain from tRNA at the P site. The completed peptide chain is now released. Finally 80S ribosome dissociates into its component units of 60S and 40S.

### 5. Post-translational Processing

- Conversion of proinsulin to insulin by proteolytic cleavage (**Fig. 21.16**).

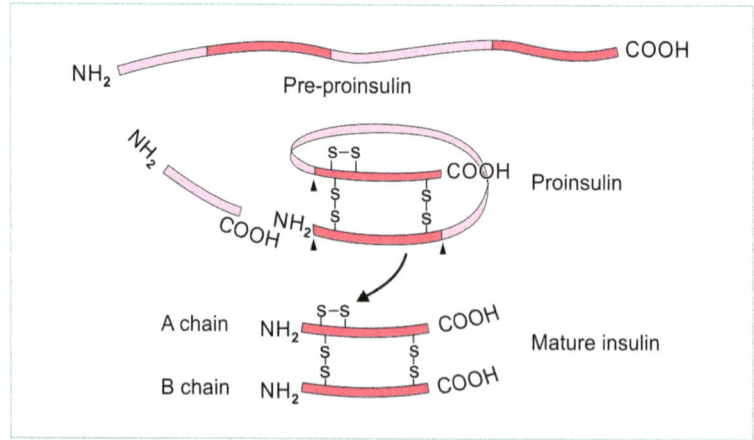

**Fig. 21.16:** Posttranslational processing of insulin by proteolytic cleavage.

- Gamma carboxylation of glutamic acid residues of prothrombin under the influence of vitamin K.
- Hydroxylation of proline and lysine in collagen with the help of vitamin C.
- Carbohydrates are attached to serine or threonine residues by a process called glycosylation.

## Inhibitors of Protein Synthesis

The modern medical practice is heavily dependent on the use of **antibiotics**. They generally act only on bacteria and are nontoxic to human beings. This is because mammalian cells have 80S ribosomes, while bacteria have 70S ribosomes. Examples of such antibiotics are tetracycline, chloramphenicol, erythromycin, and streptomycin. **Inhibitors of transcription** (described previously) will also in turn inhibit the translation process.

## Genomics and Proteomics

**Genome** means all the DNA contained in an organism or a cell, which includes both the chromosomes within the nucleus and the DNA in mitochondria. Thus, the genome of an organism is the totality of genes making up its hereditary constitution. **Genomics** is the study of the genome and its actions.

**Proteome** is the sum of all proteins expressed by the genome of an organism, thus involving the identification of the proteins in the body and determination of their role in physiological and pathological functions. While the genome remains largely unchanged, the proteins of a particular cell change dramatically as genes are turned on and off in response to the environment. **Proteomics** directly addresses the protein complement of the genome. The study of all proteins by a cell type or an organism is called "proteomics."

## SUMMARY OF THE CHAPTER

- A nucleotide is composed of a nitrogenous base, a pentose sugar, and phosphate groups esterified to the sugar.
- Purine bases in both DNA and RNA are the same.
- Purine salvage pathway is mediated by two enzymes—adenine phosphoribosyltransferase (APRTase) and hypoxanthine guanine phosphoribosyltransferase (HGPRTase).
- Committed step of de novo purine synthesis is the reaction catalyzed by amidotransferase.
- Xanthine oxidase is a metalloflavoprotein containing FAD, molybdenum, and iron. Allopurinol inhibits it, which is a good example of "suicide inhibition."
- End product of purine catabolism is uric acid. Its normal serum level is 3-7 mg/dL. However, its level increases in gout.
- Uric acid crystals deposited in the cooler areas of the body cause tophi.
- DNA sequence is always written from the 5′ end to 3′ end. This is called polarity of the DNA chain.
- Chargaff's rule states that the number of purines is equal to the number of pyrimidines.
- The two strands run antiparallel to each other.
- DNA replication in vivo is semiconservative.
- Template strand of the DNA is transcribed to mRNA.
- RNA polymerase is the enzyme synthesizing RNA in mammals.
- A transcription bubble contains RNA polymerase, DNA, and nascent NA.
- Termination of transcription can be dependent on the rho factor.
- Posttranscriptional processing of the primary RNA transcript includes polyA tailing at 3′ end, capping at 5′ end, methylation, and intron splicing.
- DNA replicates before cell division, so that each daughter DNA molecule gets an exact copy of the parent cell.

- Transfer RNA (tRNA) or soluble RNA (sRNA) is the adapter molecule between transcription and translation. Each amino acid has a specific tRNA.
- The triplet sequence on the anticodon arm of the tRNA is complementary to the codon triplet on the mRNA.
- Four high-energy phosphate bonds are required for the formation of one peptide bond, two for initial activation, one for EF-1 step and one for EF-2 step.
- Post-translational processing of proteins includes gamma carboxylations, subunit aggregation, and phosphorylations.
- Clinically useful protein synthesis inhibitors are streptomycin, chloramphenicol, tetracycline, and erythromycin.

## QUESTIONS FOR SELF-ASSESSMENT

1. Mention the difference between nucleotides and nucleosides.
2. What are functions of nucleic acids?
3. Mention the major differences between DNA and RNA.
4. Enumerate the steps of replication and mention its importance.
5. Mention the different types of RNA.
6. Mention the transcription process.
7. What are post-transcriptional modifications?
8. Mention the steps of protein synthesis.
9. Add a note on genomics.
10. Inhibitors of protein biosynthesis.

## MULTIPLE CHOICE QUESTIONS (MCQs)

21-1. The main difference between RNA and DNA is:
   a. RNA is a double helix
   b. RNA contains the sugar ribose and DNA contains deoxyribose
   c. DNA contains the base uracil but not RNA
   d. There is an RNA polymerase but no DNA polymerase.

21-2. Purine and pyrimidine bases are bound by hydrogen bonds in the DNA double helix. The typical base-pair combinations are:
   a. Adenine to cytosine
   b. Adenine to guanine
   c. Adenine to thymine
   d. Thymine to cytosine

21-3. The place where transcription of DNA to a molecule of messenger RNA occurs is:
   a. On the ribosomes
   b. In the nucleus
   c. Only during cell division
   d. When amino acids are made available by transfer RNA

21-4. Translation requires the presence of:
   a. mRNA, tRNA, and ribosomes
   b. mRNA, ribosomes, and RNA polymerase
   c. DNA, mRNA, and RNA polymerase
   d. Free nucleotide bases, amino acids, and ribosomes

21-5. The characteristics of codons are:
   a. Triplet sequences of nucleotide bases in mRNA
   b. Triplet sequences of nucleotide bases in DNA
   c. Triplet sequences of amino acids in polypeptide chains
   d. Triplet sequences of deoxyribose sugars in DNA

21-6. Because DNA strands are only made in the 5′ to 3′ direction, the lagging strand has shorter pieces of DNA known as:
   a. Leading fragment
   b. Continuous fragment
   c. Okazaki fragment
   d. Primer fragment

21-7. All the bases are found in mRNA, *except*:
   a. Adenine
   b. Guanine
   c. Uracil
   d. Thymine

21-8. **Intron is the portion of:**
 a. DNA that is cleaved off during replication
 b. Portion of mRNA that is removed after transcription
 c. tRNA that is added on after its synthesis
 d. Protein that is removed after translation

21-9. **Transcription is terminated when:**
 a. The codon on the mRNA is AUG
 b. Rho factor binds to mRNA
 c. The codon on the mRNA is UGA
 d. The ribosome reaches the polyA tail on mRNA

21-10. **DNA synthesis in eukaryotes is affected by the following, *except*:**
 a. Cytosine arabinoside
 b. 5-Fluorouracil
 c. Rifampicin
 d. 6-Mercaptopurine

## ANSWER KEYS TO MCQs

21-1. (b)  21-2. (c)  21-3. (b)  21-4. (a)  21-5. (a)  21-6. (c)  21-7. (d)  21-8. (b)
21-9. (b)  c21-10. (c)

# CHAPTER 22

# Biophysical and Biomedical Techniques

At the completion of this chapter, the reader will be able to answer questions on the following topics:

- Osmosis, osmotic pressure
- Reverse osmosis
- Dialysis
- Filtration
- Electrophoresis
- Paper electrophoresis
- Immunoelectrophoresis
- Gel electrophoresis
- Chromatography
- Microscopes; light, electron microscope
- Recombinant DNA technology
- Restriction endonucleases
- Gene therapy
- Hybridization and blot techniques
- Polymerase chain reaction

## OSMOSIS

Osmosis is an important property of particles in solution. It is the spontaneous flow of a solvent into solution when a semipermeable membrane separates the two. It is also the flow of a solvent from a more dilute solution to a relatively concentrated solution through a semipermeable membrane. Osmosis refers strictly to the flow of solvent, i.e., passive movement.

## Experiment

A beaker is partitioned by a semipermeable membrane. Both sides are filled with equal volume of water **(Fig. 22.1A)**. A known quantity of solute is dissolved in the left compartment. Water molecules are tend to passively move toward the left compartment. Once the equilibrium is reached, the water level in the left compartment increases **(Fig. 22.1B)**. Eventually, the force of the column of water on the hypertonic side of the semipermeable membrane will equal the force of diffusion on the hypotonic (the side with a lesser concentration) side, thereby creating equilibrium. When the equilibrium is reached, water continues to flow, but it flows both ways in equal amount and force, therefore stabilizing the solution.

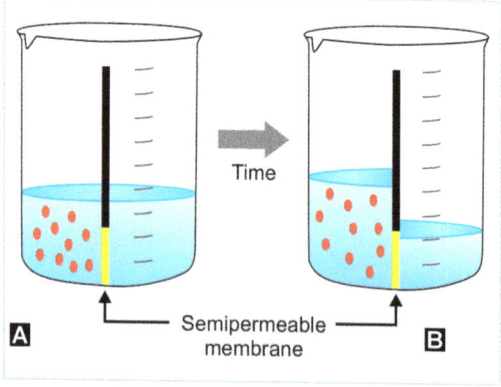

**Figs. 22.1A and B:** Osmosis.

## Osmotic Pressure

Osmotic pressure is defined as the excess pressure applied to the solution to prevent the passage of the solvent through the semipermeable membrane separating the two solutions. The osmotic pressure of a solution is directly proportional to the concentration of the solute and in turn depends on the number of molecules or ions contained in the solution. Although dependent on the number of molecules of the solute, osmotic pressure is independent of the nature of the solute. A substance of lower molecular weight will have more molecules per unit volume than a substance of higher molecular weight and will therefore exert greater osmotic pressure, e.g., albumin has lower molecular weight than globulin. Hence, albumin exerts greater osmotic pressure than globulin.

## Tonicity

Living cells have selectively permeable membrane that are permeable to water and some low-molecular-weight solutes but much less permeable to other substances. This lead to the inflow of some solute molecule along with the solvent, while others are retained. Under these conditions, the solution exhibits only that fraction of its total osmotic pressure, due to the solutes, which are retained. This fraction of total osmotic pressure of a solution is termed as its tonicity.

### Isotonic Solution

A pair of solutions, which produce no net flow through a semipermeable membrane, is said to be isotonic. If the cell is kept in hypotonic solution, the plasma membrane will allow water to pass through it and will setup an excess pressure in the interior of the cell causing the cell wall to be tight. This condition is known as "**turgor**" and the cell is said to be **turgid**. If the cell is immersed in a concentrated solution (hypertonic solution), water will pass out of the cell. The cytoplasm will then shrink and detach itself from the cell wall. This phenomenon is said to be "**lysis**".

## Role of Osmosis in Physiological Processes

- Regulation of blood volume: The osmotic pressure of plasma proteins regulates water to flow from the interstitial fluid to the blood vessels. The osmotic pressure of protein is referred to as oncotic pressure.
- The cells in the body are isotonic with the tissue fluid and blood plasma. [Living red blood cells (RBCs) if suspended in 0.92% sodium chloride solution, they neither gain nor lose water, i.e., intracellular fluid of red cell is isotonic with 0.92% NaCl solution.] An appreciable osmotic dilution of plasma would create a dangerous hydrostatic pressure in the RBCs and cells, which would take in water to achieve osmotic equilibrium. This does not occur in the body because water or salt are excreted by the kidneys, so as to keep the blood isotonic with the cells.
- Absorption from the gastrointestinal tract.
- Fluid interchange in various compartments of the body follows the principle of osmosis. Water moves spontaneously depending on the osmotic pressure.
- Destruction of RBC by **hemolysis**: When RBCs are placed in a hypotonic solution, the RBCs swell, owing to water passing into the cell osmotically (**endosmosis**). If the solution is sufficiently hypotonic, the cell may even rupture with the cell content diffusing in to the surrounding fluid. This is called hemolysis.
- If the RBCs are placed in hypertonic solution, the water passes out of the cell (exosmosis) and the RBCs will shrink. This process is called **crenation**.

## Reverse Osmosis

Suppose a solution is separated from pure solvent by a semipermeable membrane and if the pressure applied on the solution is more than the osmotic pressure, the solvent will start flowing from the solution toward the pure solvent. This phenomenon is known as reverse osmosis. This principle is being employed for getting drinking water from seawater.

## DIALYSIS

The process of separating crystalloids from colloids by means of diffusion of crystalloid through a membrane is called dialysis and the apparatus used for separation is called **dialyzer**. This method is based on the fact that the pores of the membrane are very small and allows the free passage of only the ions of the solution, but the colloidal particles are too big to pass through these pores and hence they are retained inside the bag.

Dialysis consists of parchment or cellulose acetate bag immersed in water. The impure solution (mixture of crystalloids and colloids) is placed in the dialyzer and the outside water is continuously renewed. The crystalloids slowly pass freely through the membrane into the outside water, while the colloidal solution is left inside the dialyzer. Dialysis is a very slow process. The membrane used for dialysis is semipermeable; therefore, it allows selectively the passage of only crystalloids and not the colloidal particle.

### Electrodialysis

Applying a potential difference across the membrane can accelerate the process of diffusion of crystalloid impurities from colloidal solution. Under the influence of electric field, the crystalloids migrate toward oppositely charged electrodes at a much faster rate. This process is called electrodialysis.

### Biomedical Importance of Dialysis

- Separation of proteins from small solutes.
- Formation of biological ultrafiltrates such as interstitial fluid, cerebrospinal fluid, and glomerular filtrate.
- Dialysis by artificial kidney.

Acute or chronic loss of function of kidney is a threat to life and requires removal of toxic waste products and restoration of body fluid volume and composition toward normal. This can be accomplished by dialysis with an artificial kidney. In acute renal failure, an artificial kidney may be used to tide the patient over until the kidney resumes its normal function.

Dialysis cannot maintain completely normal body fluid composition and cannot replace all the multiple functions performed by the kidney. Thus, better treatment of permanent loss of kidney function is to restore functional kidney tissues by means of kidney transplantation.

### Principle of Artificial Kidney

The basic principle of artificial kidney is to pass blood through minute blood channels bounded by a thin membrane. On the other side of the membrane is a dialyzing fluid into which the unwanted substances in the blood pass by diffusion.

## FILTRATION

Filtration is exclusion of particle having a larger diameter than the pores of the separation medium. It has often been used for the separation of solid from liquid. Two types of filtration are used to separate particle from liquid, i.e., surface filtration and depth filtration.

**Surface filtration (screen filtration)** is performed with filter paper or membranes. Surface filtration retains the solid material on the filter medium and filtrate (the liquid that passes through the filter paper) to be collected or discarded as required.

**Depth filtration:** In this the separation medium allows the particle to be retained by the body of the filter medium as well as on the surface. Depth filters are usually made up of cotton, fiberglass, asbestos, etc. The depth filter is used as a prefilter to trap larger particles, thus lengthening the life of the secondary surface filter. Combining depth and surface filters into a single filter is useful for purifying the solutions.

Filtration may be conducted under gravity, pressure, or in vacuum. The **vacuum filtration** is used to accelerate the filtration rate. Many filtrations in the clinical laboratory are carried out with filter paper. Different types of filter paper include low-ash or ash-less filter paper as well as various grades related to thickness. The membrane filters are made from homogeneous polymeric materials such as cellulose ester, cellulose acetate, polyvinyl chloride (PVC), etc.

**Ultrafiltration** is a technique for removing dissolved particles using an extremely fine filter.

## MICROSCOPES

The human eyes have limited power to distinguish small objects. The capacity of the eye to distinguish two closely placed points is the resolving power of the eye. The resolving power of the human eye is about 100 microns, i.e., two points closer than 100 microns cannot be perceived by human eye as two separate points but as a single point. So naturally cells remain invisible to our naked eye and study of the cell cannot be possible without the aid of microscope, which is having a higher resolving power.

### Compound Light Microscope

An ordinary compound microscope uses visible light to illuminate the object. It consists of lenses that magnify and resolve the image of the object and focus the light on the retina of the observer's eye. In its simplest form, the compound microscope consists of two lenses, one at each end of a hollow tube. The lens closer to the eye of the observer is called the **eyepiece** and that closer to the object is called the **objective**. The object mounted on a glass slide is placed on a stage beneath the objective lens. The light source is beneath the object and often a third lens called the **condenser** is placed between the object and the light source to focus the light on the object. In modern compound microscope, instead of a single lens, a combination of a number of lenses is used for each objective and eyepiece.

When the light rays pass through the object, different regions of the object will refract the light rays at different angles producing contrast between the structures. The objective lens and the eyepiece lens form a magnified image of the object which falls on the retina of the eye. If the object is very small in comparison with the wavelength of the light beam, the object cannot be resolved. The limit of resolution of the compound microscope is 0.2 micron.

Different types of light microscopes with increased resolving power include dark-field microscopes, phase-contrast microscopes, polarization microscopes, interference microscopes, and fluorescence microscopes.

### Electron Microscope

Instead of using visible light to illuminate the object, the electron microscope uses a beam of accelerated electors. A metal filament heated in a vacuum emits electrons which travel in a straight course similar to light rays. This electron beam is focused by electromagnetic lenses. The electromagnets are coils of wire enclosed in a soft iron casing. The beam then passes thorough the object and are deflected by objective lens which again is a magnet. The deflected rays are then projected by projector lens to a florescent screen or a photographic plate. The photographs obtained by the electron microscope are known as electron micrographs. Electron microscope has a high-resolving power. Theoretically objects less than 0.025 Å can be resolved.

Electrons are easily scattered and if the sections are not extremely thin most of the electrons would be scattered and the image would be uniformly dark. In fact, the electron beam may be scattered by even a gas molecule and that is why a vacuum field is used for electron transmission. Materials for observation under electron microscope are prepared with electron stains such as osmium tetroxide which fixes and stains the cell. One drawback of electron microscope is that living cells cannot be observed under electron microscope.

The major differences between light microscope and electron microscope are given in **Table 22.1**.

## ELECTROPHORESIS

Electrophoresis is the migration of charged particle or molecule in a medium under the influence of an applied electric field. The usual purpose of electrophoresis is to determine the number, amount, and mobility of component in a given sample or to separate them. Tiselius first performed electrophoresis.

The positively charged particles migrate toward cathode and the negatively charged particles migrate toward anode. The rate of migration depends on the number of charges each particle carries. As a result of different rates of migration, a mixture of substance can be separated in to number of fractions upon suspension in an aqueous solvent. Many important biomolecules such as nucleotides, nucleic acid, amino acid, peptides, proteins, etc., acquire either positive or negative charges.

Even typically uncharged molecules such as carbohydrates can be made to wear weak charges by derivatization as borates and sometimes as phosphates.

### Factors Affecting Electrophoretic Mobility

- **Sample:** Charge/mass ratio of the sample dictates its electrophoretic mobility. The mass depends on the size and shape of the molecule.
  - **Charge:** The higher the charge of the molecule, the greater is the electrophoretic mobility. The charge, however, is dependent on the pH of the medium.
  - **Size:** The bigger the molecule, greater are the fractional and electrostatic forces exerted up on it by the medium. So the larger particles have a smaller electrophoretic mobility compared to the smaller particles.
- **Electric field:** An increase in potential gradient increases the rate of migration. The current in the solution placed between two electrodes is carried mainly by the buffer ions. An increase in the potential difference increases the current. Resistance to flow of current is inversely proportional to the rate of migration. The power pack can control the current to a constant level without the production of heat.

**Table 22.1:** Comparison of light and electron microscopes.

| Light microscope | Electron microscope |
| --- | --- |
| The object is illuminated by sunlight or electrical lamp (visible light) | Accelerated electrons are employed instead of light |
| Glass lenses are used | Electromagnetic lenses are used |
| Image can be directly viewed by the eye | Image falls on a florescent screen or on a photographic plate |
| Image is viewed through the eyepiece | Image is projected by the projection lens |
| Light rays pass through air column | Electron beams pass through vacuum chamber |
| Resolving power comparatively low (0.2 micron) (2 Å) | Resolving power high (0.025 Å) |
| Live specimens may be observed | Only dead specimens can be observed |
| Ordinary vital or nonvital stains are used | Only electron stains can be used |
| Focusing of the image by mechanical movement of the body tube | By controlling the current through the electromagnets |

- **Support medium:** An inert supporting medium is chosen for electrophoresis. But even this inert supporting medium can exert adsorption or molecular sieving effect on the particle, thereby influencing the rate of migration.
- **Buffer:** Apart from maintaining the pH of the supporting medium, the buffer can affect the electrophoretic mobility of the sample. They are:
  - **Composition of buffer:** Commonly used buffers are phosphate, citrate, acetate, tris, and barbitone. The choice of buffer depends on the type of sample being electrophoresed. The buffer can affect the electrophoretic mobility, if it is able to bind to component of the sample being separated.
  - **Ionic strength of buffer:** Increased ionic strength of the buffer means a larger share of the current being carried by the buffer ions and or smaller proportion by the sample ions. This will lead to slower migration of sample component. Decreased ionic strength of the buffer means a larger share of the current being carried out by the sample ions, leading to faster separation. But this will cause diffusion of the smaller component that tends to be high with concomitant loss of resolution. The ionic strength used is usually between 0.05 and 0.1 M.
  - **pH of buffer:** The direction and extent of migration of sample particles are dependent on pH. Buffers ranging from pH 1 to 11 can be used to produce the required separation.

## Types of Electrophoresis

### Free Electrophoresis or Electrophoresis Without Stabilizing Media

**Microelectrophoresis:**

This technique involves the observation of motion of small particles (such as RBCs, neutrophils, bacteria, etc.) in an electric field under a microscope. The suspension is contained in a closed system composed of a thin-walled section for optical observations with suitable electrodes.

### Moving boundary electrophoresis:

This is the prototype of all modern methods of electrophoresis. In moving boundary electrophoresis, a buffered solution of macromolecules is placed under a layer of pure buffer solution contained in a U-shaped observation cell. The whole cell may then be immersed in a constant temperature bath insulated from vibrations. The power is switched on to generate an electric field between the electrodes. The macromolecule bearing a net negative charge and will move toward anode, while the macromolecule bearing a net positive charge will move toward the cathode.

### Zone Electrophoresis or Electrophoresis in Stabilizing Media

Here the molecules are immobilized by fixation in different zones. The molecules are then detected by staining them on the supporting medium. The other methods to detect the separated molecules are as follows:
- Visualization by ultraviolet (UV) light
- Detection by enzymatic reaction
- Detection by radioactivity, if the molecules are radiolabeled.
- The separated component can be eluted from the medium for further studies.

The zone separation uses stabilizing media, such as paper, or gels, such as starch, agar, agarose, polyacrylamide, etc. In zone electrophoresis, the molecules in the sample, during their migration, are separated into many different zones, as the sample contains different migrating components.

### Paper Electrophoresis

**Filter paper:** It contains 95% of cellulose and possesses only a slight adsorption capacity.

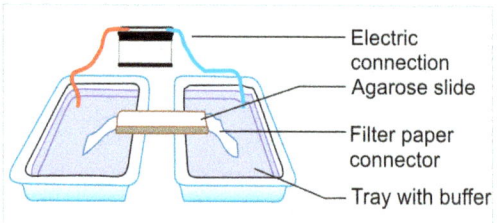

**Fig. 22.2:** Horizontal paper electrophoresis.

**Apparatus:** The equipment required for electrophoresis consists basically of a power pack and an electrophoresis cell. The power pack provides a stabilized direct current and can be programmed to give either constant voltage or constant current. The electrophoretic cell contains the electrodes, buffer reservoirs, a support for a paper, and a transparent insulating cover. The electrodes are usually made of platinum **(Fig. 22.2)**.

Two buffer reservoirs are usually used and each is partitioned into two interconnected sections. One containing the electrode and the other in contact with the supporting medium. Separate compartments are necessary, so that any change in pH occurring at the electrodes does not affect the buffer in contact with the supporting medium. Wicks are used to establish contact between the buffer reservoir and the supporting medium **(Fig. 22.2)**. Wicks are made of several layers of filter paper or gauze. The paper is placed on an insulating material usually a Perspex sheet. The whole electrophoretic unit is then covered with an insulating material to minimize evaporation during run and to assure electrical insulation. The fitter strips are commonly arranged, either horizontally or vertically.

**Sample application:** The sample may be applied as a spot (about 0.5 cm in diameter) or as a narrow uniform streak. Special devices are available commercially for this purpose.

**Electrophoretic run:** The sample is applied to the paper and the paper has been equilibrated with the buffer. Then current is switched on. Devices providing stable voltage or current are available. Overheating of the equipment can be avoided by placing the entire equipment in a cold room. The process does not usually take longer than 2 hours. The power is switched off after the run and before the paper is removed. Once removed, the paper is dried in a vacuum oven at 110°C.

## Detection

- **Fluorescence:** Staining with ethidium bromide and subsequent visualization of the electrophoretogram under UV light makes DNA and RNA fluoresce and thus facilitates their detection.
- **Ultraviolet adsorption:** Proteins, peptides, and nucleic acid absorb in the range of 260–280 nm.
- **Staining:** Different dyes are used for detecting different components. For example, bromophenol blue or Coomassie brilliant blue is used for proteins and methylene blue or toluidine blue for nucleic acid **(Fig. 22.3, right side)**.

## Quantitative Estimation

The same may be passed through a scanner to produce a graphic representation, from which the area under each peak can be noted; this will be proportional to the quantity of protein present in that peak **(Fig. 22.3, left side)**. Alternatively, the compound is eluted from the electrophoretogram The zone is excised

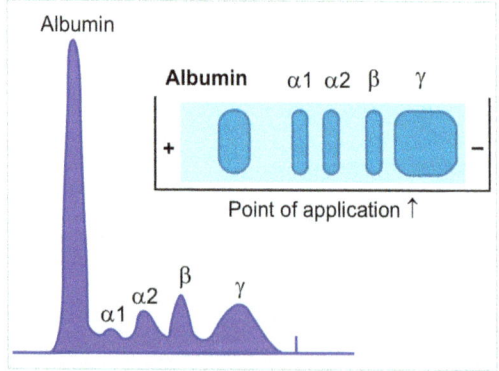

**Fig. 22.3:** Electrophoresis of normal human serum.

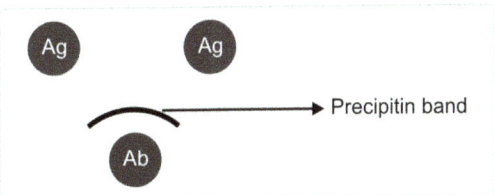

**Fig. 22.4:** Ouchterlony double diffusion.

from the paper and placed in a suitable solvent and estimated.

## Ouchterlony Technique

The individual antigen and antibody molecules filled in adjacent wells cut into the gel can diffuse freely till they come into contact. On contact, the specific antibody and the antigen form complexes with larger molecular weight. These complexes are aggregated to form a visible precipitin band **(Fig. 22.4)**.

## Immunoelectrophoresis

The technique exploits the specificity of reaction between an antigen and antibody as well as the molecular sieving of the gel. This technique was developed by Grabar and Williams. It is the modification of Ouchterlony double-diffusion technique. Immonoelectrophoresis is performed using a double-diffusion chamber. The chambers may be prepared by coating a glass microscope slide with agarose/purified agar and punching appropriate holes in the supporting medium **(Fig. 22.5)**.

The antigen is filled in the small round well. The current is switched on and the electrophoresis is allowed to continue for 1-2 hours. Immediately after disconnecting the voltage supply, the rectangular well is filled with appropriate antiserum and the gel incubated overnight at room temperature to permit diffusion of antigen and antibody toward each other. This leads to the formation of precipitin bands at the site lateral to the position where the desired component has separated during electrophoresis **(Fig.**

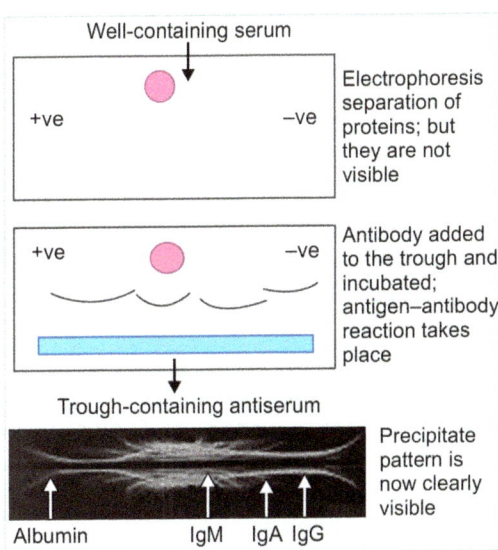

**Fig. 22.5:** Immunoelectrophoresis.

**22.5)**. The major advantage of this method is its increased resolving ability due to the combination of electrophoresis with immune specificity. The technique can be made quantitative and has been used to detect particular antigens in sera, tissue, or cell extracts and culture filtrates. It has also been used to determine the purity of a given antigen.

## Gel Electrophoresis

In paper electrophoresis, charge of the molecule was the major determinant for its electrophoretic mobility. The gels, however, are porous and the size of the molecule determines whether the molecule will enter the pore and be retarded or will bypass it. The separation thus depends on not only the charge of the molecule but also its size.

### Types of Gels

1. **Starch gel:** Potato starch is hydrolyzed in acidified acetone at 37°C. The suspension is then neutralized with sodium acetate and washed with large amount of distilled water and dried with acetone. This hydrolyzed starch when heated and cooled in an

appropriate buffer sets as a gel. The pore size in starch gel cannot be controlled and this is the biggest drawback.

2. **Agar:** Agar consists of two galactose-based polymers named as agarose and agaropectin. A solution of agar at 80°C is mixed with an equal volume of 40% polyethylene glycol which precipitates the agarose. The latter is collected, washed with distilled water, and dried with acetone. Agar solubilizes in aqueous buffers above 40°C and sets to form a gel at about 38°C. Agar is being used to separate high-molecular-weight macromolecules like proteins and nuclear acids. **Figure 22.2** shows agarose gel electrophoresis.

3. **Polyacrylamide:** It is a known neurotoxin and thus care has to be taken while preparing the gel. Acrylamide monomers are polymerized under specified conditions, when the gel will have known porosity. The running of the electrophoresis is done as discussed above.

## CHROMATOGRAPHY

Chromatography is a method of separation and purification of compounds. It is first used for separating pigments and hence the name.

## Partition Chromatography

Partition chromatography is commonly done for separation and identification of amino acids, sugar, etc., in body fluids. These compounds are soluble in both water and organic solvents and can be readily separated by this technique. The compound is partitioned between two immiscible solvents. One solvent, water, is held on the stationary supporting phase which is in the form of a thin film of inert material. The other phase consists of a mobile organic liquid that flows through the stationary phase. The components of the sample are separated depending on their partition coefficient.

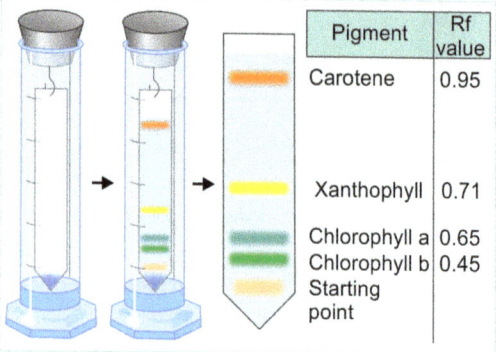

**Fig. 22.6:** Paper chromatography.

## Paper Chromatography

In paper chromatography, the mixture is spotted onto the filter paper and dried, then the solvent is allowed to flow along the paper either by gravity (descending) or by capillary action (ascending). The solvent front is marked and then the paper is dried, the position of the compound (solute) present in the mixture is visualized by a suitable staining reaction. The ratio of the distance moved by the solute and that moved by the solvent (Rf value) is more or less constant for a particular solvent system **(Fig. 22.6)**. Hence, compounds can be identified by running pure standards along with unknown mixture.

## Types of Chromatography

1. Adsorption chromatography
2. Thin-layer chromatography
3. Affinity chromatography
4. Ion exchange chromatography
5. Gel filtration chromatography
6. Gas–liquid chromatography (GLC)
7. High-performance liquid chromatography (HPLC)

### Adsorption Chromatography

In this technique, the separation is based on differences in adsorption at the surface of a solid stationary medium. The common adsorbing substances used are alumina,

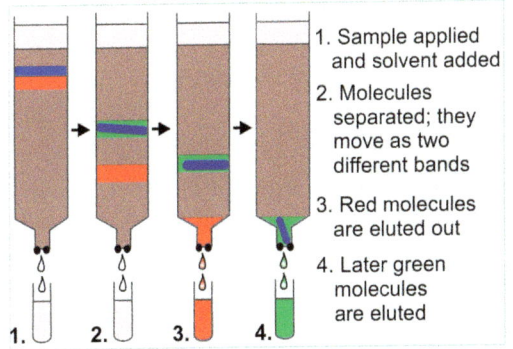

**Fig. 22.7:** Adsorption chromatography.

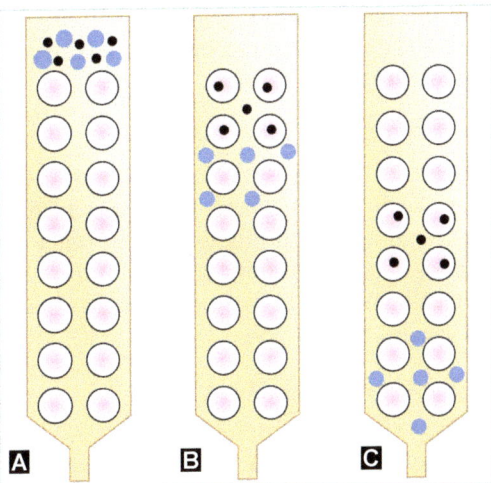

**Figs. 22.8A to C:** Sephadex (gel filtration) chromatography. (A) Protein solution is added on the top of the column; (B) Small proteins get inside the beads and so takes a longer time to reach the bottom; (C) Larger molecules cannot enter into the beads, so quickly travel and reach the bottom faster.

silicates, or silica gel. These are packed into columns and the mixture of proteins to be separated is applied in a solvent on the top of the column. The components get adsorbed on the column of adsorbent with different affinity. The fractions slowly move down; the most weakly held fraction moves fastest; followed by others, according to the order of tightness in adsorption. The eluent from the column is collected as small equal fractions and the concentration of each is measured, in each fraction (**Fig. 22.7**).

*Gel Filtration (Size Exclusion) Chromatography*

It is also called the molecular sieving. Hydrophilic cross-linked gels such as acrylamide (Sephacryl), agarose (Sepharose), and dextran (Sephadex) are used for separation of molecules based on their size. The gel is packed in a column. The gel particles are porous in nature. These pores will allow small molecules (e.g., insulin, mol.wt., 5700 D) enter into the gel. But larger molecules (e.g., immunoglobulin, mol. wt., 150,000 D) could not enter into the pores of the gel and so are excluded. Suppose a mixture of insulin plus immunoglobulin is passed through the column. The small molecules can enter the gel particles then come out and re-enter into another particle. Thus, insulin molecule has to travel a long distance inside the gels. Thus, small molecules are held back. But the large immunoglobulin molecules cannot enter the pores and sidetrack the gel particles, so they move in the column rapidly (**Figs. 22.8A to C**). In short, **larger molecules will come out first, while smaller molecules are retained in the column.**

## RECOMBINANT DNA TECHNOLOGY

**Biotechnology** may be defined as "the method by which a living organism or its parts are used to change or to incorporate a particular character to another living organism."

**Genetic recombination** is the exchange of information between two DNA segments. This is a common occurrence within the same species. But by artificial means, when a gene of one species is transferred to another living organism, it is called recombinant DNA technology. In common parlance, this is known as genetic engineering.

### Applications of Recombinant Technology

*Quantitative Preparation of Biomolecules*

If molecules are isolated from higher organisms, the availability will be greatly limited. For example, growth hormone. By means of recombinant technology, large-scale availability is now assured.

### Risk of Contamination is Eliminated

It is absolutely essential to ensure that the preparations of vaccines or clotting factors are free from contaminants such as hepatitis B particles. Recombinant DNA technology provides the answer to produce safe antigens for vaccine production.

### Specific Probes for Diagnosis of Diseases

Specific probes are useful for:
- Antenatal diagnosis of genetic diseases.
- To identify viral particles or bacterial DNA in suspected blood and tissue samples.
- To demonstrate virus integration in transformed cells.
- To detect activation of oncogenes in cancer.
- To pinpoint the location of a gene in a chromosome.
- To identify mutations in genes.

### Gene Therapy

An important application of recombinant technology is in gene therapy. Normal genes could be introduced into the patient, so that genetic diseases can be cured. These techniques are described later.

### Restriction Endonucleases

In order to transfer a gene, it is to be selectively split first from the parent DNA. This is usually achieved by restriction endonucleases (REs) which are referred to as "molecular scissors" **(Fig. 22.9)**. Certain enzymes of bacteria restrict the entry of phages into host bacteria. Hence, the name REs.

## GENE THERAPY

Gene therapy was once considered a fantasy. However, thousands of individuals have already undergone human clinical trials. A great leap in medical science has taken place on the September 14, 1990, when a girl suffering from adenosine deaminase deficiency (severe immunodeficiency) was treated by transferring the normal gene for adenosine deaminase. Gene therapy is intracellular delivery of genes to correct an existing abnormality.

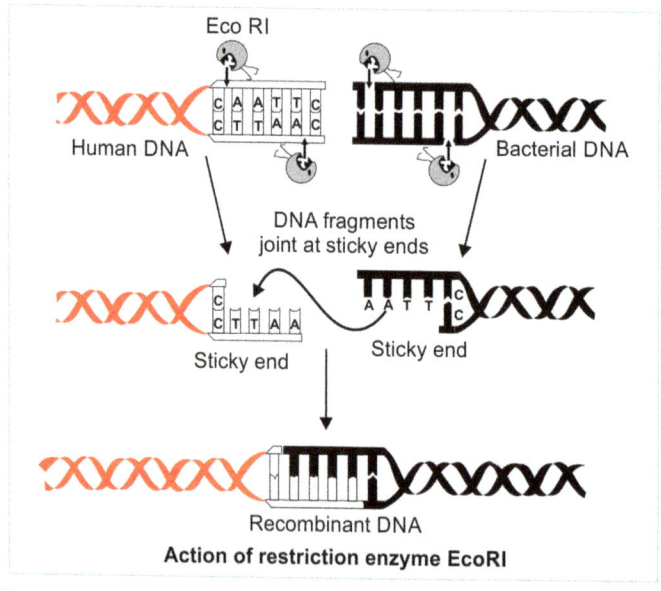

**Action of restriction enzyme EcoRI**

**Fig. 22.9:** EcoRI enzyme cuts the bonds marked with arrow. This results in the sticky ends.

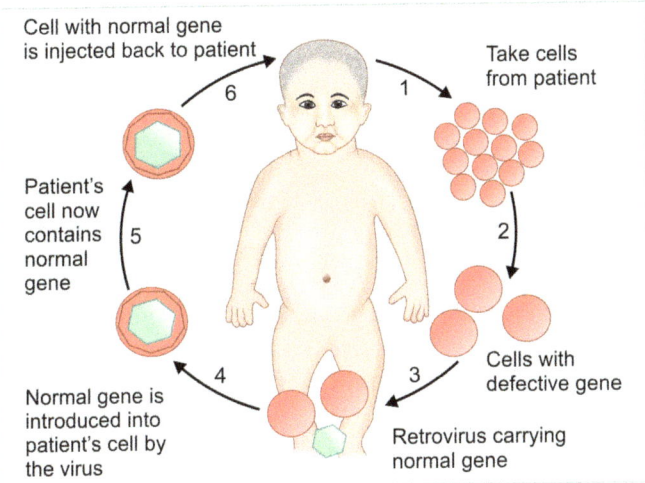

**Fig. 22.10:** Ex vivo gene therapy.

## Summary of the Gene Therapy Procedure

- Isolate the healthy gene along with the sequence, controlling its expression.
- Incorporate this gene on a carrier or vector as an expression cassette.
- Finally deliver the vector to the target cells **(Fig. 22.10)**.

## Southern Blot Technique

The method was developed by EM Southern in 1975. This is used to detect a specific segment of DNA in the whole genome. It is based on the specific base-pairing properties of complementary nucleic acid strands. This technique is also called DNA hybridization.

In Southern blot technique, the DNA is isolated from the suspected tissue. It is then fragmented by REs. The cut pieces are electrophoresed on agarose gel. It is then treated with NaOH to denature the DNA, so that the pieces become single stranded. This is then blotted over to a nitrocellulose membrane. The single-stranded DNA is adsorbed in the nitrocellulose membrane. An exact replica of the pattern in the gel is reproduced on the membrane. The DNA is then fixed on the membrane by baking at 80°C. There will be many DNA fragments on the membrane, but only one or two pieces contain the specific gene DNA. The radioactive virus probe is placed over the membrane. The probe will detect the complementary nucleotide sequence in the host DNA. So the probe is hybridized on the particular pieces of host DNA. An X-ray plate is placed over the membrane at dark for a few days. The radiation from the fixed probe will produce its mark on the X-ray plate. This is called autoradiography **(Fig. 22.11)**. Abnormal genes such as *HbS* gene or virus integration can also be identified by this method.

## Northern Blotting for Identifying RNA

The Northern blot is used to demonstrate specific RNA. The total RNA is isolated from the cell, electrophoresed, and then blotted on to a membrane. This is then probed with radioactive complementary DNA (cDNA). It will involve RNA–DNA hybridization. This is used to detect gene expression in a tissue.

## Western Blot Analysis for Proteins

In this technique, proteins (not nucleic acids) are identified. The proteins are isolated from the tissue and electrophoresis is done. The separated proteins are then transferred on

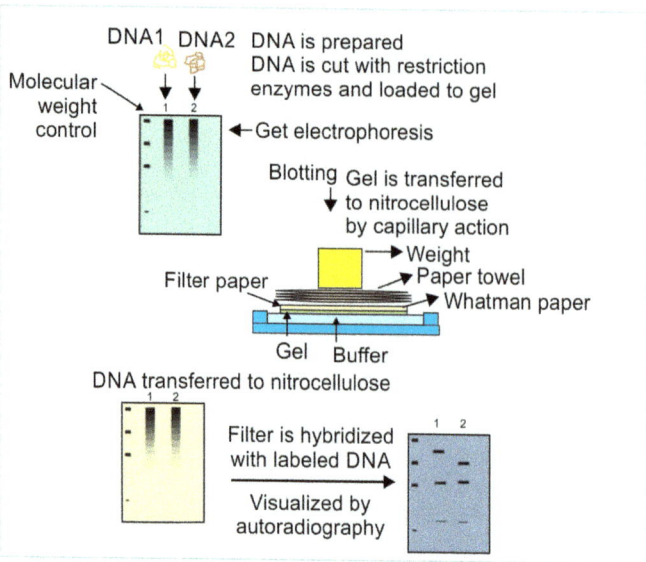

**Fig. 22.11:** Southern blot technique.

to a nitrocellulose membrane. After fixation, it is probed with radioactive antibody and autoradiographed. This technique is very useful to identify the specific protein in a tissue, thereby showing the expression of a particular gene.

## POLYMERASE CHAIN REACTION

Polymerase chain reaction (PCR) is an in vitro DNA amplification procedure in which millions of copies of a particular sequence of DNA can be produced within a few hours. It is like a photocopy machine for gene copying.

The flanking sequences of the gene of interest should be known. Two DNA primers of about 20–30 nucleotides with complementary sequence of the flanking region can be synthesized. The reaction cycle has the following steps.

**Step 1:** Separation (denaturation): DNA strands are separated (melted) by heating at 95°C for 15 seconds to 2 minutes.

**Step 2:** Priming (annealing): The primers are annealed by cooling to 50°C for 0.5 to 2 minutes. The primers hybridize with their complementary single-stranded DNA produced in the first step.

**Step 3:** Polymerization: New DNA strands are synthesized by Taq polymerase. This enzyme is derived from bacteria *Thermus aquaticus* found in hot springs. Therefore, the enzyme is not denatured at high temperature. The PCR is allowed to take place at 72°C for 30 seconds in presence of dNTPs (all four deoxy ribonucleotide triphosphates). Both strands of DNA are now duplicated (**Fig. 22.12**).

**Step 4:** The steps of 1, 2, and 3 are repeated. In each cycle, the DNA strands are doubled. Thus, 20 cycles provide for 1 million times amplifications. These cycles are generally repeated by automated instrument called Tempcycler.

**Step 5:** After the amplification procedure, DNA hybridization technique or Southern blot analysis with a suitable probe shows the presence of the DNA in the sample tissue (**Fig. 22.13**).

Chapter 22 Biophysical and Biomedical Techniques

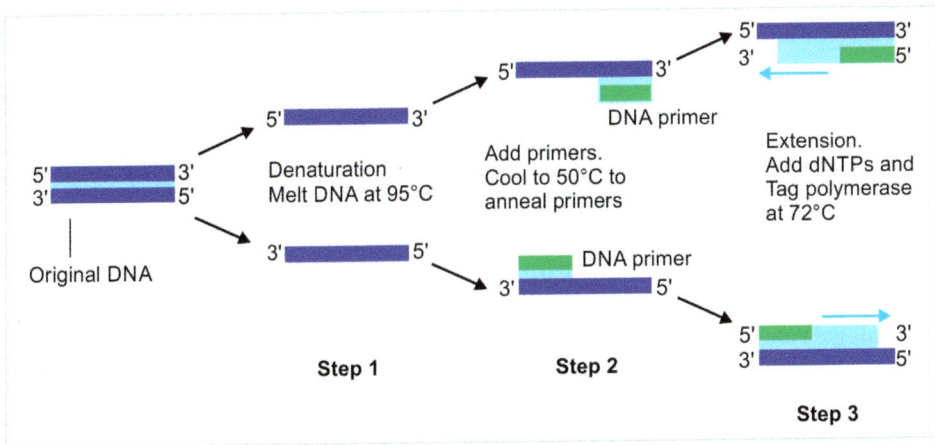

**Fig. 22.12:** Polymerase chain reaction (PCR).

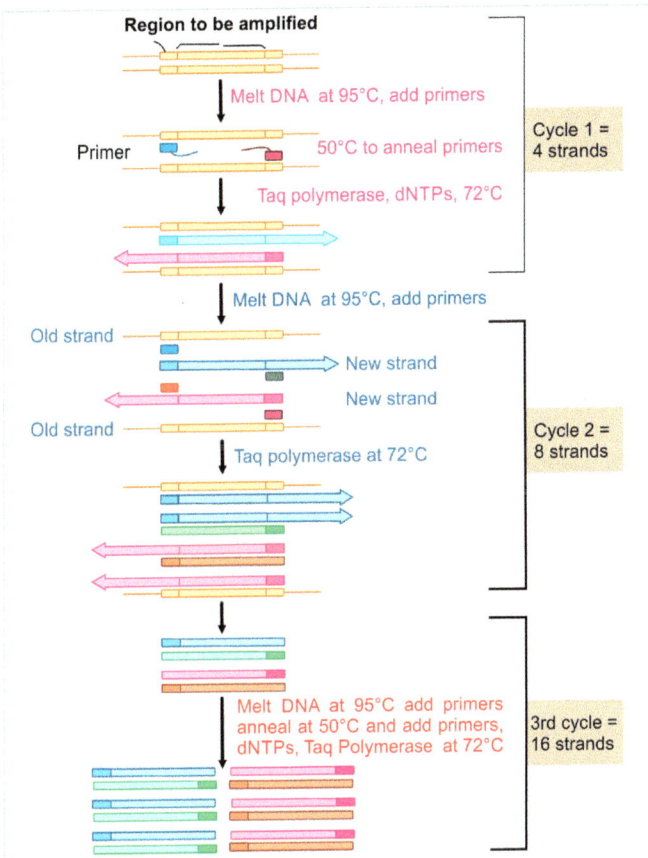

**Fig. 22.13:** Polymerase chain reaction, three cycles are shown.

## Clinical Applications of Polymerase Chain Reaction

- **Diagnosis of bacterial and viral diseases**: The PCR can detect every single bacillus present in the specimen. This technique is widely used in the diagnosis of bacterial diseases such as tuberculosis and viral infections such as hepatitis C, cytomegalovirus, and HIV.
- **Medicolegal cases**: The restriction analysis of DNA of the hair follicle obtained from the crime scene is done after PCR amplification. This pattern is then compared with the restriction analysis of DNA samples obtained from various suspects. This is highly useful in forensic medicine to identify the criminal.
- **Diagnosis of genetic disorders**: The PCR technology has been widely used to amplify the gene segments that contain known mutations for diagnosis of inherited diseases such as sickle cell anemia, beta thalassemia, cystic fibrosis, etc.
- The PCR is especially useful for prenatal **diagnosis of inherited diseases**, where cells obtained from fetus by amniocentesis are very few.
- **Cancer detection**: PCR is widely used to identify mutations in oncosuppressor genes such as p53, retinoblastoma gene, etc. (oncogenes and oncosuppressor genes).

## SUMMARY OF THE CHAPTER

- Osmosis is an important property of particles in solution. It is the spontaneous flow of a solvent into solution when a semipermeable membrane separates the two medium.
- Immunoelctrophoresis is an important technique in immunoassays.
- Genetic recombination involves the exchange of information between two segments of DNA. When a gene of one species is transferred into another under laboratory conditions, the technique is called recombinant DNA technology or genetic engineering.
- Restriction endonucleases (REs), also known as "molecular scissors" cut at sequences that are palindromes.
- Each RE is characterized by a specific "restriction site." Plasmids are commonly used vectors.
- A vector carrying a foreign DNA is called "chimeric DNA."
- The process of introducing a plasmid into a host is called transfection.
- Gene therapy involves delivering genes to generate a therapeutic effect by correcting an existing abnormality.
- Southern blot technique is used to detect specific segment of DNA in the whole genome.
- Northern blot is to identify RNA.
- Western blot analysis is done to identify proteins.

## QUESTIONS FOR SELF-ASSESSMENT

1. Write a note on osmosis and add its clinical importance.
2. What is the importance of microscopy?
3. Write a note on immunoelectrophoresis.
4. Write a note on gene therapy.
5. Name the different blot techniques with their applications.
6. Mention the importance of Western blot.
7. Write down the steps of PCR. Mention its importance.
8. Write down the importance of chromatography.
9. Mention the applications of recombinant DNA technology.
10. Mention the importance of plasmids in molecular biology

## MULTIPLE CHOICE QUESTIONS (MCQs)

22-1. The pH of the buffer maintained in serum protein electrophoresis on agar gel is:
 a. 8.6
 b. 7.6

c. 9.6
d. 6.6

22-2. **All the following are the applications of electrophoresis, *except*:**
a. Separating serum proteins
b. Separating cellular organelles
c. Hemoglobin separation
d. Lipoprotein separation and identification

22-3. **The main principle of electrophoresis depends on:**
a. Charge/mass ratio
b. Weight
c. Number of ions
d. Volume of the buffer

22-4. **Regarding chromatography, one of the following is incorrect statement:**
a. Separation of molecules is based on their volume
b. Separation of molecules is based on their molecular size of the substance
c. Separation of molecules is based on the shape of the substance
d. Separation of molecules is based on their molecular weight of the substance

22-5. **The polymerase chain reaction or PCR is a technique that:**
a. Is used to demonstrate DNA as the genetic material
b. Uses DNA polymerase to replicate specific DNA sequences in vitro.
c. Measures the ribosome transfer rate during translation
d. Detects the level of polymerases involved in replication

22-6. **DNA ligase is:**
a. An enzyme that joins fragments in normal DNA replication
b. An enzyme involved in protein synthesis
c. An enzyme of bacterial origin, which cuts DNA at defined base sequences
d. An enzyme that facilitates transcription of specific genes

22-7. **PCR is mainly used in the following, *aspect*:**
a. Digesting DNA
b. Copying plasmids
c. Amplifying DNA
d. mplifying proteins

22-8. **Which enzyme is used for preparing a recombinant DNA molecule?**
a. Restriction endonuclease
b. RNA polymerase
c. DNA polymerase
d. Topoisomerase

22-9. **Which is not amenable to gene therapy?**
a. Hemophilia
b. Severe combined immunodeficiency
c. Cystic fibrosis
d. Infective hepatitis

22-10. **The enzyme deficiency that was first corrected by gene therapy is:**
a. Adenosine deaminase
b. HGPRTase
c. Glucose-6-phosphatase
d. APRTase

## ANSWER KEYS TO MCQs

22-1. (b)  22-2. (b)  22-3. (a)  22-4. (a)  22-5. (b)  22-6. (a)  22-7. (c)  22-8. (a)
22-9. (d)  2-10. (a)

# Appendices

## APPENDIX A:
### Normal Values (Reference Values)

| Analyte | Sample | Reference value | Units |
|---|---|---|---|
| Alanine aminotransferase (ALT/SGPT) | S | 10–35 | IU/L |
| Albumin | S | 3.5–5 | g/dL |
| Alkaline phosphatase (ALP) | S | 40–125 | IU/L |
| Aspartate aminotransferase (AST/SGOT) | S | 8–20 | IU/L |
| Bicarbonate ($HCO_3^-$) | S | 22–26 | mEq/L |
| Bilirubin, total | S | 0.2–1 | mg/dL |
| Calcium | S | 9–11 | mg/dL |
| Ceruloplasmin | S | 25–50 | mg/dL |
| Chloride | S/P | 96–106 | mEq/L |
| Chloride | CSF | 120–130 | mEq/L |
| Cholesterol, total | S/P | 150–200 | mg/dL |
| Cholesterol (HDL fraction) | | | |
| Male | S | 30–60 | mg/dL |
| Female | S | 35–75 | mg/dL |
| Cholesterol (LDL fraction) | | | |
| 20–29 yr | S | 60–150 | mg/dL |
| 30–39 yr | S | 80–175 | mg/dL |
| Creatine | S | 0.2–0.4 | mg/dL |
| Creatine kinase (CK) | | | |
| Male | S | 15–100 | U/L |
| Female | S | 10–80 | U/L |
| Creatinine | S | 0.7–1.4 | mg/dL |
| Creatinine | U | 15–25 | mg/kg/d |
| Globulins | S | 2.5–3.5 | g/dL |
| Glucose (fasting) | P | 70–110 | mg/dL |

| | | | |
|---|---|---|---|
| Glucose | CSF | 50–70 | mg/dL |
| Hemoglobin | | | |
| Male | B | 14–16 | g/dL |
| Female | B | 13–15 | g/dL |
| HbA1C (glycohemoglobin) | B | 4–5.4 | % of total |
| Immunoglobulin G (IgG) | S | 800–1,200 | mg/dL |
| Immunoglobulin M (IgM) | S | 50–200 | mg/dL |
| Immunoglobulin A (IgA) | S | 150–300 | mg/dL |
| Iron | S | 100–150 | µg/dL |
| Lipids, total | S | 400–600 | mg/dL |
| Nonesterified fatty acids | P | 10–20 | mg/dL |
| pCO2, arterial | B | 35–45 | mmHg |
| pH | B | 7.4 | |
| Phosphate | S | 3–4 | mg/dL |
| Phosphate | U | 1 | g/day |
| Phospholipids | S | 150–200 | mg/dL |
| PO2 arterial | B | 90–100 | mmHg |
| Potassium | S | 3.5–5 | mEq/L |
| Proteins, total | S | 6–8 | g/dL |
| Prothrombin | P | 10–15 | mg/dL |
| Sodium | S | 136–145 | mEq/L |
| Triglycerides, fasting | S | 50–200 | mg/dL |
| Urea | S | 20–40 | mg/dL |
| Uric acid | | | |
| Male | S/P | 3.5–7 | mg/dL |
| Female | S/P | 2.5–5 | mg/dL |

(P: plasma; B: blood; S: serum; E: erythrocyte; U: urine; CSF: cerebrospinal fluid; pg: picogram; ng: nanogram; ug: microgram; mg: milligram; d: day)

# APPENDIX B:
## Recommended Daily Allowance (RDA) of Essential Nutrients

| Nutrient | Requirement per day |
|---|---|
| **1. Proteins** | |
| *Adult* | |
| Males | 1 g/kg |
| Females | 1 g/kg |
| *Children* | |
| Infants | 2.4 g/kg |
| Up to 10 years | 1.75 g/kg |
| Boys | 1.6 g/kg |
| Girls | 1.4 g/kg |
| *Pregnancy and lactation* | |
| Pregnancy | 2 g/kg |
| Lactation | 2.5 g/kg |
| **2. Fat-soluble vitamins** | |
| *Vitamin A* | |
| Adult | 750 µg |
| Children | 400–600 µg |
| Pregnancy | 1,000 µg |
| Lactation | 1,200 µg |
| *Vitamin D* | |
| Adult | 5 µg |
| Children (preschool) | 10 µg |
| Pregnancy and lactation | 1,200 µg |
| *Vitamin E* | |
| Adult males | 10 mg |
| Females | 8 mg |
| Old age | 10 mg |
| Pregnancy | 11 mg |
| *Vitamin K* | |
| Adult | 50–100 µg |
| Children | 1 µg/kg |
| **3. Water-soluble vitamins** | |
| Thiamine (B1) | 1–1.5 mg |
| Riboflavin (B2) | 1.5 mg |
| *Niacin* | |
| Adult | 20 mg |

| | |
|---|---|
| Pregnancy | 22 mg |
| Lactation | 25 mg |
| *Pyridoxine (B$_6$)* | |
| Adult | 2 mg |
| Pregnancy | 2.5 mg |
| Pantothenic acid | 10 mg |
| Biotin | 30–50 µg |
| *Folic acid* | |
| Adult | 300 µg |
| Pregnancy | 500 µg |
| Lactation | 500 µg |
| *B$_{12}$* | |
| Adult | 1–3 µg |
| Pregnancy and lactation | 3 µg |
| *Ascorbic acid* | |
| Adult | 70 mg |
| Pregnancy and lactation | 100 mg |
| **4. Minerals** | |
| *Calcium* | |
| Adult | 0.5 g |
| Children | 1 g |
| Pregnancy and lactation | 1.5 g |
| Phosphorus | 500 mg |
| Magnesium | 400 mg |
| Manganese | 5–6 mg |
| Sodium | 5–10 g |
| Potassium | 3–4 g |
| *Iron* | |
| Males | 15–20 mg |
| Females | 20–25 mg |
| Pregnancy | 40–50 mg |
| Copper | 1.5–3 mg |
| Iodine | 150–200 µg |
| Zinc | 8–10 mg |
| Selenium | 50–100 µg |

(ng: nanogram; µg: microgram; mg: milligram)

## APPENDIX C:
## Nutritional Value of Food Items

| Food | Per 100 g of edible portion ||||||
|---|---|---|---|---|---|---|
| | Protein (g) | Fat (g) | Carbohydrate (g) | Energy (kcal) | Calcium (mg) | Iron (mg) |
| Cereals (wheat, rice, etc.) | 10 | 1 | 65 | 300 | 20 | 5 |
| Pulses (bengal gram, etc.) | 20 | 5 | 55 | 300 | 50 | 10 |
| Tubers (potato, etc.) | 1 | 0 | 25 | 100 | 0 | 0 |
| Green leafy vegetables | 2 | 0 | 4 | 20 | 20 | 3 |
| Fruits (banana, etc.) | 2 | 0 | 10 | 50 | 10 | 1 |
| Nuts and oilseeds | 20 | 50 | 20 | 600 | 50 | 5 |
| Milk and curd | 3 | 4 | 5 | 60 | 200 | 0 |
| Egg | 13 | 13 | 0 | 170 | 50 | 0 |
| Meat | 20 | 3 | 0 | 100 | 150 | 3 |
| Fish | 20 | 10 | 0 | 170 | 20 | 1 |
| Oils and ghee | 0 | 100 | 0 | 900 | 0 | 0 |

(g: gram; mg: milligram)

# Index

Page numbers followed by *f* refer to figure and *t* refer to table.

## A

Abetalipoproteinemia 102
Absorbed fat, fate of 82
Absorption 183
Acetone 59
  smell of 60
Acetyl-coenzyme A
  carboxylase 85, 86, 193
  complete oxidation of 145
  transportation of 86*f*
Acetylcholinesterase 26
Achlorhydria 247
Acid phosphatase 25
Acid-base 223
  balance 223
  disturbances, types of 230*t*
  imbalance 227
  status, assessment of 230
Acidic amino acids 111
Acidosis 88, 225, 230
Acne 185
Acquired hyperammonemia 123
Acquired porphyrias 172
Actin 168
Active immunity 163
Acyl carrier protein 85
Addison's disease 231, 232
Addisonian pernicious anemia 196
Adenine 251
Adenosine
  diphosphate 43, 44, 53, 83, 105, 126*f*, 148*f*, 246, 253
  monophosphate 83, 253
  triphosphate 6, 14, 23, 43-45, 53, 54, 82, 83, 105, 126*f*, 148*f*, 149, 149*f*, 246, 253, 254
Adipokines 86
Adiponectin 88

Adipose tissue-derived hormones 86
Adrenal insufficiency 232
Adrenaline 131
Adrenoleukodystrophy 141
Adsorption chromatography 276, 277*f*
Aerobic glycolysis, energetics of 44
Agar 276
A-keto acid dehydrogenase 130
Alanine 127, 133
  amino transferase 25, 114*f*
  glyoxylate transaminase 138
  metabolic pathway of 128*f*
Albinism 137
Albumin 118, 156, 157, 243
  functions of 158
Albuminuria 243
Alcohol dehydrogenase 26
Aldolase 26
Aldonic acids 34
Aldosterone 230
Alimentary glucosuria 63
Alkali, action of 33
Alkaline phosphatase 18, 25, 241
Alkalosis 225, 233
Alkaptonuria 137, 138*f*
Allopurinol 256
Allosteric activation 23
Allosteric enzymes 23*t*
  action of 23*f*
Allosteric inhibition 23
Alpha thalassemia 178
Alpha-1-acid glycoprotein 158
Alpha-1-antitrypsin 158
Alpha-amino acid 108*f*
Alpha-fetoprotein 158
Alpha-globulins 158
Alpha-tocopherol 152

Alzheimer's dementia 152
Amide formation 113
Amino acid 2, 18, 108, 112, 112*f*, 113*f*, 114*f*, 123*f*, 133, 133*t*, 218*t*
  absorption of 122
  activation of 263
  and proteins 108, 121
  basic 111
  carbon skeletons of 132
  classification of 110, 112*t*
  color reactions of 114*t*
  correct alignment of 21*f*
  decarboxylation of 113*f*
  dibasic 109*f*
  dicarboxylic 109*f*
  first 115
  functions of 114
  hydroxy 109*f*
  limiting 218
  metabolic fate of 112, 133*f*
  metabolism of 126, 136
  mirror images of 110*f*
  nonessential 111
  numbering of 115
  reactions of 113
  semi-essential 110, 129
  simple 109*f*
  sulfur-containing 109*f*
  with amide groups 109*f*
Amino aciduria 54
Amino sugars 34
Amino terminal end 115
Aminoacidopathy 141
Aminoacyl site 263, 264
Aminoacyl tRNA synthetases 263
Aminolevulinic acid 23, 172
Ammonia 123
  mechanism 228*f*
Ammonium ions, excretion of 227

Ammonium sulfate 156
Amphibolic role 153
Amphipathic pathway 146
Amylopectin 37, 37f
Amylopectinosis 51, 140
Amylose 37
Anaerobic glycolysis 45
Anaplerotic reactions 146f
Andersen's disease 51, 140
Androgens 99
Anemia 207, 256
 classification of 209t
Angina pectoris 129
Anion gap 228
 acidosis 229
Anomeric carbon 33
Anomerism 32, 32f, 33
Anserine 115
Antagonists 195
Antibacterial agents 259
Anti-beriberi factor 189
Antibiotics 265
Antibody detection 165
Anticancer agents 22, 256, 259
Antidiuretic hormone 230
Antigen 162
 detection 165, 166f
 molecular structure of 163
Antioxidants 152, 153, 188
 commercial use of 153
Antiparallel 258
Antiport system 11
Antiport transport systems 11f
Antivitamin 116
Apoenzyme 14, 26
Apolipoprotein 76, 100
Apoprotein 76
Arachidonic acid 77, 91, 92
Arginine 129, 134
Arginino succinic aciduria 124
Aromatic amino acids 110f
Arthritis 255
Artificial kidney, principle of 270
Ascites 123
Asparaginase 24
Aspartate aminotransferase 25, 241
Aspartic acid 134
Aspirin 23, 92
Asthma 93

Asymmetric carbon 39
Asymmetric carbon atom 31
Atherosclerosis 103, 152
 prevention of 104
 risk factors for 103
Atrial natriuretic peptides 231
Augmented histamine test 248
Azaserine 128

# B

B cells 162
Bacterial diseases 282
Balanced diet 219
Barfoed's test 34
Basal metabolic rate 216
Base-pairing rule 257, 258f
Bedside medicine 3
Bence-Jones proteins 243
Bence-Jones proteinuria 162
Benedict's reagent 33
Benedict's solution 35
Benedict's test 33, 60, 64
Beriberi
 dry 190
 wet 190
Beta thalassemia 178
Beta-carotene 183
Beta-globulins 158
Beta-oxidation 85, 86, 88t
 first step of 83f
 regulation of 85
Bicarbonate
 buffer system 225
 generation of 226
 reabsorption of 227f
Bile acid 77, 98
 primary 98
Bile pigments 243
Bile salts 81, 98, 243
Bilirubin 172
 clinical significance of 174
 diglucuronide 173
 direct 239
 estimation of 239
 excretion of 175f
 free 175
 indirect 239
 metabolism 239
 normal range of 239
 production of 175f
 unconjugated 54, 174, 239

Biochemical abnormalities 137
Biochemistry 1
 history of 1t
Biological oxidation 143, 148
Biological rhythms 132
Biomolecules, quantitative preparation of 277
Biotin 189
 biological functions of 193
 deficiency manifestations of 194
 sources of 193, 194
Birth defects 195
Bitot's spot 185
Blood 286
 buffer systems in 225
 clotting 188
 coagulation 159
 high potassium in 60
 parameters 246t
 reference values in 230
Blood glucose
 estimation of 63
 fasting 56
 level 56
  factors maintaining 56
 random 56
 regulation of 56, 57f
Body fluid compartments, electrolyte concentration of 231t
Body mass index 219
Body, buffer systems in 226t
Bone diseases 26
Branched glycogen molecule 37f
Branched-chain amino acid 109f, 130, 139
 catabolism of 130t
 metabolism 130
Branching enzyme 49
Brittle bones 166
Bronze diabetes 209
Buffer 224, 273
 capacity 224
 composition of 224, 273
 effective range of 225
 ionic strength of 273
 pH of 224, 273
Burning foot syndrome 193

# C

Caffeine 153
Cahill cycle 128
Calbindin 186, 201
Calcitonin 202
Calcitriol 186, 187*f*, 201, 202*f*
  generation of 186*f*
Calcium 15, 200
  absorption of 187*f*, 200
  biological role of 202
  daily dietary allowance
    requirement of 200
  deficiency 203
  homeostasis 202*f*
  intestinal absorption of 186
  level, regulation of 201
  serum level of 201
  sources of 200
  toxicity 203
Calculi 255
Calmodulin 202, 203*t*
Cancer detection 282
Carbamino compound,
    formation of 114
Carbaminohemoglobin 179
Carbohydrate 29, 38, 39, 146
  chemistry 29
  classification of 29, 30*f*
  excess 146, 232
  functions of 29
  metabolism 41
    inborn errors of 140
  reactions of 33
  simple 29
  small amounts of 7
Carbon atoms, nomenclature
    of 69*f*
Carbon dioxide 144*f*
  isohydric transport of 179
  transport of 179
Carbonic anhydrase 226, 227*f*,
    246
Carboxy terminal end 115
Carboxyhemoglobin 177
Carboxypeptidases 122
Carcinogenesis 152
Carcinoid syndrome 192
Carcinoid tumors 132
Cardiac arrest 233
Cardiac biomarkers 24, 104

Cardiac enzyme markers 24
Cardiac failure 231
Cardiac markers, classification
    of 105*t*
Cardiac troponins 105
  elevation of 106*f*
Cardiotonic drug digoxin
    inhibits 10
Cardiovascular diseases 101
Carnitine 82
  acyltransferases 82
Carnosine 115
Carpopedal spasm 203*f*
Cartilages 38
Castle's intrinsic factor 247
Catabolism 130
Catalase 152
Catalytic proteins 115
Cataract 60, 152, 168
Cell 4, 5, 10
  membrane 7, 98
    functions of 8
  powerhouse of 6
  theory 4
Cell-mediated immunity 162,
    163
Cellular components, activities
    of 7*t*
Cellulose 38
Central nervous system 123,
    211
Cephalin 74, 74*f*
Ceramide 75
Cerebrosides 76, 76*t*
Cerebrospinal fluid 248, 286
  chloride 249
  glucose 249
  protein 249
  sample, collection of 248
Ceruloplasmin 153, 158, 159,
    207, 209
Chemical bond energy 148
Chemiosmotic theory 149
Chloride 205, 233
  biological role of 205
  deficiency 205
  functions of 205
  in plasma, normal level of
    205
  sources of 205
Cholecalciferol 185

Cholesterol 12, 67, 96, 98, 217
  bad 101
  biosynthesis of 96, 97
  excretion of 99, 99*f*
  formation of 97, 98*f*
  functions of 97
  level, clinical significance of
    103
  pool 99*f*
  structure of 96, 96*f*
Choline, deficiency of 91
Cholinesterase 25
Chondroitin sulfate 38
Chromatin 258
Chromatography, types of 276
Chromogenic substrate 63
Chromosomes 258
Chylomicrons 81, 101
  function of 101
Chyluria 82
Chymotrypsin 121, 122
Cigarette smoking 104
Cirrhosis 50, 90, 158
Cistrons 258
Citrate synthase 144*f*
Citric acid cycle 43, 57, 143, 148*f*
  energetics of 143
  functions of 143
  regulation of 147
Citrullinemia 124
Clonal selection 162, 163
*Clostridium botulinum* 116
Cobalamin 195
Coenzyme 14, 14*t*
  A, structure of 83*f*
Cohn's fractionation 156
Collagen 166
  abnormalities in 166
  functions of 166
  structure of 166
Color blindness 184
Coma 60
Common monosaccharides 30*t*
Competitive inhibition 21, 21*f*
Compound carbohydrates 30
Compound light microscope
    271
Compound lipids 68, 72
Cones 184
Congenital cataract 54
Congestive cardiac failure 232

Conjugated bilirubin 174, 175, 239
   formation of 174f
Conopsin 184
Convulsions 192
Cooley's anemia 177
Copper 15, 209
   biological role of 209
   deficiency anemia 210
   dietary sources of 209
   metabolic abnormalities of 209
   requirement of 209
Cori's cycle 48, 48f
Cori's disease 51, 140
Cornea 38
Coronary artery disease 103, 219
Cotransport system 11
C-reactive protein 159
Creatine 129
   phosphate, generation of 126f
   synthesis of 126
Creatine kinase 24, 26, 105, 106
   isoenzymes of 106
   reaction 105f
Creatinine 126
Creatinine clearance test 245
   disadvantages of 245
   procedure for 245
Cretinism 210
Crigler-Najjar syndrome 176
Cristae 6
Crystallins 168
Cushing's syndrome 233
Cyanocobalamin 189
Cyclic monophosphate 253
Cyclooxygenase pathway 92, 92f
Cycloserine 128
Cysteine 133
   storage disease 138
Cystinosis 138
Cytidine monophosphate 253
Cytochrome 149f
   oxidase 22
Cytosine 252
   diphosphate 253
   triphosphate 253
Cytoskeleton 8

Cytosol 6
Cytosolic pathway 85

## D

De novo synthesis, reactions of 85
Debranching enzyme 48
Decarboxylation 113, 125, 192
Dehydration 60
Dehydrogenations 148
Deoxy sugar 34, 35f
Deoxy-hemoglobin 178
Deoxyribonucleic acid 5, 6, 187f, 251, 262
   replication 259
Deoxyribonucleosides 252
Depth filtration 270
Derived lipids 68, 76
Dermatan sulfate 38
Desmosine 167
Dextrin 38
D-fructose 30
D-galactose 30
D-glucose 30
Diabetes
   biochemical explanation of 58
   maturity onset 58
Diabetes mellitus 1, 33, 42, 56, 58, 88, 103, 104, 169, 219, 242
   chronic complications of 60
   clinical presentations in 59
   complications of 59
   diagnostic criteria for 61
   laboratory investigations in 61
   management of 64
   metabolic derangements in 59f
   type 1 58
   type 2 58
   insulin resistance in 58f
Diabetic ketoacidosis 59
Diagnostic enzymes 24, 25
Dialysis 270
   biomedical importance of 270
Diarrhea 231-233
Dickens-Horecker pathway 50

Dicoumarol 22
Dietary carbohydrates 217
Dietary deficiency 194
Dietary fats 217
Dietary fibers 217
Diffusion potentials 9
Digestion 41
Digestive enzymes 24
Dipeptide 115
Dipstick method 244t
Disaccharide 34, 41
   reducing 36f
D-mannose 30
Dry epithelium 185
Dubin-Johnson syndrome 176

## E

Edema 123, 231, 232
Ehrlich's test 175
Eicosanoids 91
Elastin 166
Electric field 272
Electrodialysis 270
Electrolyte
   balance 223, 230
   diffusion of 9
Electron flow 149f
Electron transport chain 148, 148f, 150f
   organization of 149, 149f
Electrophoresis 272, 273, 274f
   free 273
   types of 273
Electrophoretic mobility 272
Embden-Meyerhof pathway 42, 44f
Endopeptidases 121
Endoplasmic reticulum 5-7, 12
Endosmosis 269
End-product inhibition 23
Energy
   density 215
   requirements 216, 264
   yield 46t
Enterohepatic circulation 174
Enzymatic method 63
Enzyme 14, 24
   active center of 20, 21f
   activity, factors influencing 16

classification of 15, 16t
concentration 17, 17f
  effect of 18f
containing metals 15t
inhibition 21
profiles 26t
reaction, mechanisms of 19
regulatory 21
therapeutic use of 24t
Enzyme-linked immunosorbent assay test 165
Enzyme-substrate complex 20f, 27
Epimerism 32, 32f
Epinephrine 131
Epitopes 162
Ergocalciferol 185
Ergosterol 185
Erythrocyte 286
*Escherichia coli* 262
Essential amino acids 110, 111
Essential fatty acids 71, 91
Essential nutrients, recommended daily allowance of 287
Estrogens 99
Ethanolamine plasmalogen 75f
Eukaryotic cells 4t, 5
Ewald's test meal 247
Ex vivo gene therapy 279f
Exons 258
Exopeptidases 121
Extracellular fluid 200f
  isotonic contraction of 231
Eyepiece 271
Eyes, complications in 60

## F

Facilitated diffusion 9, 10f
Familial hyperlipoproteinemias 103
Fat
  chylomicrons, absorption of 81f
  intake, moderation in 104
  utilization of 82f
Fat-soluble vitamins 182, 189t, 287
Fatty acid 46, 69, 85
  activation of 82, 83f
  beta-oxidation of 82, 84f
classification of 68t
de novo synthesis of 87f
even-chain 68
free 100
oxidation of 71
  energetics of 83
properties of 69
synthase system 85
  functional division of 88f
synthesis of 85, 88t
  regulation of 86
Fatty liver 90, 94
Ferritin 207
Ferrous iron 207
Fibrinogen 156
Fibrous proteins 116
Filter paper 273
Fischer's template theory 19
Fischer's theory 20f
Five-carbon unit, formation of 98f
Flavin adenine dinucleotide 2, 143, 149f, 196
Flavin mononucleotide 149f, 196
Fluid
  and electrolyte balance 231
  exchange 159
  infusion of 231
Fluorescence 274
Fluoride 211
  biological role of 211
  dietary sources of 211
  inhibits 63
  requirement of 211
  toxicity of 211
Fluorosis 211
Foam cells 152
Folate deficiency, causes for 194
Folic acid 189, 194, 209
  analogue 195
  biological functions of 194
  deficiency 195f
  sources of 194
Food
  determine items of 220
  general composition of 220
  items, nutritional value of 289
  preservatives 153
Foodstuffs, oxidation of 148f
Fouchet's reagent 240
Fouchet's test 175, 243
Fractional test meal 247
Frederickson's classification 103t
Free radical 143, 150
  generation of 150
  scavenger systems 152, 152f
Fructokinase 52
Fructose 33
  intolerance 53
  metabolism of 52, 53f
  structure of 33f
Fructosuria 53, 64, 140
Fumarylacetoacetate hydroxylase, absence of 138f
Furanose form 32
Furanose ring 32

## G

Galactokinase 53
Galactose metabolism 53, 53f
Galactose-1-phosphate uridyltransferase 53, 140
Galactosemia 53, 54, 140
Galactosuria 54, 64
Gamma aminobutyric acid 147f
Gamma glutamyl transferase 25, 241
Gamma-globulins 159
Gangliosides 76
Gap junction 8
Gastrectomy 63
Gastric atrophy 196
Gastric function tests 246
Gastric secretions 247
Gastrointestinal tract 57
Gaucher's disease 76, 139, 141
Gel
  electrophoresis 275
  filtration chromatography 277, 277f
  types of 275
Gene therapy 278
  procedure, summary of 279
General immunity 162
Genetic disorders, diagnosis of 282

Genome 265
Gestational diabetes mellitus 62
Gilbert syndrome 176
Globular proteins 117
Globulin 156, 158
Glomerular filtration rate 243, 244
Glomerular proteinuria 243
Glomerulonephritis, chronic 151
Glucagon-like peptide-1 58
Glucan transferase 48
Glucocerebrosidase 139
Glucocorticoids 99
Glucogenic amino acids 112, 133f
Glucokinase 42
Glucometer 63
Gluconeogenesis 46, 47
   key enzymes of 47t
Gluconic acid 34
Glucosaccharic acid 34
Glucosazone 34f
Glucose 127f, 232, 233, 242
   absorption 42f
   alanine cycle 128
   and fructose, mixture of 35
   carbon atom of 31
   D and L forms of 31f
   intestinal absorption of 42f
   metabolism 42
   oxidase peroxidase method 63
   regulation, impaired 62
   residues 36f
   tolerance, impaired 62
   transporter type 42, 57
Glucose-6-phosphatase 47, 49, 50
   dehydrogenase 25, 51
   metabolic fate of 43f
Glucose-dependent insulinotropic polypeptide 58
Glucosuria 33, 56, 59, 64
Glucuronic acid 34
Glutamate 125f
Glutamic acid 125
Glutaminase 227
Glutathione 115, 152f
   reductase 152f

Glycated hemoglobin 63, 177
Glycemic index 221
Glyceraldehyde, D and L forms of 31f
Glycerol and fatty acids 72f
Glycerophospholipids 73
Glycine 126, 127, 133, 166
   biological importance of 126
   catabolism of 127
   cleavage of 127
   from serine, formation of 126f
   metabolism 127f
   synthesis 126f
Glycinuria 138
Glycocholate 243
Glycogen 37
   functions of 49
   metabolism, regulation of 49
   phosphorylase 48, 49, 49f
   storage diseases 50, 51t, 54, 140, 140t
   synthase 49, 49f, 50f
Glycogenesis 49
Glycogenolysis 48
   key enzyme of 49f
Glycogenosis 140
Glycohemoglobin 63
Glycolipids 76
Glycolysis 42, 44f
   key enzymes of 47t
Glycoproteins 38
Glycoside 36, 36f, 37t
   bond 33
   formation 33
Glycosphingolipids 75
Glyoxylic acid, formation of 127
Goiter 210
Goiterous belts 210
Goitrogenic substances 210
Golgi apparatus 6, 7
Gout 255
   clinical finding of 255
   primary 255
Graft, rejection of 163
Grierson-Gopalan syndrome 193
Growth 218
Guanine 251
Guanosine 5'-monophosphate 253

Guanosine
   diphosphate 47, 253, 264
   triphosphate 47, 253, 264

# H

Haptoglobin 158
Hartnup disease 139
Hays sulfur powder test 240
Hays test 243
Heart
   attacks 92
   diseases 97
   valves 38
Heat and acetic acid test 118, 158, 243
Heat coagulation 118
Heavy metal poisoning 211
Heme
   biosynthesis of 171
   catabolism of 172
   metabolism 171
   oxygenase system 174f
   structure of 171
Heme synthesis
   disorders of 172
   regulation of 172
   steps of 173f
Hemoglobin 171, 176, 177, 208
   abnormal 177
   buffer system 226
   derivatives 177
   modified 177
   oxygenated 178
Hemoglobinopathy 177
Hemoglobinuria 243
Hemolysis 194, 233, 269
Hemolytic anemia 194
   drug-induced 51
Hemolytic jaundice 174, 239, 240
Hemorrhage 123, 209
Hemosiderosis 209
Henderson-Hasselbalch equation 224
Heparin 38
Hepatic coma 123
Hepatic diseases 26
Hepatic jaundice 174, 239
Hepatitis B virus 239
Hepatocellular jaundice 174, 239, 240

## Index

Hepatomegaly 123
Hereditary fructose intolerance 140
Heteroglycans 37, 38
Hexokinase 16, 26, 42, 47
  reaction 43f
Hexose 30t
  monophosphate 6, 43, 189
  shunt pathway 50, 52f
High-sensitivity cardiac troponins 105
Hippuric acid 127
Histamine stimulation test 247
Histidase 139
Histidine and proline 110f
Histidinemia 139
Histocompatibility complex 164
Homeostasis, maintenance of 241
Homogentisic acid oxidase 137
  absence of 138f
Homoglycans 37
Hormonal functions 241
Hormones 116, 218
Hormone-sensitive lipase 90
Human body functions 2
Human genome project 2
Human immunodeficiency virus antibody, detection of 165
Human insulin gene 2
Human leukocyte antigens 164
Humoral immunity 162, 163
Hyaluronic acid 38
Hyaluronidase 24
Hydrochloric acid secretion, mechanism of 246, 246f
Hydrogen
  bonding 257
  ions 224t
    excretion of 227f
  peroxide 150
Hydrolases 16, 71
Hydrolytic enzymes 12
Hydrolytic rancidity 72
Hydroxyl ions 224t
Hydroxyl radical 150
Hydroxyphenylpyruvic acid oxidase 137
Hyperacidity, causes for 248t
Hyperaldosteronism 233

Hyperammonemia 124
Hyperbilirubinemia 174
Hypercalcemia 203
Hyperchloremia 233
Hyperchlorhydria 247
Hypergammaglobulinemia 162
Hyperglycemia 56, 232
Hyperkalemia 205, 230, 233
Hyperlipoproteinemia 102, 103t
Hypermagnesemia 204, 234
Hypernatremia 205
  causes of 232
Hyperornithinemia 124
Hyperoxaluria 138
Hypersensitivity 163
Hypertension 104, 219
  failure 231
Hyperthyroidism 63
Hypertonic contraction 231
Hypertyrosinemia 138f
Hyperuricemia 255
  secondary 255
Hypervitaminosis
  A 185
  D 187
Hypoacidity, causes for 248t
Hypoalbuminemia 157
Hypocalcemia 187, 203
Hypochloremia 205
  causes for 233
Hypochloremic alkalosis 233
Hypochlorhydria 247
Hypogammaglobulinemia 162
Hypoglycemia 54, 56, 64
  fasting 50
Hypokalemia 205, 233
Hypolipoproteinemia 102
Hypomagnesemia 204, 234
Hyponatremia 205
  causes of 232
Hypothyroidism 103
Hypotonic contraction 231
Hypoxanthine-guanine phosphoribosyltransferase 141

## I

Ibuprofen 92
Immune response 162
  primary 163f
  secondary 163, 163f

Immune thrombocytopenia purpura 47
Immunity against infections 163
Immunoelectrophoresis 275, 275f
Immunoglobulin 156, 159
  A 161, 161f
  D 161, 161f
  E 161, 161f
  G 160, 161f
  M 161, 161f
  molecule 160f
  secreting plasma cells 5
  structure of 160
Immunological functions 159
Immunological system 184
Inborn errors of metabolism
  groups of 141
  salient features of 136
Incretin hormones 58
Indole group, tryptophan with 110f
Indomethacin 92
Infections, chronic recurrent 59
Inherited diseases, diagnosis of 282
Inhibition, types of 22t
Inhibitors, effect of 19
Inorganic metal ions 14
Insulin 57, 233
  effect of 57
  injections 64
  insensitivity 58
  overdose of 64
  resistance syndrome 61
Intermittent porphyria, acute 172
Intestinal mucosal cells 101
Intestinal system 129
Intracorpuscular defects 209
Intravenous infusion 232
Introns 258
  removal of 262
Invisible fats 217
Iodine 210
  biological role of 210
  deficiency manifestations of 210
  functions of 210
  number 70

requirement of 210
sources of 210
Iron 15, 209
　absorption of 247
　atom 171
　conservation of 207, 208f
　deficiency 208f
　　anemia 196
　　causes of 207
　　manifestations 207
　　treatment for 209
　in plasma, normal level of 207
　kinetics, normal 208f
　storage of 207
　toxicity 209
　transport of 207
Irreversible inhibition 22
Isoelectric pH 113
Isoenzymes 25, 27
Isoleucine 130
Isomaltase 41
Isomerase 16, 80
Isoniazid 192
Isonicotinic acid hydrazide 22
Isonicotinylhydrazide 192
Isoprene units, formation of 97
Isotonic expansion 231
Isotonic solution 269
Isotretinoin 185

## J

Jaundice 123, 174, 237, 239, 239t
　physiological 175
　post-hepatic 174, 240
　pre-hepatic 174, 240
　types of 176f, 176t, 240t
Juvenile diabetes 58

## K

Keratan sulfate 38
Keratinization 185
Keratins 167
Keratomalacia 185
Ketoacidosis 59, 232
　consequences of 60
　management of 60
Ketogenesis 88, 89f
Ketogenic amino acids 112, 132

Ketolysis 88
　pathway of 89f
Ketone bodies 88, 242
　formation of 88, 89f
Ketonemia 59, 88
Ketonuria 59, 88
Ketosis 50, 88
　causes for 88
Key enzymes 21, 23, 24
Kidney
　disease 242
　functions of 241
Kilocalorie 215
Koshland's induced fit model theory 20, 21f
Krebs cycle 143, 144f
Kussmaul's respiration 60
Kwashiorkor 218, 219t

## L

Lactase 41
Lactate dehydrogenase 15, 24, 26, 45, 48
　reaction of 15f
Lactate production, significance of 45
Lactic acidosis 50
Lactosazone 36f
Lactose 35
　contains 35f
　free diet 54
　intolerance 53
Lactosuria 64
Lanosterol 97
Lauric acid 69
Lead poisoning 21
Lecithin 74
　functions of 74
Lens proteins 168
Lesch-Nyhan syndrome 141
Leucine 130
Leukopenia 194
Leukotrienes 93
Levorotation 35
L-gulonolactone oxidase 51
Ligases 16
Linoleic acid 91
Linolenic acid 91
Lipase 25, 80

Lipid
　absorption of 81
　chemistry 67
　classification of 68, 68t
　clinical applications of 67
　digestion of 80
　functions of 67
　in blood, transport of 99
　in intestines, digestion of 80
　metabolism 80
　　inherited disorders of 139
　simple 71
Lipoprotein 76, 96
　classes of 100t
　classification of 100
　functions of 76, 100
　high-density 99-102, 102f
　metabolism 101f
Lipotropic factors 91
Lipoxygenase pathway 92f, 93
Liver
　disease, routine markers of 238
　dysfunction, clinical manifestations of 237
　enzymes 25
　in fat metabolism, role of 90
　major functions of 238t
Liver function tests 237, 238
　classification of 238t
　indications for 238
Long-chain fatty acids, absorption of 80
Low intracellular sodium 10
Low potassium diet 233
Low-density lipoprotein 100, 101, 102f, 188
Lungs 38
Lyases 16
Lymphocytes, small 5
Lynen cycle 87f
Lysosome 6, 7
　functions of 7
　primary 7
　secondary 7

## M

Macrocytic anemia 195f
Macronutrients 200
Macrophages 129, 163
*Madhumeha* 1, 58

Magnesium 204, 233
  functions of 204, 234
  normal serum level of 204, 234
  requirement of 204
Malate-oxaloacetate shuttle 86f
Malonyl-coenzyme A, formation of 85
Maltase 41
Maltosazone
  petal-shaped crystals of 36f
  sunflower-shaped crystals of 36f
Maltose 36
Manganese 15
Maple syrup urine disease 130, 139
Marasmus 218, 219t
McArdle's disease 51, 140
Medicolegal cases 282
Medium-chain fatty acids 69
Megaloblastic anemia 194
Megaloblasts 196
Melanin 131
  synthesis 210
Melatonin 132
Membrane receptor 57
Menkes kinky hair syndrome 141, 210
Mental retardation 130, 137
  severe 54, 139, 140
Metabolic acidosis 60, 228, 233
  causes of 228
Metabolic alkalosis 229
  causes of 229
  compensatory mechanism of 229
Metabolic disorder 130, 137f
Metabolic fate 112t
Metabolic pathways, integration of 146
Metabolic processes 2
Metabolic syndrome 60, 219
Metabolic waste products, excretion of 241
Metabolism 121
  inborn errors of 136, 141t
  intermediary 2
  primary 2, 147
  secondary 2
  tertiary 2

Methemoglobin 177
Methione 264
Methionine 128, 133
  synthesis of 195, 195f
  to cysteine 129f
Methotrexate 195
Methyl transfer reactions 194
Mevalonate
  formation of 97f
  synthesis of 97
Michaelis constant 17
Michaelis-Menten constant 27
Michaelis-Menten theory 19, 27
Microalbuminuria 243
Microangiopathy 60
Microcytic hypochromic anemia 208, 208f
Microelectrophoresis 273
Microelements 200
Microscopes 271
Mineral 200, 200t, 288
  metabolism 212t
Mitochondria 6, 7
  cut section of 149f
Mitochondrion, structure of 149
Molecular biology 251
  central dogma of 260, 260f
Molecular oxygen 174f
Molisch test 33
Molybdenum 15
Monoclonal band 162
Monophosphates 251
Monosaccharides 29
  modified 34
  properties of 31
Monounsaturated fatty acid 77, 91, 217
Mucoproteins 38
Multienzyme complex 85
Muscle cells 8
  glucose transport in 43f
Muscle diseases 26, 106
Muscle proteins 168
Muscular dystrophies 106, 126
Mutarotation 33
Myeloma, multiple 157, 161
Myeloperoxidase 151, 151f
Myocardial infarction 26, 105, 126
Myosin 168
Myxedema 210

# N

Neonatal mortality 62
Nephron 241
Nephropathy 60
Nephrotic syndrome 103, 157, 158, 232
Nerve conduction 98
Neuropathy 60
Neutral aminoaciduria 139
Neutral fat, structure of 71f
Niacin 132, 189, 191
  biological functions of 191
  deficiency 132
    causes for 191
    manifestations of 192
  dietary sources of 191
  recommended dietary allowance of 191
  sources of 191
  structure of 191f
Niacinamide, structure of 191f
Nicotinamide adenine dinucleotide 2, 15, 15f, 148f, 149f, 183f, 184f, 196
  hydride 183f
  hydrogen 44, 45, 177
  phosphate 136, 151f, 186, 196
  phosphate hydrogen 174f, 177
  phosphate oxidase 151
  structure of 15
Nicotinamide group 15
Nicotinic acid 189, 191
Niemann-Pick disease 75, 139, 141
Night blindness 185
Nitric oxide 129, 150
  physiological actions of 129
  synthase 151
Nitrogen balance 217, 218, 218f
Nonapeptide 115
Noncompetitive inhibition 22, 22f
Nonoxidative deamination 125
Nonreducing disaccharide 35f
Nonsteroidal anti-inflammatory drugs 92
Norepinephrine 131
Normal pH 225

Normoglycemia 56
Nucleic acid 251
   metabolism, inborn errors of 141
Nucleoside 251, 252t, 253t
   diphosphate 153
   triphosphate 153, 251, 252, 253t
Nucleotide 251, 252, 252t, 253t
   composition of 251
Nucleus 5
Nutrients, calorific value of 215t
Nutrition 215
Nutritional importance 110
Nutritional value, classification on 117

# O

Obesity 219
   index 219
Obstructive jaundice 103, 174, 240
Obstructive liver disease, markers of 241
Ochronosis 137
Odd-chain fatty acids 68
Oils and fats, composition of 71t
Oncogenic virus 240
Opsin 183
Optical activity 31
Optical isomerism 110
Oral glucose
   load test 62
   tolerance test 61, 62f
Oral hypoglycemic agents 64
Organic acidopathies 141
Ornithine 129
   decarboxylase 27
Orosomucoid 158
Orotic aciduria 256
Osazone formation 34
Osmosis 268, 268f
   reverse 269
   role of 269
Osmotic diuresis 60
Osmotic pressure 158, 269
Osteoblasts 186
Osteocalcin 186
Osteogenesis imperfecta 166

Osteomalacia 187
Ouchterlony double diffusion 275f
Ouchterlony technique 275
Oxalates 201
Oxaluria
   explanation for 127f
   primary 127
Oxidation reactions 34
Oxidative deamination 113, 125
Oxidative decarboxylation 189
Oxidative phosphorylation 148, 149
Oxidative rancidity 72
Oxidatively decarboxylated 46
Oxidoreductases 16
Oxygen
   dissociation curve 179f
   transport of 178

# P

Palmitoleic acid 91
Pancreatic lipase 80
Pancreatin 24
Pancreatitis, acute 122
Pantothenic acid 189, 193
   recommended dietary allowance of 193
   sources of 193
Papain 24
Paper chromatography 276, 276f
Paper electrophoresis 273
Paracetamol 92
Parathyroid hormone 201
Partition chromatography 276
Passive immunity 163
Passive transport 9
Patients, diabetic 61t
Pellagra 132, 192
Pentapeptides 115
Pentose phosphate pathway 50, 54
Pentosuria 64
   essential 52
Pepsin 121, 122, 247
Peptide bond 114
   formation 263
Peptidyl site 263, 264
Peptidyl transferase 263

Peripheral blood smear 208f
Peroxidase 152f
Peroxisomes 7
pH
   effect of 18
   on enzyme velocity, effect of 18f
Phenylalanine 130
   and tyrosine, catabolism of 131f
   hydroxylase 131, 136
      absence of 138f
   metabolism, abnormalities in 131
Phenylketonuria 131, 136, 137, 138f
Phosphate 252t
   buffer system 225
   mechanism 227f
Phosphatidic acid 73, 74f
Phosphatidylcholine 74
Phosphatidylethanolamine 74, 74f
Phosphatidylinositol 75, 75f
Phosphatidylserine 75
Phosphodiester bonds 256
Phosphoenolpyruvate carboxykinase 47
Phosphofructokinase 47, 54
Phosphogluconate oxidative pathway 50
Phospholipids 7, 67, 72
   glycerol of 7
Phosphorus 203
   biological role of 204
   deficiency manifestations of 204
   distribution of 203
   requirement of 203
   serum level of 203
   sources of 203
   toxicity of 204
Phosphorylation, substrate-level 148
Phosphosphingosides 75
Photoreceptor matrix 184f
Phrynoderma 71, 91
Phytanic acid oxidase 139
Phytic acid 201
Plaque formation 60
Plasma 286

cells 162
glucose levels 61*t*
insulin 58
Plasma lipid 99
  profile 100*t*
Plasma membrane 7, 8, 12
  fluid mosaic model of 7, 8*f*
  layers of 7
Plasma protein 156
  estimation 240
  functions of 159
  normal range of 240
  separation of 156
Plasmacytoma 161
Plasmalogens 75
Platelet 25
  aggregation 93
Polyacrylamide 276
Polycystic ovary disease 61
Polydipsia 59
Polymerase chain reaction 280, 281*f*
  clinical applications of 282
Polyneuritis 190
Polynucleotide 257*f*
Polyphagia 59
Polyribosomes 6
Polysaccharide 37, 38, 38*t*, 41
Polyunsaturated fatty acids 70*f*, 91, 188, 217
  peroxidation of 151
Polyuria 59
Pompe disease 140
Porphobilinogen 172
Porphyria 171, 172
Porphyrin ring 171*f*
Postprandial blood glucose 56
Postprandial lipemia 81
Potassium 205, 232
  biological functions of 205
  deficiency of 205
  excretion 232
  in plasma, normal level of 205
  requirement of 205
  sources of 205
  supplementation, excess 233
  taurocholate 243
  toxicity of 205
Pregnancy 194, 196, 218
  complications in 60

Procollagen, maturation of 167*f*
Product concentration, effect of 19
Proenzyme 19
Progesteins 99
Pro-hormone 186
Prokaryotic cells 4*t*
Pronuclear membrane 5
Prostacyclins 93
Prostaglandins 92
  functions of 93
Prostate
  cancer 26
  cells 25
Protein 116, 159, 166, 218*t*, 287
  acute phase 159
  biosynthesis 262
  buffer system 225
  calorie malnutrition 158, 218
  classification of 115
  complete 117
  conjugated 116, 117*t*
  contractile 116
  defense 115
  deficiency 218
  denaturation of 117, 118*f*
  derived 116
  dietary 217
  digestion of 121
  energy malnutrition 218
  gastric digestion of 121
  general functions of 108
  in plasma, normal values of 156
  incomplete 117
  intestinal digestion of 122
  levels of organization of 117, 117*f*
  metabolism of 123*f*
  mutual supplementation of 218
  nutritionally rich 117
  pancreatic digestion of 122
  poor 117
  protective 115
  regulatory 116
  requirement 218
  simple 116, 116*t*
  structural 115
  synthesis 12
  tissue 156

  total 156
  transport 115
  western blot analysis for 279
Proteinuria, overflow 243
Proteome 265
Proteomics 265
Prothrombin 188
Provitamin 185
Pseudoxanthoma elasticum 167
Pulmonary hypertension 129
Purine 251
  metabolism, disorders of 255
  structure of 251*f*
  synthesis of 126
Purine nucleotides
  biosynthesis of 253
  degradation of 254
Purine synthesis 254*t*
  inhibitors 254, 256
  regulation of 254
Pyranose 32
Pyridoxal phosphate 124, 196
Pyridoxine 189, 192
  biological functions of 192
  deficiency manifestations of 193
  recommended dietary allowance of 193
  sources of 192
Pyrimidine 251, 252*f*, 256, 256*f*
  de novo synthesis of 256
  metabolism, disorders of 256
Pyruvate carboxylase 47, 194
Pyruvate dehydrogenase reaction 46
Pyruvate kinase 47

# R

Racemic mixtures 32
Rancidity 72
Rate limiting enzyme 24, 97
Reactive oxygen species 150, 151
  generation of 151*f*
Red blood cell 5, 11, 25, 63, 156, 171, 175*f*, 208*f*
Reduce dietary cholesterol 104
Refsum's disease 139

Renal failure 233
  chronic 232
Renal function 126
Renal function tests 237, 241
  classification of 242t
Renal glomerular function 242
Renal glucosuria 63
Renal mechanism 226, 229
Renal threshold 63
Renal tubules 186
Renin-angiotensin system 230
Reperfusion injury 152
Repression mechanism 172
Respiratory acidosis 229
  causes of 229
  compensatory mechanism of 229
Respiratory alkalosis 229
  causes of 229
  compensatory mechanism of 229
Respiratory burst 151
Respiratory chain 148
Respiratory diseases 151
Respiratory mechanism 226, 229
Respiratory quotient 215
Restriction endonucleases 278
Reticulocytosis 194
Retinal pigment epithelium 184f
Retinoic acid 183
Retinol 183
Retinopathy 60
Reversible noncompetitive inhibition 22
Rheumatoid arthritis 151
Rhodopsin cycle 183
Riboflavin 189, 190
Ribonucleic acid 251
  polymerase 261
  synthesis, inhibitors of 262
Ribonucleosides 252
Ribosome 6
Rickets 187
Rothera's test 59, 60, 94, 242
Rough epithelium 185

# S

Saline infusion, excess 233
Salvage pathway 254
Saponification 72, 72f
Sarcolemma 8, 168
Saturated fatty acids 69, 217
Sclerosis, multiple 152
Screen filtration 270
Scurvy 197
Selenium 211
  biological role of 211
  requirement of 211
Seliwanoff's test 33
Sephadex chromatography 277f
Serine 128, 133
  metabolism 128f
Serotonin 132
Serum 286
  calcium 201t
  cholesterol level 103
  enzyme activity 241
  triglyceride 104
  vitamin D 187
Short-chain fatty acid 69
  absorption 81
Sickle cell hemoglobin 177
Skin 38
Sodium 204, 230, 231, 243
  biological role of 205
  deficiency 205
  in plasma, normal level of 205
  loss 60
  pump 10, 11
  requirement of 204
  restriction 232
  sources of 204
  toxicity 205
Sodium-dependent glucose transporter 41
Sodium-potassium activated atpase 10
  pump 11f
Southern blot technique 279, 280f
Sphingolipids 75
Sphingomyelin 75, 75f
Sphingomyelinase 139
Sphingosine 75
Squalene, formation of 97, 98f
Starch 37, 41
  content 247
  gel 275
Stereoisomerism 31, 109

Steroid 76
  ring 76
Steroid hormones 98
  from cholesterol, synthesis of 99f
  functions of 99t
Sterols 76
Streptokinase 24
Subcellular organelles 5
  metabolic functions of 6t
Substrate concentration 17
  effect of 17f
Substrate saturation level 17
Sucrase 41
Sucrose 34, 35f
Suicidal inhibition 22, 27, 256
Sulfatides 75
Sulfolipids 76
Sulfonamides 195
Superoxide
  anion radical 150
  dismutase 151f
Surplus amino acids 123
Sweet urine 58
Symport transport systems 11f
Synthesis 254
  pathways 86
  regulation of 97

# T

T lymphocytes 162
Tangier's disease 102
TATA box 261
Temperature, effect of 18
Tetany 203
Tetrahydrofolate 14
Tetrahydrofolic acid 126, 196, 254, 254f
Tetrapyrrole 171
Thalassemia 178
  major 178
  minor 178
  syndromes 178
Therapeutic enzymes 24
Thiamine 189
  deficiency of 190
  pyrophosphate 14, 189, 196
  recommended dietary allowance of 190
Threonine 128, 264

# Index

catabolism of 128f
Thromboxanes 93
Thymidine
   diphosphate 253
   monophosphate 253
   triphosphate 253
Thymine 252
Thyroid hormones 216
Thyroxin 131, 210
Tight junction 7, 8f
Titratable acid 226
Tonicity 269
Toxic regulatory substances 144f
Toxins 116
Trace elements 200
Transamination reaction 124f
Transfer ribonucleic acid 264
Transferases 16
Transmembrane proteins 12
Transport mechanisms 9, 11t
Tremors 203
Triacylglycerols 71
   digestion of 80
Tricarboxylic acid 6, 143
Tricarboxylic acid cycle 143, 145
   anaplerotic role of 147
   inhibitors of 147
   intermediates
      efflux of 147f
      influx of 146f
   significance of 145
Triglyceride 81f, 82, 100
   breakdown of 90
   hydrolysis of 72f
   metabolism of 89
   properties of 71
   synthesis of 90
Triiodothyronine 210
Triosephosphate isomerase 26
Tripeptide 115
Tropocollagen 166
Tropomyosin 168
Troponins 168
Trypsin 121, 122
Trypsinogen 19
Tryptophan 131, 134, 139
   dietary deficiency of 191
   metabolism 132f
Tuberculosis 59
Tubular fluid 227f
Turgid 269
Turgor 269
Typical cell 5f
   structure of 5
Tyrosinase 137
Tyrosine 130, 134
   pathway, metabolic defects in 138f
   products from 131
Tyrosinemia 137

## U

Ultraviolet adsorption 274
Uniport transport systems 11f
Universal energy currency 149, 253
Unsaturated fatty acid 69, 69f, 70f
   properties of 70
Urea clearance test 246
Urea cycle 123, 123f
   defects 141
   disorders 124t
Uric acid 255
   structure of 255f
   synthesis, inhibitor of 22
Uricosuria 255
Uricosuric drugs 256
Uridine 5'-monophosphate 253
Uridine diphosphate 49, 174f, 253
Uridine triphospahate 49, 253
Urine 242, 286
   abnormal constituents of 242
   acidification of 226
   bile salt 240
   bilirubin 240
   differential diagnosis of reducing substances in 64t
   normal constituents of 242
   normal organic constituents of 242
   reducing substances in 63
   sample, observations of 242
Urobilinogen 174, 175f
Urokinase 24

## V

Vacuum filtration 271
Van den Bergh reaction 239
Van den Bergh test 175
Vascular diseases 60
Very low-density lipoprotein 82, 99-101, 102f
Viral diseases 282
Viscosity 159
Vision 183
Vitamin 182, 189
   A 183, 189
      role of 183
      sources of 185
      structure of 183f
   B complex 189t, 196t
   $B_1$ 189
      source of 189
      structure of 189
   $B_{12}$ 189, 195
      absorption and storage of 195
      biological functions of 195
      deficiency, causes of 195
      sources of 195
   $B_2$ 189, 190
      deficiency, symptoms of 191
      sources of 190
   $B_6$ 189, 192
   C 153, 196, 209
      deficiency manifestations of 197
      functions of 197
      sources of 196
   D 77, 98, 185, 186, 189, 201
      biochemical effects of 186
      deficiency 187
      effect of 186
      sources of 185
   E 152, 153, 188, 189
      biochemical role of 188
      deficiency manifestations of 188
      functions of 188
      sources of 188
   K 188, 189
      analog of 22
      antagonists of 188
      functions of 188

sources of 188
types of 182t
Vomiting 231-233
von Gierke disease 50, 140

## W

Wald's visual cycle 183, 184f
Water balance 230
Water-soluble vitamins 182, 188, 287
Watson-Crick model 257, 257f

Weak acid, action of 34
Wernicke-Korsakoff syndrome 190
White blood cell 25, 156, 171
Wilson's disease 141, 159, 209

## X

Xanthine oxidase 255
Xerophthalmia 185
Xylitol dehydrogenase 52
Xylulosuria 64

## Z

Zellweger syndrome 141
Zinc 15, 210
  biological role of 210
  deficiency 211
  dietary sources of 210
  normal serum level of 210
  requirement of 210
  toxicity of 211
Zwitter ion 112, 112f
Zymogen 19, 23

EU GSPR Authorised Reprsentative
Logos Europe, 9 rue Nicolas Poussin
1700, La Rochelle, France
Phone: +33 (0) 6 67 93 73 78
E-mail: contact@logoseurope.eu

www.ingramcontent.com/pod-product-compliance
Ingram Content Group UK Ltd.
Pitfield, Milton Keynes, MK11 3LW, UK
UKHW050429150426

5217IPUK00019B/1305